Microbes for Humankind and Applications

The Editors

Dr. Deepak Kumar Malik is Assistant Professor at University Institute of Engineering and Technology, Kurukshetra University, Kurukshetra. He has a PhD degree in Microbiology from Choudhary Charan Singh Haryana Agriculture University, Hisar in the area of PGPR. He is having 14 year of teaching and 19 years of research experience in the field of Microbiology. He has published 60 Research paper/Review artical in various disciplines of Microbiology and Nanotechnology.

Dr. Meenu Rathi is Assistant Professor at Gandhi Memorial National PG College Ambala Cantt. She has a PhD degree in Botany from Choudhary Charan Singh University Meerut in the area of phosphate solubilzation by bacteria. She is having 9 year of teaching experience and 4 years research experience in the field of plant microbe interaction.

Dr. Rajesh Kumar is Assistant Professor at University Institute of Engineering and Technology, Kurukshetra University, Kurukshetra. He has a PhD degree in Biotechnology from Kurukshetra University Kurukshetra in the area of Microbial Biotechnology. He is having 13 year of teaching profession and 16 years research experience in the field of Microbial Biotechnlogy. He has published research paper in various discipline of Enzyme Technology.

Dr. Divya Bhatia is presently working as Assistant Professor at University Institute of Engineering and Technology, Kurukshetra University, Kurukshetra. She has done M. Tech and Ph.D. in Biotechnology from Kurukshetra University, Kurukshetra. She has expertise in Microbial Biotechnology and Nanobiotechnology. She has more than four year teaching experience and has contributed 17 research papers and 3 book chapters.

Microbes for Humankind and Applications

Deepak Kumar Malik

Meenu Rathi

Rajesh Kumar

Divya Bhatia

2020

Associated Publishing Company®

A Division of

Astral International Pvt. Ltd.

New Delhi – 110 002

ISBN : 9789390371211 (Int Edition)

Publisher's Note:

Every possible effort has been made to ensure that the information contained in this book is accurate at the time of going to press, and the publisher and author cannot accept responsibility for any errors or omissions, however caused. No responsibility for loss or damage occasioned to any person acting, or refraining from action, as a result of the material in this publication can be accepted by the editor, the publisher or the author. The Publisher is not associated with any product or vendor mentioned in the book. The contents of this work are intended to further general scientific research, understanding and discussion only. Readers should consult with a specialist where appropriate.

Every effort has been made to trace the owners of copyright material used in this book, if any. The author and the publisher will be grateful for any omission brought to their notice for acknowledgement in the future editions of the book.

Published by : **Associated Publishing Company®**
A Division of
Astral International Pvt. Ltd.
– ISO 9001:2015 Certified Company –
4736/23, Ansari Road, Darya Ganj
New Delhi-110 002
Ph. 011-43549197, 23278134
E-mail: info@astralint.com
Website: www.astralint.com

Digitally Printed at : **Replika Press Pvt. Ltd.**

Preface

The healthy man has ten times as many microbial cell as human cells, so what good things of these microbes doing for us? Humans have made the use of microorganisms to carry out fermentation processes to make various fermented foods. Genetic engineering has been applied in various fields including medicine, industrial biotechnology and agriculture for the benefit of humankind. In research, GMOs are used to study gene function and expression through decrease and increase of function, tracking and expression experiments. By knocking out of some genes, it is likely to make animal model organisms of human diseases. Some soil microbes form relationships with plant roots to provide nutrients like nitrogen or phosphorus. Microbes provide many benefits including drought tolerance, heat tolerance, resistance to insects and resistance to plant diseases. However, the invented stories associated with microbes are needed to be generalized. The all review article/ research paper brings current information with latest references. This book will help all the researcher and teachers for better understanding of their relevant research fields. For that purpose, the proposed book Microbes for Humankind and Applications will be a novel and useful collection of large applications of microbes in different areas like food, pharmaceutical, diagnostic, medical, agriculture, environment and industry *etc.* The idea is to compile a reference book in various aspect of microbes application, that will help the life science researchers. Our aim is to make available a massive variety of topics related to microbes within a single book. The all chapters bring advanced information with current references. We will focus on topics that are presently challenging with applications of microbes.

Deepak Kumar Malik

Meenu Rathi

Rajesh Kumar

Divya Bhatia

Contents

Chapter 1

Development of Probiotic Functional Foods as Novel Attentiveness for Human Health and Disease Management using Fermentation Technology

Surbhi Panwar[1], Saurabh Jain[2], K Siddaardha Duggirala[3], Ashok Kumar Yadav[4] and Ashwani Kumar[3]*

[1]*Department of Genetics and Plant Breeding, Chaudhary Charan Singh University, Meerut – 250004, Uttar Pradesh*
[2]*Department of Applied Science, Jai Parkash Mukand Lal Innovative Engineering and Technology Institute, Radaur – 135 133, Yamuna Nagar, Haryana*
[3]*Department of Nutrition Biology, Central University of Haryana, Mahendergarh, Jant-Pali – 123 029, Haryana*
[4]*Centre for Molecular Biology, Central University of Jammu, Rahya Suchani Bagla, Samba – 181 143, Jammu and Kashmir*
**E-mail: ashwanindri@gmail.com*

ABSTRACT

Fermentation process uses microbes to develop flavour in food products. The microbe also synthesise bioactive component that enhance bio-functionality of food products which are known as functional or designer foods. Recently various prevalent chronic diseases like obesity, diabetes, cardiovascular diseases and cancer particularly guts related cancer are associated with poor food habits. Therefore, there is a need to screen novel good microbes for functional food production which can augment biocompatibility and functionality of foods against several life-threatening diseases. Several bioactive molecules are generated during fermentations process that can exert better health benefits for the host. Development of functional foods by supplementation of probiotic microorganisms during fermentation process can be of great significance in food industries. Therefore, optimization of fermentations process to produce large amount of microbial cell mass is also of enormous importance. Moreover, original test and flavour of function foods can be enhanced through microorganism that generates bioactive molecules.Therefore, intensive research is essential for developingthe probiotic bacteria fortified functional foods that can augment bio-functionality and flavour in the product.

Introduction

Current society is more conscious about health along with changes in life style in fast growing environment. There has been a noteworthy move in customer's perspective to treat food not purely as a nutrition source but also as an effective agent in the management of their health care and well-being. The highly awakened new generation is now looking for safe natural products as substitute to the conventional drug-based therapy for routine health care and disease management since the latter apart from being cost prohibitive also suffer from undesirable side effect. In this framework, functional food containing probiotics have recently emerged as safe, cost effective and easily affordable as prophylactics or biotherapeutics for general health promotion and specifically in the executive of the target disease particularly gut related inflammatory metabolic disorders (Ringel *et al.*, 2012).

Foods which contain novel ingredients which provide some health benefits beyond basic nutrition are called Functional foods. Functional food is also fortified with nutrient contents present in the food. The probiotic functional foods provide several health benefits that help in maintaining a good balanced gut microflora and which decrease the infection of gut pathogens.

Currently, the demand of probiotic functional foods is growing quickly due to increasing knowledge of consumers concerning the positive impact of suchfoods on health. The major challenge in developing probiotic functional food is to maintain optimal dosage of probiotic during the time of delivery because several factors interfere during processing and storing.Sensory properties of certain foods can be changed by some probiotic strains.

Probiotic

Probiotic strains are using as food ingredient to enhance flavour and bio-functionality of foods by synthesis of bioactive peptides. Several chronic diseases like diabetes, obesity, gut related cancer and cardiovascular diseases are linked with poor food habits. To overcome these life-threatening diseases, market demand to develop novel probiotic functional food that can enhance bio-functionality of foods is growing rapidly. The LAB (lactic acid bacteria) contains large group of genus lactobacillus. The genus lactobacillus contains large number of species belonging to human, animal and plant foods. The *lactobacilli* show phenotypic, biochemical and physiological variability (Popova *et al.*, 2012). Various foods like yoghurt, sourdough breads, silage, table olives and sauerkraut are manufactured using lactobacilli as a starter culture that ferment and also used as natural preservatives in non-fermented vegetables. The probiotic *lactobacilli* attributes health benefits to host that colonised the gastrointestinal tract (Figure 1.1). These probiotic attributes are immune-stimulation, decreases pH of the local environment, trapping of mutagenic

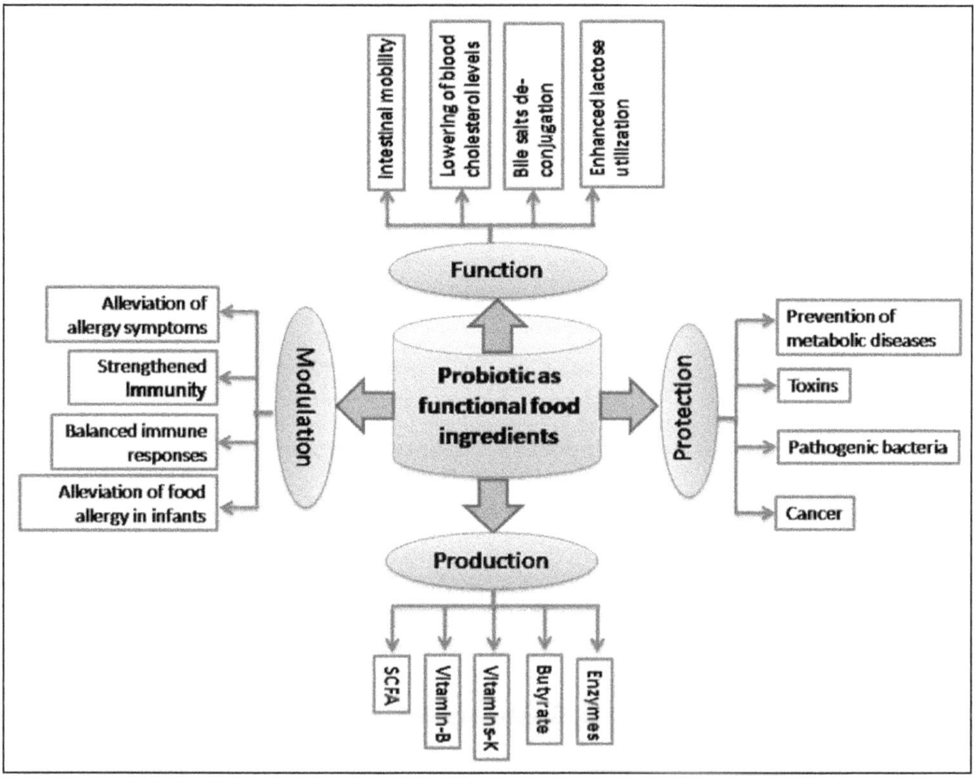

Figure 1.1: Health Benefits of Probiotic.

compounds, bacteriocins production, inhibition of pathogenic bacterial adhesion to the epithelial cell and hypocholesteraemic effects (Sanders 1993). The *Lactobacillus* species, generally recognised as safe (GRAS), use in future will possibly include the screening of strains that have the ability of producing specific biomolecules and enzymes and other epitopes displaying on bacterial surface for the purpose of delivery. The production of viable culture at large-amounts will depend on the cost-effective methods (Wyszyñska *et al.*, 2015; Walker and Girgis, 1999). All these applications exposed the *lactobacilli* to various environmental stresses, which in turn may affect cells in both the physiological status and its properties. Therefore, proteomics would be helpful in generating valuable knowledge about this environment stress of the cell.

The Genus *Lactobacillus*

Lactobacillus genus contains approximately over 100 species. The detailed properties are mentioned in Figure 1.2 which is collected from www.ncbi.nl.nih.gov/taxonomy. Their normal habitats are plants, dairy products, fermented foods, sourdough breads, meat products *etc.* Since, most of them are acid tolerant and anaerobic or aero-tolerant they easily survive in normal microflora of humans. They can be found in oral cavities, gastro intestinal tract and Vagina.

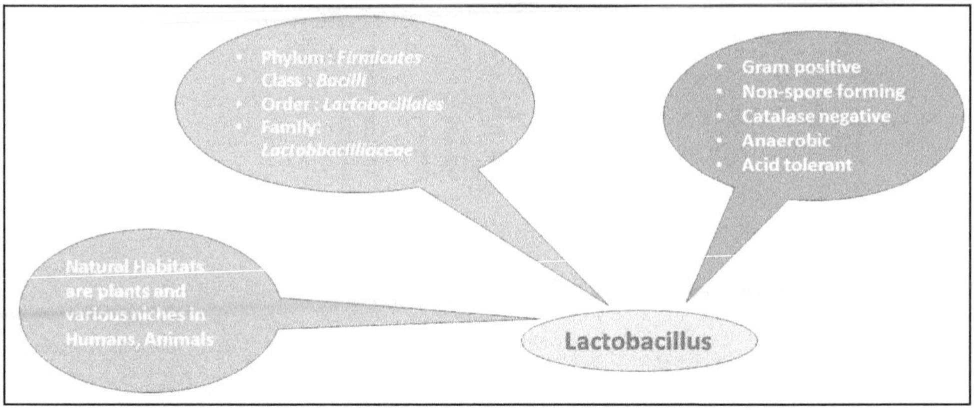

Figure 1.2: Properties of Genus Lactobacillus.

Establishment and Colonization of Probiotic Lactobacilli

The organisms can colonize in the Gastrointestinal tract of humans within a few days just after their birth. Factors such as dietary constituents, type of delivery and gestational age influence the course of colonization.Approximately over two-week period the initial phase of colonization occurs. Low colonization rates of lactobacilli in infant GI has been reported by some authors (Balmer and Wharton, 1989; Ellis-Pegler *et al.*, 1975; Matsumiya *et al.*, 2002), whereas, others found significantly higher quantities (10^{7-9} CFU/g) of lactobacilli infaeces of infants.The Table 1.3 shows the colonization potential of probiotic *Lactobacilli.*

Table 1.3: Colonization Potential of Probiotic *Lactobacilli* in GI tract of Humans

AGE	SAMPLE	QUANTITY	REMARKS	REFERENCES
2 months	Stools	10^{7-9} CFU/g	*L.rhamnosus* and *L.gasseri* were most common.	Yadav et al., 2013; Ahrne *et al.*, 2005; Hall *et al.*, 1990; Kleessen *et al.*, 1995
6 months	Stools	10^{7-9} CFU/g	*L.rhamnosus* was most abundant followed by *L.paracasei*, *L.plantarum*, *L.fermentum* and *L.gasseri*.	Yadav et al., 2013; Ahrne *et al.*, 2005; Hall *et al.*, 1990; Kleessen *et al.*, 1995
34 days – 7 months	Oral Cavity	-	Lactobacilli are present in oral cavity in 47% of the infants.	Plonka *et al.*, 2012
2-3 years	Saliva	$2x10^3$ to $4x10^4$ CFU/mL	Lactobacilli are in present in saliva in 50% of the children.	Carlsson *et al.*, 1975; Köhler *et al.*, 1984
Adults	Intestine & Oral	-	In adults, *lactobacilli* form a minor portion of both the intestina and oral microbiota.	Dal Bello *et al.*, 2003; Kimura *et al.*, 1997; Lebeer *et al.*, 2008

Lactobacillus as a Probiotic Strain

L.acidophilus, L.bulgaricus along with *L.rhamnosus etc.* are having probiotic characteristics and are commercially used in many milk and milk products. *L. delbrueckii subsp. bulgaricus* is used in many fermented milk products and also used in milk preservation which produced antibacterial compound. Another strain of probiotic bacteria is *L.rhamnosus* GG, isolated from healthy human gut. The strain *L rhamnosus* GG includes several prominent health benefits like adhesion to human intestinal mucosal cells, bacteriocin production, immunomodulation and inhibition of diarrhoea *etc.* Probiotic strains of lactobacilli obtained from the human origins includes *L.johnsonii* NCC 533, *L. acidophilus* strain LB, *L. casei* DN-114 001, *L. acidophilus* NCFM, *L. reuteri, L. Plantarum* 299v, *L. fermentum* ME-3 and *L. acidophilus* CL1285.

Probiotic Attributes of *Lactobacilli*

The genus *lactobacilli* have several health promoting functions like cholesterol lowering, lactose intolerance, immunomodulation to cure IBD and diarrhoea. Studies revealed that the *L. rhamnosus* GG improve IL-10 production, which helps in alleviation of diarrhoea in small children (Duary*et al.*, 2012; Pessi*et al.*, 2000, de Roos and Katan *et al.*, 2000). Similarly, Harmful faecal enzyme activity was reduced when yoghurt with *Lactobacillus* GG was administered. Many probiotic *lactobacillus* strains canreduce the levels of serum cholesterol as confirmed by different investigators. Cholesterol level has been reduced when yoghurt containing BSH positive *L. plantarum* was consumed. (De Smet *et al.*, 1994). In a different study, intake of *L. reuteri* also helpful in serum cholesterol levels reduction up to 22-23 per cent. Similarly, implantation of microencapsulated genetically engineered probiotic bacterium (*L. plantarum* 80) where BSH was over expressed revealeddecrease in serum cholesterol (Jones *et al.*, 2004).

Beyond such implications on health, lactic acid bacteria are also vastly used in food industry,industrial fermentations, nutraceuticals and pharmaceuticals. Hence, there is lot of opportunity for isolation and identification of novel strains of lactobacilli and new methodologies which are rapid and specific for extensive applications in food and dairy industry. Increasing consumer demand and growing market involved in commercialization and booming of healthy functional foods. Accurate identification of probiotic strain has become tremendously important to exclude the prospect of false health claims and to resolve arguments which concerns their identity as probiotic strains.

Bile Salt Hydrolase BSH: A Probiotic Marker

Bile salt hydrolase BSH also known as Cholyglycine hydrolase (Classification E.C.3.5.1.24). It belongs to family of Cholyglycine hydrolase subfamily (Pfam: PF02275). This enzyme is prevalent among gastrointestinal microflora and other bacteria in the archaea kingdom. Newly synthesized Bile acid is secreted from liver into duodenum directly. When these Conjugated Bile salts pass through intestinal tract, they get deconjugated, dehydroxylated, dehydrogenated and desulphated by BSH produced by gut microbes. Many bacterial genera present in GI tract can express BSH. Interestingly activity of BSH was only observed in stationary phase cells of lactobacillusisolates like *L. salivarius* AK 21 and *L. aviarus* AK 311. When BSH activity was investigated in mouse gut the results confirmed that *lactobacilli* are predominant producers of BSH approximately 86 per cent of total BSH activity was found in ileum and 74 per cent in cecum (Kumar *et al.*, 2011a, b)

Bile Salt Hydrolase's (BSH) and its Characterisation

Bacterial species which can deconjugate Bile salts may use electron acceptors such as amino acid taurine (Tannock *et al.,* 1997). BSH protein produced by gut microbes in some extent help in survival and colonization of bacteria in intestine since it can protect bacteria from bile acid shock (Moser *et al.*, 2001). It brings us to an assumption that BSH activity can be asurvival method for bacteria because it involves in detoxification of conjugated bile salts. It is observed that free bile acids interfere with growth of common gut microbes but not with conjugated bile acids (Kumar *et al.*, 2012). BSH gene in *L. Johnsonii* is encoded by two separate loci, of which one encodes for cbsH? which is a part of operon of three genes encoding for Bile hydrolysis and for its transporter, this locus in conserved in other Lactic acid bacteria species. The other locus encodes for cbsHβ which is not arranged in tandem. Lactobacilli with in the species of *L. acidophilus, L. brevis, L. buchnerii, L. fermentum, L. gasseri*and *L. plantarum* can express a variable BSH phenotype.

Significance of BSH on Health

Hydrolysis of bile salts performs a gate keeping function. Synthesis of bile acid takes place in liver from cholesterol during regular enterohepatic circulation (EHC). About 4.0 g bile salts is secreted into the duodenum two times after each meal which helps in digestion of lipids. Sterol transformation can occur only after deconjugation of Bile salts (Park *et al.*, 2007; Kumar *et al.*, 2011). The conjugated bile salts amphipathic in nature, capture cholesterol and fats; and ease and enhance their absorption by the intestinal epithelium. Sodium dependent transporters actively transport bile salts through the epithelium into the blood. Later, they reach liver to complete the enterohepatic cycle. Majority of secondary bile salts get absorbed in the lower intestinal tract during intestinal transit and reached to liver through the hepatic portal vein. The bile salt pool in liver is refilled or replaced with synthesis of new bile from serum cholesterol. Factors such as dosage, mode of therapeutic administration, frequency, composition of microcapsule membraneandpotential effects of byproducts must be considered in Cholesterol therapy (Tannock *et al.*, 1994). The mechanism of BSH activity in the GI tract is shown in Figure 1.3.

Deconjugation of Bile Salts: A Way to Lower Cholesterol by Probiotic LAB

Certain species of *lactobacilli* which reside in the intestinal tract have the ability to deconjugate both glycocholic and taurocholic acids in anaerobic conditions. Deconjugated bile acids are not efficient in capturing cholesterol. Hence, the conjugated bile acids are more efficient in intestinal absorption of cholesterol (Kumar *et al.*, 2011). Enhanced bile acids excretion incites the synthesis of substitute bile acids from cholesterol; therefore, cholesterol levels in the body going to reduce. If lactobacilli cells can incorporate the deconjugated bile salts and cholesterol in their cell membrane during growth in intestine, then it can increase the potential to

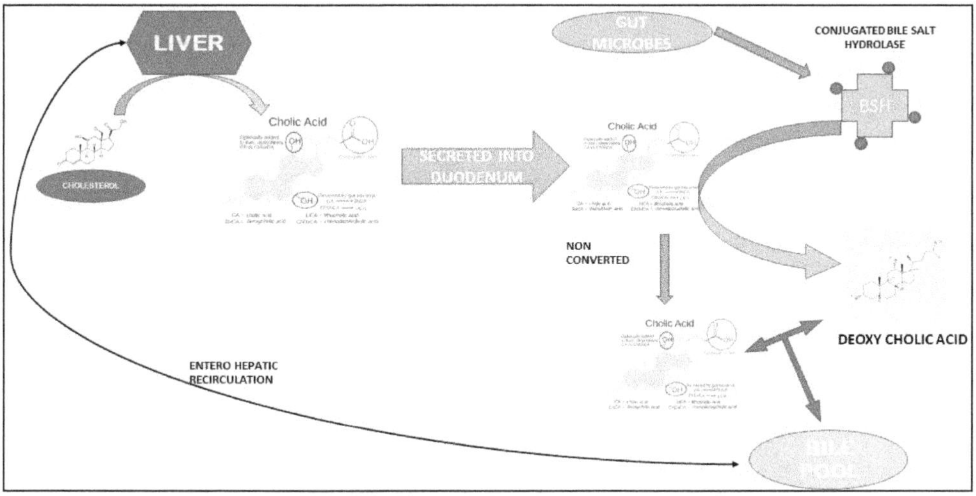

Figure 1.3: Mechanism of BSH Activity in the GI Tract.

decrease the serum cholesterol. Absorption on the surface of cells makes cholesterol unavailable for absorption into the blood.

Lactobacillus acidophilus is predominantly present in the Gastrointestinal tract of healthy humans. Bile tolerance is a prerequisite characteristic of *lactobacilli* strains. *Lactobacillus acidophilus* has capability to reduce the levels of serum cholesterol which is one of the important beneficial health attributes. *Lactobacillus plantarum* 80 which is microencapsulated can efficiently break-down the bile slats in *in vitro* physiological condition.

Commercial Uses of Probiotic *Lactobacillus*

Several probiotic lactobacilli are commercially used in many fermented dairy and non-dairy foods. The *Lactobacillus casei* Shirota (YIT9029), *L. acidophilus*, *Lactobacillus brevi*and *L. fermentum* can change the texture, flavour of the final product.They are also found in other fermented products mostly in milk and milk products and helps in the preservation. *Lactobacillus rhamnosus* GG (ATCC 53103) (Valio Ltd. Finland)isolated in 1983 from the intestinal tract of a healthy human being produces bacteriocins (Simova*et al.*, 2008) and therefore, acts as a bacteriocide. (Lievin and Servin, 2014; Ljungh and Wadström, 2006; Auclair*et al.*, 2015). Some commercial therapeutics like Neolactoflorene use lab bacteria in curing dismicrobism and inflammatory bowel disease. Bifilact (Fidia Nutrceutical S.p.A, padova Italy) a nutraceutical consist of *L.acidophilus*, *L. debreuckiisubsps* and *L. sporogenes* which usually help in curing diarrhoea, gastroduodentis.

Lactobacilli as Delivery Agents for Vaccines or Therapeutics

Lactobacilli generally regarded as safe, are generally consumed through many fermented food products, mostly milk and milk products which are consumed in huge amounts and popular among consumers. Some bacterial species of *Lactobacilli* are considered as attractive delivery vehicles for therapeutic products or vaccines (Liu *et al.*, 2014). Foreign therapeutic molecules can easily be expressed on surface of lactobacilli. Some studies showed that when mice were administered with recombinant *L. casei* cells which are expressing FliC a flaggelin protein from *Salmonella entericaserovar Enteriditis* were bestowing defensive immunity against consequent Salmonella infection (Kajikawa *et al.*, 2007). Genome sequencing along with expression refinement has been carried out to optimise the vehicle, with certain view to develop microbicides for mitigation of transmission of HIV in Heterosexuals and viral pathogens which are transmitted mucosally (Liu *etal.*, 2008). Figure 1.4 depicts some properties of *Lactobacillus* as delivery agents for therapeutics.

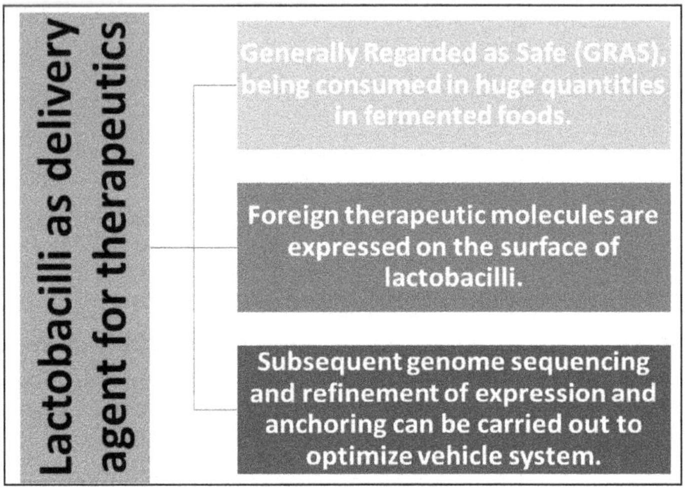

Figure 1.4: Lactobacillus as Delivery Agent for Therapeutics.

Conclusion

Probiotics bacteria can be considered as food ingredient in functional foods or nutraceuticals because it provides health benefits along with the basic nutrition. Currently, most of the probiotic functional foods products are of genus *Lactobacilli* and *Bifidobacterium* belong to thelactic acid bacteria. Probiotic microbes showed several beneficial effects that maintained gut microflora, which promotes wellbeing and defend from the invasive potential pathogens present in the gut. It would be used in therapeutic treatment or prophylactic treatment for several types of diseases like lactose intolerance, diarrhoea, prevention of colon cancer, reduction of blood cholesterol level, inflammatory bowel disease, cancer and inhibition of *Helicobacter pylori* infection. Improvement of the properties of imbalanced indigenous microbiota forms the rationale of probiotic therapy. Nevertheless, the probiotic functional foods can be used as an alternative to the conventionaldrug-based therapy andfor routine health care and disease management. The probiotic functional foods are safe natural products, being cost prohibitive and there is no undesirable side effect. In this context, probiotics have recently emerged as safe, cost effective and easily affordable as prophylactics or biotherapeutics for general health promotion and specifically in the management of the target disease particularly gut related inflammatory metabolic disorders.

REFERENCES

Ahrne, S., Lonnermark, E., Wold, A.E., N. Aberg, Hesselmar, B., Saalman, R., Strannegard, I.L., Molin, G. and Adlerberth, I. 2005. *Lactobacilli* in the intestinal microbiota of Swedish infants, *Microbes and infection/Institut Pasteur* 7, 1256-1262.

Auclair, J., Frappier, M., and Millette, M. 2015. Lactobacillus acidophilus CL1285, *Lactobacillus*casei LBC80R, and *Lactobacillus*rhamnosus CLR2 (Bio-K+): characterization, manufacture, mechanisms of action, and quality control of a specific probiotic combination for primary prevention of Clostridium difficile infection. *Clin Infect Dis*; 60(suppl 2): S135–43.

Balmer, S.E., and Wharton, B.A. 1989. Diet and faecal flora in the newborn: breast milk and infant formula. *Arch Dis Child*, 64, 1672-1677.

Carlsson, J., Grahnen, H., and Jonsson, G. 1975. *Lactobacilli* and *streptococci* in the mouth of children. *Caries Res*, 9, 333-339.

Dal Bello, F., Walter, J., Hammes, W.P. and Hertel, C. 2003. Increased complexity of the species composition of lactic acid bacteria in human feces revealed by alternative incubation condition. *Microb Ecol*, 45, 455-463.

De Roos, N.M. and Katan, M.B. 2000. Effects of probiotic bacteria on diarrhea, lipid metabolism, and carcinogenesis: a review of papers published between 1988 and 1998. *American J Clin. Nutrition*, 71: 405-11.

De Smet, I., Van Hoorde, L., De Saeyer, N., VandeWoestyne, M. and Verstraete, W. 1994. In vitro study of bile salt hydrolase (BSH) activity of BSH isogenic *Lactobacillus plantarum* 80 strains and estimation of cholesterol lowering through enhanced BSH activity. Microbial Ecol. *Health Disease*, 7: 315-29.

Duary, R.K., Bhausaheb, M.A., Batish, V.K., and Grover, S. 2012. Anti-inflammatory and immunomodulatory efficacy of indigenous probiotic *Lactobacillus plantarum* Lp91 in colitis mouse model. *MolBiol Rep*, 39(4): 4765–75.

Ellis-Pegler, R.B., Crabtree, C., and Lambert, H.P. 1975. The faecal flora of children in the United Kingdom. *J Hyg (Lond)* 75, 135-142.

Hall, M.A., Cole, C.B. Smith, S.L., Fuller, R., and Rolle, C.J. 1990. Factors influencing the presence of faecal lactobacilli in early infancy. *Arch Dis Child*, 65, 185-188.

Jones, M.L., Chen, H., Ouyang, W., Metz, T. and Prakash, S. 2004. Microencapsulated genetically engineered *Lactobacillus plantarum* 80 (pCBH1) for bile acid deconjugation and its implication in lowering cholesterol. *J. Biomed Biotechnology*, 1: 61-9.

Kajikawa, A., Satoh E. *et al.*, 2007. Intragastric immunization with recombinant *Lactobacillus casei* expressing flagellar antigen confers antibody-independent protective immunity against Salmonella enterica serovar enteritidis. *Vaccine* 25(18): 3599–3605.

Kimura, K., McCartney, A.L., McConnell M.A., and Tannock, G.W. 1997. Analysis of fecal populations of *bifidobacteria* and *lactobacilli* and investigation of the immunological responses of their human hosts to the predominant strains. *Appl Environ Microbiol*, 63, 3394-3398.

Kleessen, B., Bunke, H., Tovar, K., Noack, J. and Sawatzki, G. 1995. Influence of two infant formulas and human milk on the development of the faecal flora in newborn infants. *ActaPaediatr*, 84, 1347-1356.

Kohler, B., Andréen, I. and Jonsson, B. 1984. The effect of cariespreventive measures in mothers on dental caries and the oral presence of the bacteria *Streptococcus*mutans and lactobacilli in their children. *Arch Oral Biol* 29, 879-883.

Kumar, M., Verma, V., Nagpal, R., Kumar, A., Behare, P.V., Singh, B. and Aggarwal, P.K. 2011a. Anticarcinogenic effect of probiotic fermented milk and Chlorophyllin on aflatoxin-B1 induced liver carcinogenesis in rats. Brit J Nutr (doi: 10.1017/S0007114511003953).

Kumar, M., Verma, V., Nagpal, R., Kumar, A., Gautam, S.K., Behare, P.V., Grover, C.R. and Aggarwal, P.K. 2011b. Effect of probiotic fermented milk and chlorophyllin on gene expressions and genotoxicity during AFB1-induced hepatocellular carcinoma. *Gene*, 490: 54-59.

Kumar, R., Grover, S. and Batish, V.K. 2011. Hypocholesterolaemic effect of dietary inclusion of two putative probiotic bile salt hydrolaseproducing*Lactobacillus plantarum* strains in Sprague-Dawley rats. *Br J Nutri*, 105: 561–73.

Kumar, R., Grover, S. and Batish, V.K. 2012. Bile salt hydrolase (Bsh) activity screening of *Lactobacilli*: in vitro selection of indigenous *Lactobacillus* strains with potential bile salt hydrolysing and cholesterol-lowering ability. *ProbiotAntimicrob Proteins*, 4(3): 162–72.

Lebeer, S., Vanderleyden, J., and De Keersmaecker, S.C. 2008. Genes and molecules of *lactobacilli* supporting probiotic action. *MicrobiolMolBiol Rev*, 72, 728-764.

Lievin-Le Moal, V. and Servin, A.L. 2014. Anti-infective activities of lactobacillus strains in the human intestinal microbiota: from probiotics to gastrointestinal anti-infectious biotherapeutic agents. *ClinMicrobiol Rev*, 27: 167–99.

Liu, J.K., Wei, C.H., Hou, X.L. and Yu, L.Y. 2014. Passive protection of mice pups through oral or intranasal immunization of dams with recombinant *Lactobacillus casei* vaccine against ETEC F41. *Res Vet Sci.*, 96(2): 283-87.

Liu, X., Lagenaur, L.A. *et al.*, 2008. Engineering of a human vaginal *Lactobacillus* strain for surface expression of two-domain CD4 molecules. *Appl Environ Microbiol*, 74(15): 4626–4635.

Ljungh, A. and Wadström, T. 2006. Lactic acid bacteria as probiotics. *Curr Issues IntestMicrobiol*, 7: 73–89.

Matsumiya, Y., Kato, N., Watanabe, K., and Kato, H. 2002. Molecular epidemiological study of vertical transmission of vaginal *Lactobacillus* species from mothers to newborn infants in Japanese, by arbitrarily primed polymerase chain reaction.

Journal of infection and chemotherapy: *official journal of the Japan Society of Chemotherapy*, 8, 43-49.

Moser, S.A. and Savage, D.C. 2001. Bile salt hydrolase activity and resistance to toxicity of conjugated bile salts are unrelated properties in *lactobacilli*. *Appl. Environ. Microbiol.*, 67(8): 3476-80.

Park, Y.H., Jong, G.K., Young, W.S., Sae, H.K. and Kwang, Y.W. 2007. Effect of dietary inclusion of *Lactobacillus acidophilus* ATCC 43121 on cholesterol metabolism in rats. J MicrobiolBiotechnol 17: 655–662.Pessi, T., Sütas, Y., Hurme, M. and Isolauri, E. 2000. Interleukin-10 generation in atopic children following oral *Lactobacillus rhamnosus* GG. *Clin Exp Allergy*, 30: 1804-1808.

Plonka, K., Pukallus, M., Barnett, A., Walsh, L., Holcombe, T. and Seow, W. 2012. Mutants*streptococci* and *lactobacilli* colonization in predentate children from the neonatal period to seven months of age. *CariesRes*, 46, 213- 220.

Popova, M., Molimard, P., Courau, S., Crociani, J., Dufour, C., Vacon, F.L. and Carton, T. 2012. Beneficial effects of probiotics in upper respiratory tract infections and their mechanical actions to antagonize pathogens. Journal of Applied Microbiology, 10: 1365-2672.

Ringel, Y., Quigley, E., and Lin, H. 2012. Probiotics and gastrointestinal disorders. *Am J GastroenterolSuppl*; 1: 34–40.

Sanders, M.E. 1993. Summary of conclusions from a consensus panel of experts on health attributes of lactic cultures: significance to fluid milk products containing cultures. *J. Dairy Sci.*, 76, 1819–1828.

Simova, E.D., Beshkova, D.M., Angelov, M.P. and Dimitrov, Z.P. 2008. Bacteriocin production by strain *Lactobacillus delbrueckiissp. bulgaricus* BB18 during continuous prefermentation of yogurt starter culture and subsequent batch coagulation of milk. *J IndMicrobiolBiotechnol*, 35: 559–67.

Tannock, G.W., Bateup, J.M., and Jenkinson, H.F. 1997. Effect of sodium taurocholate on the in vitro growth of *lactobacilli*. *Microbiology Ecology*, 33, 163-167.

Tannock, G.W., Tangerman, A., Van Scack, A. and McConnell, M.A. 1994. Deconjugation of bile acids by *lactobacilli* in the mouse small bowel. *Appl. Environ. Microbiol.*, 60: 3419–3420.

Walker, D.C., Girgis, H.S., and Klaenhammer, T.R. 1999. *Appl. Environ. Microbiol.*, 65, 3033–3041.

Wyszyñska, A., Kobierecka, P., Bardowski, J., and Jagusztyn-Krynicka, E.K. 2015. Lactic acid bacteria-20 years exploring their potential as live vectors for mucosal vaccination. *ApplMicrobiolBiotechnol*, 99(7): 2967-7.

Yadav, A.K., Tyagi, T., Kaushika, J.K., Saklanib, A.C., Grovera, S., Batish. V.K. 2013. Role of surface layer collagen binding protein from indigenous *Lactobacillus plantarum* 91 in adhesion and its anti-adhesion potential against gut pathogen. *Microbiol. Res.*, 168(10): 639-645.

Chapter 2

Microbial Enzymes and its Applications: Bacterial Calpain as a Possible Member of GRAS Family in Meat Industry

Prachi Gupta[1,2] *and Ashwani Mittal*[1*]

[1]*Skeletal Muscle Lab, Department of Biochemistry, Institute of Integrated and Honors Studies, Kurukshetra University, Kurukshetra, Haryana*

[2]*Department of Biotechnology, Govt College for Girls, Sector 14, Gurugram, Haryana*

**E-mail: mittala@kuk.ac.in*

ABSTRACT

Enzymes are biological catalysts coordinating the intricate meshwork of biochemical reactions. After realizing the vast applications of isolated animal and plant enzymes in the various commercial human industries, the competitive race started for the discovery of some better, less expensive and more readily available substitute sources for the enzymes. In this search race the culturing of microbial diversity has attracted the attention, as it may serve as the pool for natural and human designed enzymes. Enzymes from microbes (including bacteria, moulds, fungi, yeasts) have offered high scope of their commercialization attributed to their vast availability, easiness in the production, high purification rate and complete inactivation after using simple processes. The journey of cultured microbial enzymes is full of number of challenges beginning from the identification of correct species and locating the target enzyme in the ocean of millions of enzymes and ending with the commercialized packing of enzymes without adverse reactions in final trials. Genetic manipulation, recombinant DNA technology and codon optimization have completely changed the scenario as now it is possible to create, produce and extract commercial important artificial enzymes from microbial world. Immense research and field trials have finally deepen the roots of microbial enzymes in every possible sector of humanization ranging from clothing, food, and medicine up to future science and technology. Another hunt was to discover homologue of those human enzymes/proteins in bacterial world which are playing critical role in treating diseases. Extensive research requires a large quantity of such protein from animal sources which is unlikely present in limited amount or where extraction is an issue. Calpain is one such important proteolytic enzyme involved in number of physiological and pathophysiological conditions related to muscle. The involvement of calpain system in muscle, make it important in the meat industry. Till the recent discovery, it was thought to be widely distributed in eukaryotes only. But now a bacterial strain *Porphyromonas gingivalis* has also been discovered carrying and excreting a calpain homologue. This discovery has opened the future possibility of bacterial calpain research in meat industry.

Keywords: *Microbial enzyme, Eukaryotic calpain, Bacterial calpain homologue, Meat industry.*

Introduction

Enzymes are biological catalysts ubiquitously present in the biosphere making life possible on this earth. Enzymes are actually *"a revert action of nature's impatience"* as their presence in concerted fashion perfectly coordinate the intricate meshwork of biochemical reactions. Microbial world occupy the largest domain hence serve as the biggest factory outlet for the enzymes. Diversity due to versatility in terms of chemical, physical and biological features of enzymes have made one species of microorganism different from other and even different strains of same species. Enzymes are also light spot catcher due to their specificity in action, particularly in terms of substrate and type of reactions they commence. Chemically enzymes are complex macromolecules made up of protein with or without non-protein cofactor group (co-enzyme or prosthetic or metal ions).

Besides playing important role in maintaining life, microbial enzymes are constantly being exploited by humans long before the actual role of enzymes was understood. Brewing where barley malt was used for starch conversion and leather making where animal dung was used for bating the hides, are some perfect examples when the world was unaware of the actual causative agents (Underkofler, Barton, and Rennert, 1958). After realizing the merits of enzymes from animal and plant sources, there arises a competitive era of producing better, less expensive and more readily available sources. It was then decoded that microorganisms possesses that potential and led to the development of microbial enzymes on large scale. Dr. Jokichi Takamine (1894, 1914) and Boidin and Effront (1917) were the pioneers who have realized the possibility to introduce the cultivated enzymes with the technology at industry level. Inauguration of commercial enzymes from microorganisms has replaced enzymes extraction from animal and plant sources at number of places. Microbial enzymes can now be credited for the human modernization and industrialization evolution. Easiness in the production and purification of enzymes from microbes (including bacteria, moulds, fungi, yeasts) and also, their availability have offered high scope of their commercialization. Complete inactivation of enzyme using simple processes adds crown on its controlled utilization so to prevent unfavorable alterations (Saxena, 2015).

Though production of microbial enzymes sounds easy, but it required hard core research identifying the location of the target enzyme. After locating the target enzyme or particular strain of microbe, the next challenge was to isolate and extract from the pool of microbial enzymes. Every single microbe produces diversified list of enzymes of different nature including hydrolyzing, reducing or oxidizing and so on. Also, some enzymes are intracellular and some are secreted out for various reasons. Technically, a single strategy for extraction is not going to work with every enzyme. Depending upon the strain of the microorganism (*i.e.* fungus, bacteria, mold *etc.*) and characteristic of enzyme, different methods have been evolved and are still evolving. Surface culture method was the most primitive method where solid substrates were used for cultivation of mold (fungus). Later, submerged culture methods have gained a lot extensive use for bacteria as well as mold. Various modifications regarding culture flasks, inlet for sterile air, vigorous agitation, and fine aeration were implemented for better commercialization. There exist merits and demerits in each method, and thus choice of method depends on the culture type, plant equipment, relative yields, space, convenience and applications. In this direction membrane filtration technique has augmented downstream processing resulting in the isolation of purified enzymes from the culture plants (Vittaladevaram, 2017). Now-a-days, microbial enzymes have conquered majorities of industries including food, textiles, diagnostics and therapeutics (Sanchez and Demain, 2011).

Microbial Enzymes in Textiles

Textile industry is one of the largest industries worldwide providing the third most important basic necessities of life *i.e.* clothing. Textiles fibers include fabric from natural origin such as cotton, silk, wool *etc.* and from artificial origin such as nylon, rayon, or mixed fiber. This growing industry has traditionally been using considerably amounts of energy source, water and harsh chemicals. Textile industry is a major chemical processing sector contributing a huge share in the global pollution and hence this industry is always been in high environmental risk alert zone. But after the evolution of white biotechnology, new products and processes using renewable resources and green technologies are known for low energy consumption and eco-friendly waste products. This white/industrial biotechnology is nothing but the use of enzymes at large commercial level to reduce the energy burden and time consumption. Enzyme involvement in textile industry has thus provided positive benefits in terms of environmental impact and product quality (Nierstrasz and Warmoeskerken, 2003). Hydrolases (including amylases, cellualses, proteases, pectinases, lipases, esterases) and oxidoreductases (including catalase, laccase) are principle enzymes empowering textile industry. Microorganisms and fungi serve as a large factory for these enzymes. Microbial enzymes are replacing harsh chemicals and excess water waste in number of textile processing such as removal of starch, denim finishing, degradation of hydrogen peroxide, softening, bio-stoning, textile bleaching, decolourization of textile effluents and many more (Mojsov, 2012).

Microbial Enzymes in Pharmaceuticals

In majority of medical ailments, mutation, overproduction, underproduction or deletion of some critical enzymes have been found responsible. And as natural the treatment resides in prescribing external enzyme or its inhibitor as the medication or boosting internal machinery for regaining the normal state. Microbial enzymes are fixing their roots in pharmaceutical industry more strongly than plant and animal

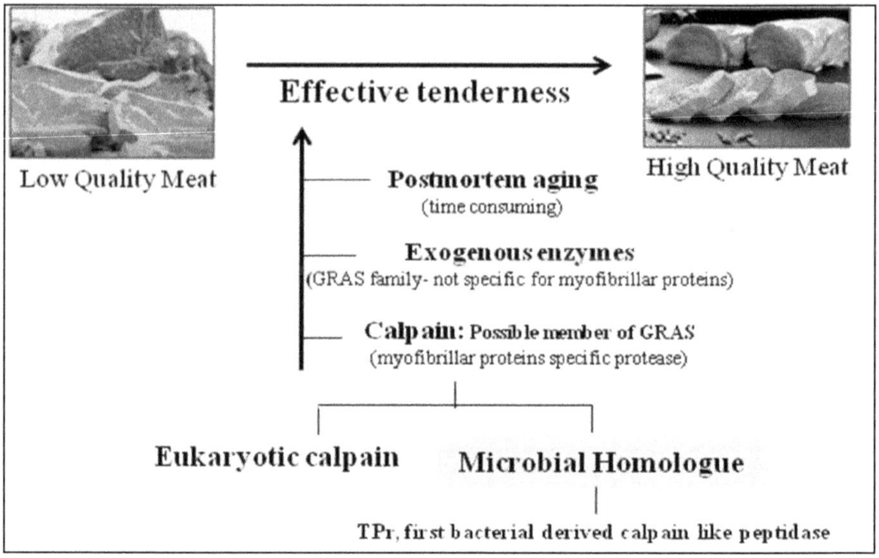

Figure 2.1: Key Factors Behind Effective Meat Tenderness.

enzymes (Gonzalez and Isaacs, 1999). The obvious reasons may be that microbial enzymes are more stable and active, and also microorganisms can be cultured for enzyme production in short time by fermentation at larger scale. Biochemical diversity and ease in the genetic manipulation has offered the production of desired therapeutic enzyme, so that there are minimal side effects. Microbial enzymes have served as a therapeutic replacement strategy for some genetic deficiencies in 1960s (Vellard, 2003). Capitalization of enzyme as drugs begins in the late 19th century when crude proteolytic enzymes were prescribed to rectify gastrointestinal disorders. This triggers the story and became the milestone covering nuclease of *Bacillus pyocyaneus* used for killing anthrax bacilli. Now enzymes in individual or in different combination with other enzymes are available as pills, capsules, injections and powder form. Microbial enzymes are gaining success in different pathological conditions such as- (a) removing cytotoxic agents from blood circulation; (b) treating oncolytics and thrombolytics; (c) acting as anti-coagulants; (d) replacements for metabolic deficiencies; (e) promising in genetic diseases including Gaucher, SCID, CF, PFU, *etc.*; (f) acting as anti-neoplastic in treating cancer; (g) treating protozoal, fungal and bacterial infectious diseases (Kaur and Sekhon, 2012; Mane Prajakta and Vidya, 2015; Sabu, 2003). Profound examples are asparginase for acute lymphocytic leukemia (Gurung, Ray, Bose, and Rai, 2013); chitinase as potent antimicrobial (Fusetti *et al.*, 2002); collagenase healing burns and dermal ulcers (Ostlie *et al.*, 2012); lipase as digestive aids (Matsumae, Furui, and Shibatani, 1993); nattokinase reducing blood clotting and cardiovascular disorders (Selvarajan and Bhatnagar, 2017); serratiopeptidase useful in pain and inflammation and others thrombolytic drugs (Banerjee, Chisti, and Banerjee, 2004) *etc.*

Microbial Enzymes in Food Industry

Microbial enzymes have gained a lot of importance in food industry by improving the characteristics of final products including flavor, texture, odor, shelf life and more (Raveendran *et al.*, 2018). Also, enzymes can be used for optimizing the process, improving the bioavailability of minerals, subsiding unfavorable changes and more. In brief, enzymes fulfill three major roles *i.e.* (a) additives; (b) ingredients and (c) favoring in food processing industry including dairy, meat, baking, beverages and fermented products (Fraatz, Ruhl, and Zorn, 2014; Mieszczakowska-Frąc, Markowski, Zbrzeźniak, and Płocharski, 2012; Teh *et al.*, 2014). Widely used enzymes in dairy industry for removing lactose, cheese ripening, clarification *etc.* are lactase, lipase, protease, pectinase, asparaginase and transglutaminase. Also, enzymes like phytase, laccase, and xylanase-cellulase complex are attracting attention for enhancing bioavailability of minerals, reducing off-odours and production of nutraceuticals.

After dairy industry, meat industry represents the major market value throughout the world. As per UN Food and Agriculture Organization (FAO) database 2017, there is an increasing trend in the per capita meat consumption which indicates much faster increasing rate of total meat production as compared to population growth (Ritchie and Roser, 2019). Besides juiciness and flavor, meat tenderness is the most important palatability trait of meat quality that attracts consumers and money (Miller, Carr, Ramsey, Crockett, and Hoover, 2001). However, variability in meat tenderness is one of the major issues in the meat industry that influences their profit. Studies show that tenderness commonly depends upon three main factors *i.e.* connective tissue (collagen content), sarcomere length and proteolytic degradation of structural myofibrillar proteins (Arshad *et al.*, 2016). As in other food industries, enzymes here also reported to play crucial roles in converting muscle-to-meat by inducing complex physico-chemical changes in muscle through inducing proteins degradations. Thus a large number proteases were explored which are in use in meat industry isolated from various sources including plants and bacteria. Till date there are five exogenous protein degrading enzymes (*i.e.* papain, ficin, bromelain, *Aspergillus oryzae* protease and *Bacillus subtilis* protease) that have been considered as GRAS (Generally Recognized as Safe) by Food Safety Inspection services and are added to meat for improving tenderness (Bekhit, Hopkins, Geesink, Bekhit, and Franks, 2014) even then variation in tenderness is still a challenging issue for meat industry. Available reports also suggest that the most primitive method for having effective tenderness is postmortem aging where another proteolytic enzyme *i.e.* calpain is the only protease responsible for degradation of key myofibrillar proteins (including titin, nebulin, filamin, troponin-T and desmin) within muscles fibers (Bhat, Morton, Mason, and Bekhit, 2018; Coria, Carranza, and Palma, 2018; Lian, Wang, and Liu, 2013). Postmortem proteolysis is a time consuming event that may offer unwanted changes or attract negative entities, hence is associated with flaws.

Thus a smarter tenderization strategy should target the isolation of calpains and its application as an exogenous enzyme for stronger economic impact or increasing consumer's purchase. Therefore, the next part of this chapter covers the brief description about eukaryotic calpain and also regarding its discovery as bacterial homologue.

Eukaryotic Calpains and Related Challenges

Calpains are cytosolic Ca^{2+}-dependent papain-like proteases involved in a variety of biological phenomenon including embryo development, maintenance of cellular architecture and structure, cell development and migration *etc.* (Sorimachi and Suzuki, 2001). Being non-lysosomal cysteine proteases, calpain is known for their limited proteolysis and thus required for modulating several proteins in the cell (Ono, Ojima, Shinkai-Ouchi, Hata, and Sorimachi, 2016). Calpains are widely distributed in almost all eukaryotes but are reported in only few bacteria (Sorimachi, Hata, and Ono, 2011). Calpains are also famous for their ~15 homologues in eukaryotes specifically in humans. Among these human calpains, eight are ubiquitously present in all cells whereas six have been restricted to specific organ or tissue. There are many substrates of calpains like protein kinase C, transcription factors like c-fos, c-jun, NF-κB, p53, enzymes like tyrosine phosphatases, Ca-ATPase, and many others (Han, Weinman, Boldogh, Walker, and Brasier, 1999; Kubbutat and Vousden, 1997; Shoshan-Barmatz, Weil, Meyer, Varsanyi, and Heilmeyer, 1994). Calpain also present in muscles and are also associated with diseases conditions like muscular dystrophy, muscle atrophy and many muscular deformities. Calpains digest a large number of myo-cytoskeletal proteins such as vimentin, myelin basic protein, tropomyosin alpha-1 chain, microtubule-associated protein tau, beta crystallins *etc.* (Banik, Chou, Deibler, Krutzch, and Hogan, 1994; Fischer, Vandekerckhove, Ampe, Traub, and Weber, 1986; Liu *et al.*, 2011; Wu *et al.*, 2007). Out of 14 isoforms, skeletal muscle expresses mainly three types of calpains *i.e.* calpain 1, calpain 2 and calpain 3 (p64)(Sorimachi *et al.*, 1989). Among these three isoforms, calpain 1 and calpain 2 are "conventional or classical" calpains as they are majorly studied and universally expressed. Both of these calpains are calcium-activated proteases and require Ca^{2+} in micro- and millimolar concentrations for activation, hence termed as μ-calpain and m-calpain respectively (Goll, Thompson, Li, Wei, and Cong, 2003). These heterodimeric proteins structurally consist of two domains, larger subunit of 80 kDa assist catalytic function and smaller subunit of 30 kDa. In literature there are more than a dozen genes related to calpains, and half of these genes are showing tissue or organ specific expression (Ono *et al.*, 2016). Calpain 3 is expressed only in muscle but due to their extremely unstable and autolytic nature, this protein has not been successfully purified (Kinbara *et al.*, 1998). Recently, we have reviewed and published the challenges faced by different research group in studying eukaryotic calpains and developed two zymographic protocols which can be used to study calpains (at < 10 ug protein) in all type of muscles (skeletal and cardiac) at both *in vitro* and *in vivo* levels (Dutt *et al.*, 2017). Requirement of high protein concentrations, more time consuming process and lack of clearly defined protocol have placed such a powerful and potential approach (zymography) on least priority for the study of calpain. Though calpain systems are present in almost all eukaryotes but reports related to microorganisms specifically bacteria are very scanty (Bhat *et al.*, 2018). Thus finding the microbial source for extracting calpain and its use as an exogenous enzyme could be a breakthrough achievement for meat research industry.

Bacterial Calpain Homologue: Interesting Outbreak

The quest for more development begins as soon as man realizes that there are many hidden microorganisms that can be exploited as an alternate source for various tedious eukaryotic proteins or enzymes. Also, microorganism can be grown even in harsh conditions and under such conditions they may produce totally different microbial enzymes (Anbu, Gopinath, Chaulagain, Tang, and Citartan, 2015; Sardessai and Bhosle, 2004). This quest initiated isolation and production of new bacterial and fungal strains from extreme conditions either by changing their environment such as pH, temperature, salinity, heavy metals, and organic solvents or by the means of biotechnology tools. The major motive was to increase consortium of microbial enzymes that can be exploited for therapeutic or industrial purposes or clearing toxic xenobiotics from environment. Also, advancement in biotechnology in the present era has even made possible to insert our target genes in microbes for the production of enzymes that are not produced by that microbe naturally. There are number of articles highlighting the applications and importance of biotechnologically and industrially valuable microbial enzymes besides naturally occurring microbial enzymes (Anbu *et al.*, 2015; Anbu, Gopinath, Chaulagain, and Lakshmipriya, 2017). Genetic manipulation has thus increases the quantity production and diversity of microbial enzymes from microbial world (Demain and Vaishnav, 2009). The current scenario of designing heterologous expression system *i.e.* codon optimization (guided mutagenesis) has made possible such gene expression system that can be utilized for mankind ranging from industrial, agricultural to medicinal.

Till date only one bacterial calpain homologue has been characterized biochemically from the *Porphyromonas gingivalis*, a peptide-fermenting periodontal pathogen incapable of breaking carbohydrates for energy. Genome analysis of *P. gingivalis* reveals that besides gingipains there are several genes encoding cysteine peptidases and out of those, *PG1055* gene encoding a protein similar to the cysteine peptidases of the calpain and papain families. This protein was termed as thiol protease (TPr) which is expressed as inactive zymogen with molecular weight ~55kDa. This protease is responsible for the generation of peptides essential for the survival of *P. gingivalis*. This study also documented that like eukaryotic calpain, protein structure and activity of TPr is dependent on calcium concentration (> 1mM). As per report, TPr represents the first biochemically characterized bacterial-derived calpain-like peptidase. Also, Stanlec and his coworker, 2015 has successfully standardized the culture conditions for the growth of this bacterial strain and procedure for isolation of TPr (Staniec *et al.*, 2015). In the same year (2015), another research group reported that calpain homologues are rare and occur in a minority of bacterial phyla. From the analysis of a phylogenetic tree, they also illustrated that calpains in eukaryotes could be derived from genes horizontally transferred from a bacterium (Rawlings, 2015). This enzyme homologue connection between eukaryotes and bacteria has opened numerous paths for subsiding calpain isolation with bacterial sources and its use in food industry.

Conclusions

This chapter highlights the comparative applicability of microbial enzymes over animal and plant enzymes in the current commercialized world including in textile, pharmaceutical, and food industry. In this hunt, bacterial calpain is the biggest achievement as it was thought earlier that it is found in eukaryotic cells only. Calpain is in limelight as it is playing a key role in muscle tenderness being its involvement naturally in postmartem aging. Since postmortem aging is a time consuming process while use of GRAS enzymes exogenously is not sufficient for high quality meat because of non-specificity of these enzymes against myofibriller proteins, thus this chapter has elaborated only one example where bacterial calpain homologue has the best potential to be used as an exogenous enzyme either directly or as a new member of GRAS family in meat industry (Figure 2.1). Although much has to be done in the area of calpain homologue, it is just the beginning of new chapter.

REFERENCES

Anbu, P., Gopinath, S. C., Chaulagain, B. P., Tang, T. H., and Citartan, M. (2015). Microbial Enzymes and Their Applications in Industries and Medicine 2014. *Biomed Res Int, 2015,* 816419. doi: 10.1155/2015/816419.

Anbu, P., Gopinath, S. C. B., Chaulagain, B. P., and Lakshmipriya, T. (2017). Microbial Enzymes and Their Applications in Industries and Medicine 2016. *Biomed Res Int, 2017,* 2195808. doi: 10.1155/2017/2195808.

Arshad, M. S., Kwon, J.-H., Imran, M., Sohaib, M., Aslam, A., Nawaz, I.,. Yildiz, F. (2016). Plant and bacterial proteases: A key towards improving meat tenderization, a mini review. *Cogent Food and Agriculture,* 2(1). doi: 10.1080/23311932.2016.1261780.

Banerjee, A., Chisti, Y., and Banerjee, U. C. (2004). Streptokinase—a clinically useful thrombolytic agent. *Biotechnol Adv, 22*(4), 287-307.

Banik, N. L., Chou, C. H., Deibler, G. E., Krutzch, H. C., and Hogan, E. L. (1994). Peptide bond specificity of calpain: proteolysis of human myelin basic protein. *J Neurosci Res, 37*(4), 489-496. doi: 10.1002/jnr.490370408.

Bekhit, A. A., Hopkins, D. L., Geesink, G., Bekhit, A. A., and Franks, P. (2014). Exogenous proteases for meat tenderization. *Crit Rev Food Sci Nutr, 54*(8), 1012-1031. doi: 10.1080/10408398.2011.623247.

Bhat, Z. F., Morton, J. D., Mason, S. L., and Bekhit, A. E.-D. A. (2018). Role of calpain system in meat tenderness: A review. *Food Science and Human Wellness, 7*(3), 196-204. doi: 10.1016/j.fshw.2018.08.002.

Coria, M. S., Carranza, P. G., and Palma, G. A. (2018). Calpain System in meat tenderization: A molecular approach. *Revista MVZ Cordoba,* 23(1), 6523. doi: 10.21897/rmvz.1247.

Demain, A. L., and Vaishnav, P. (2009). Production of recombinant proteins by microbes and higher organisms. *Biotechnol Adv,* 27(3), 297-306. doi: 10.1016/j.biotechadv.2009.01.008.

Dutt, V., Kaur, N., Gupta, P., Gujar, R., Grewal, A., Gupta, S.,. Mittal, A. (2017). Efficient and modified protocol for zymography to detect muscle specific calpain activity. *Biocatalysis and Agricultural Biotechnology, 10*, 96-103. doi: http: //dx.doi.org/10.1016/j.bcab.2017.02.011.

Fischer, S., Vandekerckhove, J., Ampe, C., Traub, P., and Weber, K. (1986). Protein-chemical identification of the major cleavage sites of the Ca2+ proteinase on murine vimentin, the mesenchymal intermediate filament protein. *Biol Chem Hoppe Seyler, 367*(11), 1147-1152.

Fraatz, M. A., Ruhl, M., and Zorn, H. (2014). Food and feed enzymes. *Adv Biochem Eng Biotechnol, 143*, 229-256. doi: 10.1007/10_2013_235.

Fusetti, F., von Moeller, H., Houston, D., Rozeboom, H. J., Dijkstra, B. W., Boot, R. G.,. van Aalten, D. M. (2002). Structure of human chitotriosidase. Implications for specific inhibitor design and function of mammalian chitinase-like lectins. *J Biol Chem*, 277(28), 25537-25544. doi: 10.1074/jbc.M201636200.

Goll, D. E., Thompson, V. F., Li, H., Wei, W., and Cong, J. (2003). The calpain system. *Physiol Rev, 83*(3), 731-801. doi: 10.1152/physrev.00029.2002.

Gonzalez, N. J., and Isaacs, L. L. (1999). Evaluation of pancreatic proteolytic enzyme treatment of adenocarcinoma of the pancreas, with nutrition and detoxification support. *Nutr Cancer, 33*(2), 117-124. doi: 10.1207/S15327914NC330201.

Gurung, N., Ray, S., Bose, S., and Rai, V. (2013). A broader view: microbial enzymes and their relevance in industries, medicine, and beyond. *Biomed Res Int, 2013*, 329121. doi: 10.1155/2013/329121.

Han, Y., Weinman, S., Boldogh, I., Walker, R. K., and Brasier, A. R. (1999). Tumor Necrosis Factor-α-inducible IκBα Proteolysis Mediated by Cytosolic m-Calpain. *Journal of Biological Chemistry, 274*(2), 787-794. doi: 10.1074/jbc.274.2.787.

Kaur, R., and Sekhon, B. S. (2012). Enzymes as drugs: an overview. *J. Pharm. Educ. Res., 3*(2), 29-41.

Kinbara, K., Ishiura, S., Tomioka, S., Sorimachi, H., Jeong, S. Y., Amano, S.,. Suzuki, K. (1998). Purification of native p94, a muscle-specific calpain, and characterization of its autolysis. *Biochem J, 335 (Pt 3)*, 589-596.

Kubbutat, M. H., and Vousden, K. H. (1997). Proteolytic cleavage of human p53 by calpain: a potential regulator of protein stability. *Mol Cell Biol, 17*(1), 460-468.

Lian, T., Wang, L., and Liu, Y. (2013). A New Insight into the Role of Calpains in Post-mortem Meat Tenderization in Domestic Animals: A review. *Asian-Australas J Anim Sci, 26*(3), 443-454. doi: 10.5713/ajas.2012.12365.

Liu, M. C., Kobeissy, F., Zheng, W., Zhang, Z., Hayes, R. L., and Wang, K. K. (2011). Dual vulnerability of tau to calpains and caspase-3 proteolysis under neurotoxic and neurodegenerative conditions. *ASN Neuro, 3*(1), e00051. doi: 10.1042/AN20100012.

Mane Prajakta, and Vidya, T. (2015). Overview of Microbial Therapeutic Enzymes. *Int. J. Curr. Microbiol. App. Sci 4*(4), 17-26.

Matsumae, H., Furui, M., and Shibatani, T. (1993). Lipase-catalyzed asymmetric hydrolysis of 3-phenylglycidic acid ester, the key intermediate in the synthesis of diltiazem hydrochloride. *Journal of Fermentation and Bioengineering, 75*(2), 93-98. doi: 10.1016/0922-338x(93)90216-u.

Mieszczakowska-Frąc, M., Markowski, J., Zbrzeźniak, M., and Płocharski, W. (2012). Impact of enzyme on quality of blackcurrant and plum juices. *LWT - Food Science and Technology, 49*(2), 251-256. doi: 10.1016/j.lwt.2011.12.034.

Miller, M. F., Carr, M. A., Ramsey, C. B., Crockett, K. L., and Hoover, L. C. (2001). Consumer thresholds for establishing the value of beef tenderness. *J Anim Sci, 79*(12), 3062-3068.

Mojsov, K. (2012). ENZYME APPLICATIONS IN TEXTILE PREPARATORY PROCESS: A REVIEW. *International Journal of Management, IT and Engineering (IJMIE), 2*, 272-295.

Nierstrasz, V. A., and Warmoeskerken, M. M. C. G. (2003). *"Textile Processing with Enzymes"*. Cambridge, England: Woodhead Publishing Ltd.,.

Ono, Y., Ojima, K., Shinkai-Ouchi, F., Hata, S., and Sorimachi, H. (2016). An eccentric calpain, CAPN3/p94/calpain-3. *Biochimie, 122*, 169-187. doi: 10.1016/j.biochi.2015.09.010.

Ostlie, D. J., Juang, D., Aguayo, P., Pettiford-Cunningham, J. P., Erkmann, E. A., Rash, D. E.,. St Peter, S. D. (2012). Topical silver sulfadiazine vs collagenase ointment for the treatment of partial thickness burns in children: a prospective randomized trial. *J Pediatr Surg, 47*(6), 1204-1207. doi: 10.1016/j.jpedsurg.2012.03.028.

Raveendran, S., Parameswaran, B., Ummalyma, S. B., Abraham, A., Mathew, A. K., Madhavan, A.,. Pandey, A. (2018). Applications of Microbial Enzymes in Food Industry. *Food Technol Biotechnol, 56*(1), 16-30. doi: 10.17113/ftb.56.01.18.5491.

Rawlings, N. D. (2015). Bacterial calpains and the evolution of the calpain (C2) family of peptidases. *Biol Direct, 10*, 66. doi: 10.1186/s13062-015-0095-0.

Ritchie, H., and Roser, M. (2019). Meat and Seafood Production and Consumption. *Our World in Data*.

Sabu, A. (2003). Sources, properties and applications of microbial therapeutic enzymes. *Indian J. Biotech., 2*, 334- 341.

Sanchez, S., and Demain, A. L. (2011). Enzymes and Bioconversions of Industrial, Pharmaceutical, and Biotechnological Significance. *Organic Process Research and Development, 15*(1), 224-230. doi: 10.1021/op100302x.

Sardessai, Y. N., and Bhosle, S. (2004). Industrial potential of organic solvent tolerant bacteria. *Biotechnol Prog, 20*(3), 655-660. doi: 10.1021/bp0200595.

Saxena, S. (2015). Microbial Enzymes and Their Industrial Applications. 121-154. doi: 10.1007/978-81-322-2259-0_9.

Selvarajan, E., and Bhatnagar, N. (2017). Nattokinase: an updated critical review on challenges and perspectives. *Cardiovasc Hematol Agents Med Chem*. doi: 10.2174/1871525716666171207153332.

Shoshan-Barmatz, V., Weil, S., Meyer, H., Varsanyi, M., and Heilmeyer, L. M. (1994). Endogenous, Ca(2+)-dependent cysteine-protease cleaves specifically the ryanodine receptor/Ca2+ release channel in skeletal muscle. *J Membr Biol, 142*(3), 281-288.

Sorimachi, H., Hata, S., and Ono, Y. (2011). Calpain chronicle—an enzyme family under multidisciplinary characterization. *Proc Jpn Acad Ser B Phys Biol Sci, 87*(6), 287-327.

Sorimachi, H., Imajoh-Ohmi, S., Emori, Y., Kawasaki, H., Ohno, S., Minami, Y., and Suzuki, K. (1989). Molecular cloning of a novel mammalian calcium-dependent protease distinct from both m- and mu-types. Specific expression of the mRNA in skeletal muscle. *J Biol Chem, 264*(33), 20106-20111.

Sorimachi, H., and Suzuki, K. (2001). The structure of calpain. *J Biochem, 129*(5), 653-664.

Staniec, D., Ksiazek, M., Thogersen, I. B., Enghild, J. J., Sroka, A., Bryzek, D.,. Potempa, J. (2015). Calcium Regulates the Activity and Structural Stability of Tpr, a Bacterial Calpain-like Peptidase. *J Biol Chem, 290*(45), 27248-27260. doi: 10.1074/jbc.M115.648782.

Teh, K. H., Flint, S., Palmer, J., Andrewes, P., Bremer, P., and Lindsay, D. (2014). Biofilm – An unrecognised source of spoilage enzymes in dairy products? *International Dairy Journal, 34*(1), 32-40. doi: 10.1016/j.idairyj.2013.07.002.

Underkofler, L. A., Barton, R. R., and Rennert, S. S. (1958). Production of microbial enzymes and their applications. *Appl Microbiol, 6*(3), 212-221.

Vellard, M. (2003). The enzyme as drug: application of enzymes as pharmaceuticals. *Curr Opin Biotechnol, 14*(4), 444-450.

Vittaladevaram, V. (2017). Fermentative Production of Microbial Enzymes and their Applications: Present status and future prospects. *Journal of Applied Biology and Biotechnology, 5*, 90-94. doi: 10.7324/jabb.2017.50414.

Wu, Y., Liang, S., Oda, Y., Ohmori, I., Nishiki, T., Takei, K.,. Tomizawa, K. (2007). Truncations of amphiphysin I by calpain inhibit vesicle endocytosis during neural hyperexcitation. *EMBO J, 26*(12), 2981-2990. doi: 10.1038/sj.emboj.7601741.

Chapter 3

Emerging Environmental Pollutant: Classification, Detection and Risk Assessment

Dharmender Kumar* and Sunil Chopra

Department of Biotechnology, Deenbandhu Chhotu Ram University of Science and Technology, Murthal – 131 039 Sonepat, Haryana
****E-mail: dkbiology@gmal.com***

ABSTRACT

The emerging environmental pollutants like antibiotic, pesticides, metals, pharmaceuticals and personal care products (PPCPs, used for personal health and cosmetics, drugs like endocrine disrupters, illicit drugs, steroids *etc.*), surfactants and food additives presence, transport and their effects is one of the emerging issues around the world. They are released from various industries, household use. These pollutants were divided into various classes on the basic of their physiological properties. Presently, many organization and governments are focused on the removal of such pollutant from the environment. The technology enhancement plays the major role in the detection of such pollutants. The techniques like GC/MS, LC-ESI-MS/MS, and HPLC–MS/MS make possible to detect such pollutants from the various environmental sites up to ng/l. The common method for screening of organic compound is through bio-filtration. The coating of filters has been done with support materials like nanomaterials and organic materials. Biodegradation is one of the techniques used for removal of such pollutant with the help of micro-organisms. Microbial attached-growth technologies are getting alternative to the conventional microbial technology. The biodegradation capability of Bio-filters increases with the passage of time and also due their self-regenerating capability.

Keywords: *Emerging pollutants, GC/MS, LC-ESI-MS/MS, HPLC–MS/MS, PPCPs, Bio-filters, Attached-growth technologies.*

Introduction

Pollutants commonly known as those substances which are polluting water, air, land and various environmental bodies and making them inappropriate for consumption. This process get complexed, due to undesirable substances, pathogenic microorganisms *etc.* The term "contamination" is sometime used for pollution. Chapman (2007) proposed clarification as: *Contamination is simply the presence of a substance where it should not be or at concentrations above background. Pollution is contamination that results in, or can result in, adverse biological effects to resident communities. All pollutants are contaminants, but not all contaminants are pollutants."* The contaminants are distributed into three main groups namely: chemical, physical, and microbiological. Some of these contaminates caused various chronic toxicity, adverse effect on aquatic wildlife, endocrine disruption in humans and developed bacterial pathogenic resistance. In India various government incentives were taken for identification and control of such contaminants in various systems. The dictionary meaning of emerging is *"become apparent or prominent"*. The combination of these two terms makes "emerging pollutants". Emerging pollutants (EPs) defined by UNESCO-IHP International Initiative on Water Quality (IIWQ) (2014-2017) as *"Any synthetic or naturally-occurring chemical or any microorganism that is not commonly monitored or regulated in the environment with potentially known or suspected adverse ecological and human health effects"*. As per USEPA, pollutants are having two groups: Priority and Emerging pollutants. Priority pollutants are those compounds which are selected on the bases of their carcinogenicity, teratogenicity, and acute toxicity defined by well analytical tests.

European aquatic environment published report on more than 700 EPs, their metabolites and the transformed products (*www.norman-network.net*). In some cases the EPs presence in environment by the continuous production and consumption of these products by human population. Many EPs may not recognize previously until advanced detection methods were developed. A specific and targeted quality monitoring system is required for concentration identification of most EPs detected in water bodies; for risk assessment across river basins (Geissen, *et al.*, 2015). These industries plays major role in world's economy, considered as a strategic sector and significant for the GDP of any country. Both advantages and disadvantages were associated for the use of such products whose cause may be known or unknown due to their effect on to the living system. Such pollutants includes various chemicals released from various industries, household and natural products such as: pharmaceuticals and personal care products (PPCPs), agricultural pesticides, industrial chemicals, household waste products, metals oxides, various type of surfactants, various type of industrial additives and solvents. There are almost twenty million tons PPCPs compounds produced every year (Wang *et al.*, 2016) and their traces are present in many water bodies throughout the world. Many of these products released directly or indirectly in environment and detected continuously in various ecosystem (Table 3.1). The quantity may vary from different place to place and product to product.

Chemical Pollutants

The chemical pollutants are classified as organic and inorganic compounds based on their physical properties. They are directly or indirectly comes into the environment by manmade activities. This substance leads to various chemical reactions, photochemical reactions and metabolic transformations in the environment. These reactions gives the birth of various primary pollutants. The secondary pollutants are produced due to transformation of primary products. The latter has the possibility of transformed products are difficult to handle, after metabolized in living organisms. After transformation these contaminants produced more toxic compounds than former toxic or potentially toxic substances. The original molecule and its metabolite are considered as a pollutant. The organic and inorganic substances data exceeds more than 127 million compounds and most of them are organic compounds. Therefore, Organic compounds are considered as most pollutants. According to environmental chemistry view the organic pollutants are further classified into VOCs (volatile organic compounds) and POPs (persistent organic pollutants) (Jones *et al.*, 2009).

VOCs having small length carbons structures (not more than 10 or 12 Carbons), low boiling points and vapor-pressure. These compounds evaporated readily and present mainly in atmosphere. But these compounds also detected in various soils, ground waters and surface waters. POPs are the semi-volatile molecules and having low volatility that having significant toxicity. The most appropriates example of such compounds are tri-halomethanes or formaldehyde. This compound resists chemical and biological degradation having half-life of a year or more. It takes more than decades in soil than in water while in atmospheric forms exist for several days (Jones *et al.*, 2009).

Environmental Pollutants

Environmental pollutants discharged may be "natural" or "human-made". Natural pollutants are those substances which appeared in environment naturally (without human introduction). The most appropriate examples are trace metals. These compounds generally found in moderate concentration in the environment; do not pose any health threats and having anthropogenic origins. Due to human activities high amount of inorganic compounds are released into the environment. These contaminants not only toxic but their concentrations make them more toxic. Some of these compounds possess: chemical stability, intrinsic nature, solubility in water and vapor pressure. But other compounds effect on various conditions, like partition-coefficient in octanol/water, air/water sorption and bio-concentration. These chemical compounds transformed in environment by chemical, photochemical, or microbiological processes or by combination of these processes. The reaction processed by various chemical compounds occurs via mechanisms like: acid–base transformations, redox reactions, hydrolysis, elimination, substitution, precipitation and complexation.

Classification of EPs

EPs are classified on the basis of their used, origin and their potential effects on environment. Some of the major groups for EPs are: pharmaceutical compounds and veterinary products (PCVP), disinfectant, illicit drugs (ID), personal care products and other lifestyle products, food additives, industrial chemicals, biological toxins, waterborne pathogens and nanomaterials. Some other categories are also present on the basis of their nature, such as surfactants (used in detergents, removal of grease and as emulsifier in cosmetics), synthetic hormones *etc.* Unfortunately, sometimes EPs categories overlapped each other, this creates some confusion and there is no standard categorization method of these EPs. They are classified into different categories (Table 3.1).

Table 3.1: The Classification of different Categories of Emerging Pollutants (EPs) and Chemical Compounds Acting as Environmental Pollutant

Sl.No.	*Category of EPs*	*Chemical Compound*
1	Antibiotics (Human and Veterinary)	Ampicillin, amoxicillin, azithromycin, doxycycline, erythromycin, lincomycin, sulfamethoxazole, trimethoprim
2	Analgetics, anti-inflammatory drugs and NSAIDs	Acetaminophen, aspirin, codeine, diclofenac, fenoprofen, ibuprofen, dipyrone metabolites
3	Psychiatric drugs	Bromazepam, carbamazepine, diazepam, lorazepam
4	Lipid regulators	Amlodipine, atorvastatin, bezafibrate, clofibric acid, cilazapril, fenofibric acid, enalapril, simvastatin,
5	dippa-Blockers	Bisoprolol, metoprolol, propranolol, timolol,
6	Steroids and hormones	Diethylstilbestrol, estradiol, estrone, estriol
7	Illicit drugs	Dihydrocodeine, cocaine, morphine
8	Sun-screen agents, insect repellents	Benzophenone, 3-(4-methylbenzylidene)camphor, N,Ndiethyl-3-methyl-benz-amide
9	Fragrances	Nitro, macrocyclic and polycyclic musks
10	Detergents	2-(2-(4-Nonylphenoxy)ethoxy)ethanol, 2-(2-(4-octylphenoxy) ethoxy)ethanol
11	Food additives	Sucralose, triacetin
12	Antioxidants	2,6-Di-tert-butylphenol
13	Pesticides	Organophosphorus pesticides, thiocarbamates, 2-aminobenzimidazole
14	Nanoparticles	Limestone, titanium dioxide
15	Algal toxins	Microcystins (microcystin-LR)
16	Bio-terrorism/sabotage agents	Chloropicrin
17	Perfluorinated compounds	Perfluoroctane sulfonates (PFOS), perfluoroctanoic acid
18	Anti-corrosives	1H-Benzotriazole,
19	Antifouling compounds	Organotins (dibutyltin and triphenyltin ions), cybutryne
20	Plasticisers	Bisphenol-A
21	Metals	Arsenic, cadmium, lead, mercury, methylmercury
22	Pharmaceuticals and personal care products	Benzylpenicillin, carazolol, derquantel, emamectin, flumequine, triclabendazole
23	Endocrine disruptors	Testosterone, progesterone, E2, PFOS, PFOA

Detection of these Pollutants

EPs are found in low quaintly (ng/L) in several systems, released by n natural or anthropogenic activity. These pollutants were detected in air, wastewater, soil, and tissues and milk of living organisms. The complete knowledge of these compounds quantitatively and qualitatively is one of the problems. Advances in analytical technologies make enabled to detection of such molecules such a low concentration (PPB/PPT in environmental and biological samples). These low concentration detection by modern analytical methods attracts the scientific community to study the toxicity levels that leads to toxic effect on the environment. The first unexpected pollutants detected in Swedish water in1966 after sampling marine animals were PCBs and DDTs (Jensen *et al.*, 1969).

Protocols used for detection of EPs which are accepted by international organizations and many institutions. These are the valuable tool for study of regulations effect for such pollutants. Gas Chromatography (GC) is one of the most fundamental techniques used for the analysis of the environmental sample, which allows considering the organic compounds as pollutants. The commonly used GC-techniques have limitation to analysis due to the uncharged non-polar compounds exists as environmental compounds. While, the polar compounds were analyzed through GC/MS and final detection is based on chemical transformation of these compounds into volatile form. The liquid chromatography/mass spectrometry (LC/MS) have been used for the effectively identification of both polar and nonpolar organic compounds.

Mass spectroscopy provides information about the structure determination of the chemical pollutants present in the environment. This is one of very sensitive and selective technique. The chromatogram used peaks library match and MS used to separate organic compounds present in the complex mixture. Consequently by combining together these techniques helps to analyze such compounds qualitatively and quantitatively. There combines techniques were used in many studies, such as LC-ESI-MS/MS, GC/MS/MS, and HPLC–MS/MS (Richardson, *et al.*, 2014; Baker and Kasprzyk-Hordern, 2013; Wille, *et al.*, 2012). The toxicity and risk assessment of PPCPs has been described in terms of environmental fate and transport through the food chain and food web in the ecosystem (Chopra and Kumar, 2018).

Biodegradation of EPs

There are the two way by which degradation path of EPs follows 1) In natural environment by biodegradation 2) photo-degradation. The biodegradation is the one of the most effective and economical clean-up technologies. It having potential to degrade pollutants partially or completely. In biodegradation different microorganisms are used like bacteria and fungus present in soil and water that having potential to uptake the targeted EPs as growth supplement and leading for elimination (Stasinakis, 2012; Yang *et al.*, 2013). Microbial attached-growth technologies are getting alternative to the conventional microbial technology for the treatment of wastewater. A biofilm is attached and covered with the supported packed material in an attached-growth technology. The biofilm is made by the microorganisms, specific materials and some extracellular polymers (Loupasaki and

Diamadopoulos, 2013). The attached-growth technologies classified into two type: fixed bed and moving bed type of bioreactors. The microbial growth mecanisms having some major advantages over the conventional microbial processes (Guo *et al.*, 2012) and has the following implications:

1. The attached-growth processes improved oxygen transfer (treatment), increased nitrification rate and a biomass concentration gets higher.
2. These techniques are more actual in removal of organic pollutants in wastewater and having high higher organic-loading rates with shorter hydraulic retention time.
3. These methods allow microorganisms to develop with low specific-growth rates.
4. Less space are required due to the smaller size of reactor.
5. Economical because operational low costs.

A general procedure for removal of organic contaminants is bio-filtration: in this the waste water (effluent) is passed through a column or a filter (bio-filter) coated with support materials having microorganisms onto the surface. Such bio-filters are able to degrade/transform the organic pollutant present in effluent into simple forms. Various specific strains of microorganisms has been identified involved in remediation of toxic compounds. This technique having notable advantages over the conventional techniques (activated carbon absorbers). Bio-filters capability was increased over the time due to its self-regenerating capability of microbial grow than their adsorption capacity gets increased with increased of time. The emerging contaminants are probably removed by the bio-filters with biological transformation process. So the bio-filtration appears as a promising technique for EPs degradation (Reungoat *et al.*, 2011).

Conclusion

The removal/management of various pollutants present in the environment require more sophisticated studies, research, regulation and concern. Biodegradation processes are more economical as compare to other processes Emerging contaminants are present in our ecosystem from a long time but the advances in detection techniques lead to identification of such contaminants which together helps us to develop the future degradation strategies. These contaminants have potential to cause new and serious challenges for ecosystems and human health. The advancement in research and technologies leads the future regulations, based on occurrence, bioremediation targets and their potential environmental and health risks.

REFERENCES

Baker, D.R. and Kasprzyk-Hordern, B. 2013. Spatial and temporal occurrence of pharmaceuticals and illicit drugs in the aqueous environment and during wastewater treatment: new developments. *Sci. Total Environ.*, 454–455, 442–456. doi: 10.1016/j.scitotenv.2013.03.043.

Chapman, P.M. 2007. Determining when contamination is pollution – weight of evidence determinations for sediments and effluents. *Environ. Int.*, 33 (4), 492–501. doi: 10.1016/j.envint.2006.09.001.

Chopra, S and Kumar, D. 2018. Pharmaceuticals and Personal Care Products (PPCPs) as Emerging Environmental Pollutants: Toxicity and Risk Assessment. In: *Advances in Animal Biotechnology and its Applications.*Editors S. K. Gahlawat,J. S. Duhan, R.K. Salar, P. Siwach, S. Kumar, P. Kaur (Springer). Pp. 337-353

Geissen, V., Mol, H., Klumpp, E., Umlauf, G., Nadal, M., van der Ploeg, M., van de Zee, S.E.A.T.M. and Ritsema, C.J. 2015. Emerging pollutants in the environment: A challenge for water resource management. *Int. Soil Water Conserv. Res.* (3), 57–65. https: //doi.org/10.1016/J.ISWCR.2015.03.002

Guo, W., Ngo, H. and Vigneswaran, S. 2012. Enhancement of membrane processes with attached growth media, in Membrane Technology and Environmental Applications (eds T. Zhang, R. Surampalli, S. Vigneswaran, R. Tyagi, S. Ong, and C. Kao), *American Society of Civil Engineers, Reston, VA*, pp. 603–634. doi: 10.1061/9780784412275.ch20.

Jensen, S., Johnels, A.G., Olsson, M., and Otterlind, G. 1969. DDT and PCB in marine animals from Swedish waters. *Nature,* 224 (5216), 247–250. doi: 10.1038/224247a0.

Jones, K.C. and De Voogt, P. 1999. Persistent organic pollutants (POPs): state of the science. *Environ. Pollut.*, 100 (1-3), 209–221. doi: 10.1016/S0269-7491(99)00098-6.

Loupasaki, E. and Diamadopoulos, E. 2013. Attached growth systems for wastewater treatment in small and rural communities: a review. *J. Chem. Technol. Biotechnol.*, 88 (2), 190–204. doi: 10.1002/jctb.3967.

Richardson, S.D. and Ternes, T.A. 2014. Water analysis: emerging contaminants and current issues. Anal. Chem., 86 (6), 2813–2848. doi: 10.1021/ac500508t.

Stasinakis, A.S. (2012) Review on the fate of emerging contaminants during sludge anaerobic digestion. *Bioresour. Technol.* 121, 432–440.

Wang, Jianlong; Wang, Shizong (2016). "Removal of pharmaceuticals and personal care products (PPCPs) from wastewater: A review". *Journal of Environmental Management*. **182**: 620–640. doi: 10.1016/j.jenvman.2016.07.049. PMID 27552641

The air (prevention and control of pollution) act, 1981 No. 14 of (1981).

UNESCO-IHP International Initiative on Water Quality (IIWQ) (2014-2017); https: //en.unesco.org/emergingpollutants

Wille, K., De Brabander, H.F., Vanhaecke, L., De Wulf, E., Van Caeter, P. and Janssen, C.R. 2012. Coupled chromatographic and mass-spectrometric techniques for the analysis of emerging pollutants in the aquatic environment. TrAC, *Trends Anal. Chem.*, 35, 87–108. doi: 10.1016/j.trac.2011.12.003.

Yang, S., Hai, F.I., Nghiem, L.D., Price, W.E., Roddick, F., Moreira, M.T. and Magram, S.F. 2013. nderstanding the factors controlling the removal of trace organic contaminants by white-rot fungi and their lignin modifying enzymes: a critical review. *Bioresour. Technol.* 141, 97–108.

Chapter 4

Microbes as a Source of Therapeutic Natural Products

Divya Bhatia[1], Ritu Saini[2], Harnek Singh Saini[1], and Deepak Kumar Malik[1]*

[1]*Department of Biotechnology Engineering, University Institute of Engineering and Technology, Kurukshetra University, Kurukshetra – 136 119, Haryana*
[2]*Indian Institute of Wheat and Barley Research, Karnal, Haryana*
**E-mail: divyabhatia228@gmail.com*

ABSTRACT

Nature is an inexhaustible source of chemical compounds for drug discovery. Since several decades, natural products have been exploited for the cure and prevention of diseases. Natural compounds are an indispensable source of novel products with therapeutic potential. Among natural sources, microorganisms have great potential as a producer of therapeutically significant products like antibiotics, antioxidants, enzymes, immune-suppressants, enzyme inhibitors, growth hormones and vitamins. Natural products have tremendous potential to be employed to combat severe diseases like cancer, anemia, HIV, diabetes *etc.*

Keywords: *Antibiotic, Natural products, Immunosuppressant, Antimicrobial.*

Introduction

Natural products, the compounds synthesized by living organism either by primary metabolic pathways or by secondary metabolism. Polysaccharides, proteins, nucleic and fatty acids essential for growth are some common primary metabolites produced by all living beings. The secondary metabolites (SMs) are low molecular weight (MW 3000), chemically and taxonomically diverse compounds with tremendous therapeutic potential. These natural products with novel structure not only possess significant biological activities but also can act as a base for the production of synthetic and semi-synthetic therapeutic products. Secondary metabolites from living organisms (microbes, plants, animals) have increased the chances to prevent and treat infections, which lead to increased human life span (Demain, 2006). Microbial secondary metabolites exhibit antimicrobial, anticancer and/or antiviral activities. This leads to great interest in the exploration of microorganisms from nature. *The greater part of the world's population is dependent upon natural medicine for their health related issues (Taylor, 2013).* In the USA, a major portion of the population has expressed interest in medicines from nature for the prevention and cure of diseases (WHO, 2013). A major proportion of drugs approved for the human diseases have been obtained from natural products (Newman *et al.*, 2012). Microorganisms being an endless source of bioactive metabolites have contributed some significant pharmaceutical products (Gupta *et al.*, 2014). Out of 170,000 reported natural products, more than 22,500 bioactive metabolites are produced by microorganisms, with about 45 per cent contributed by actinomycetes and about 38 per cent of fungi (Seyedsayamdost and Clardy, 2014; Demain and Sanchez, 2009). Previous studies had reported that natural sources are a key contributor of novel antimicrobial agents. Nearly 50 per cent antibiotics are isolated from actinomycetes, 10-15 per cent from bacteria and approx 20 per cent fungi. Most of the significant compounds from natural compounds are synthesized either by chemical reaction or by biotransformation.

Initially, emphasis was only on antimicrobial agents from natural sources. However, from the last few decades, natural products are not only explored for novel antimicrobial products but also for other various applications. Bacteria, yeast and actinomycetes have been explored for the production of antioxidants, enzymes, growth factors, vitamins and immunosuppressants (Newman and Cragg, 2007). Most significant microbial products are antimicrobial compounds, anticancer drugs, immunosuppressants, enzyme inhibitors, antiparasitic agents, herbicides, growth regulators, pesticides and insecticides. Other important applications of natural products that are seeking attention include treatment of viral diseases, dermatological problems, malaria, prion related diseases, diabetes, obesity, gastric ulcers, hemochromatosis and alumina overload in kidney patients. The approved natural products include actinomycin D, anthracyclines, bleomycin, enediynes, taxol and epothilones (Gupta *et al.*, 2014).

Bacteria and Actinomycetes

Bacteria are the most significant source for the production of natural products and/or their synthetic intermediates. Among bacteria, *Bacillus*, *Pseudomonas* Myxo and Cyanobacteria species are the most common source of natural products. However, Mycobacteria, mycoplasmatales and spirotheces are usually not used for production (Raja and Prabakarana, 2011). Marine organisms produce different novel metabolites *such as polyketides, alkaloids, peptides, lipids, shikimates, glycosides and iso-prenoids which exhibit several different pharmacological properties (Jin et al., 2016).* In addition to antibiotics, other significant antimicrobial compounds like bacteriocins, antimicrobial peptide. Hydrogen peroxide, nitric oxide *etc.* are also produced by probiotic bacteria (Deshmukh and Thorat, 2014).

In the last few decades, antimicrobial peptides (AMPs) are seeking interest as an effective solution for antimicrobial drug resistance problems. AMPs are produced by all organisms; however, they have svariation in structure and function. Bacteriocins, ribosomally synthesized peptides have a narrow spectrum antimicrobial actvity. Bacteriocins are produced by all major species of bacteria and exhibit toxic effect only against related species of producer strain (Benmechernene *et al.*, 2013). Bacteriocins, a product of lactic acid bacteria (LAB) not only exert antibacterial activity but also sporostatic/sporicidal activity (Egan *et al.*, 2016). *E. faecalis* produce *Enterocins DD28 and DD93* which exhibit bactericidal activity against methicillin-resistant *S. aureus* (MRSA) and have increased activity when combined with erythromycin or kanamycin (Al Atya *et al.*, 2016). Laparaxin, a polypeptide produced by *L. paracasei* exhibit a significant antibacterial activity against Gram-positive bacteria (Liu *et al.*, 2012). Microcin, a polypeptide isolated from *E. coli* has bacteriostatic activity against *Salmonella* (Hammami *et al.*, 2015) and antibacterial activity against *Enterobacteriaceae* family (Gomaa *et al.*, 2017).

Among bacteria *Bacillus* species are preferable for commercial production because antibiotics produced are in the soluble protein structures that result in economic production of antibiotic. Some peptide antibiotics isolated from *Bacillus* species are bacitracin, polymyxin, gramicidin, tyrocidine, subtilin, bacilysin *etc.* (Trookman *et al.*, 2011). Variety of peptide antibiotics produced by *Bacillus* sp. has been approved for the food industry. Polycyclic antibacterial peptide, nisin is the first approved antimicrobial peptide to be used as a food preservative (Egan *et al.*, 2016). Nisin is also an effective against *Clostridium* and can be used to treat *C. difficile* mediated infections (Le Lay *et al.*, 2016). Enterocin DD14, colistin and nisin when used in combination were effective against biofilm cultures of *E. coli* CIP54127 (Al Atya *et al.*, 2016). Consequently, bacteriocins can be used individually during food storage and also in combination augment antibiotic treatments.

The ribosomally antimicrobial peptides are synthesized by all living organisms; however, Non-ribosomally synthesized peptides are mainly reported from bacteria (Hancock and Chapple, 1999). In addition to antimicrobial application, AMPs also express immunoregulatory function (Fjell *et al.*, 2012), which render them a suitable candidate for the development of novel therapeutics. Most AMPs are 10-50 amino acids long with an overall positive charge (+2 to +11) and mainly consist of hydrophobic amino acid (Hancock and Sahl, 2006; Pasupuleti *et al.*, 2012). In 1939, an antimicrobial peptide (AMP) extracted from a *Bacillus* species exhibited inhibitory activity against *Pneumococci* and later identified as gramicidin (Hotchkiss and Dubos, 1940). Gramicidin was used effectively for treatment of wounds and ulcers (Epps and Dubos, 2006). After that, another AMP, tyrocidine with broad spectrum activity was produced (Dubos and Hotchkiss, 1941). Inside bacterial cell, AMPs hinder some essential pathways such as DNA replication and protein synthesis required for cell survival (Brogden, 2005).

In the last few decades, interest in microbial flavors and colors from natural sources has been intensified. Carotenoids, melanin, violacein, prodigiosin, pyocyanin, actinorhodin and zeaxanthin are some important pigments produced by bacteria (Ahmad *et al.*, 2012; Venil *et al.*, 2014). Pigment producing bacteria have been isolated from different environment such as soil, rhizospheric soil, desert sand, fresh water, marine areas and also from endophytes (Zhu *et al.*, 2007; Peix *et al.*, 2005; Liu *et al.*, 2009; Asker *et al.*, 2008; Franks *et al.*, 2005; Deng *et al.*, 2011). Various species such as *S. marcescens, R. rubra, S. rubidaea, V. psychroerythrus, A. rubra, S. aureus, X. oryzae, Rhodococcus* and *Kitasatospora* produce a wide variety of pigments. A correlation between antioxidants and microorganisms has been observed (Gupta *et al.*, 2013). Carotenoids are the most important group of antioxidants that can be synthesized by microorganisms. Carotenoids have different applications such as food colorants, as feed supplements, nutraceutical, cosmetics and in medicine (Jaswir *et al.*, 2011). The most significant carotenoids are lycopene, carotenes, clutein, violaxanthin and zeaxanthin (Rymbai *et al.*, 2011). Carotenoids producing bacteria mainly belongs to *Flavobacterium* sp., *Rhodobacter* sp., *Sphingomonas* sp., *G. jacobea* and *Rhodosporium* sp. (Bhosale and Bernstein, 2004; Chen *et al.*, 2006, Aksu and Eren, 2005; Silva *et al.*, 2004; Dufosse, 2006; Panesar *et al.*, 2015). *S. thermophilus* possess strong antioxidant activity which helps to scavenge free radicals from circulation. Some examples of antioxidants include mycotrienin, trienomycin and trienomycin. An oxidant isolated from *Streptomyces* showed inhibitory effect against 5-lipoxygenase enzyme (Gupta *et al.*, 2014).

Chromobacterium violaceum and *Janthinobacterium lividum* produce a mixture of violacein and deoxybiolacein pigments (Dufosse, 2009). Violacein are used not only in dye industry but also exhibit a strong antiviral, antibacterial, antileishmanial, antitumor and enzyme transition properties (Duran *et al.*, 2007; Soliev *et al.*, 2011). Prodigiosin (red pigment) isolated from *Serratia marcescens* exhibit insecticidal, antimicrobial and anti-malarial activities (Kamble and Hiwarale, 2012). Pyocyanin, a blue pigment extracted from *Pseudomonas* possess broad spectrum antimicrobial activity (Jayaseelan *et al.*, 2014).

Melanin, an indolic polymer confers resistance to UV light and prevents photo-induced damage (Hill, 1992) and also used to produce monoclonal antibodies as a therapy for melanoma (Surwase *et al.*, 2013). *Colletotrichum* sp., *Magnaporthe* sp., *Cryptococcus* sp., *Vibrio* sp. and *Alteromonas* are main producer of melanin (Soliev *et al.*, 2011; Surwase *et al.*, 2013).

Enzyme inhibitors are considered as useful tools, not only to elucidate structures and mechanisms but also for treatment of severe diseases. Various enzyme inhibitors with huge industrial potential are produced by microbes (Adrio and Demain, 2014). Metabolic enzymes responsible for many diseases like glucosidases, lactamase, amylases, lipases, proteases and Xanthine Oxidase (XO) may be the target for inhibitors. Acarbose, isolated from *Actinoplanes* sp. inhibits intestinal α-glucosidase and leads to starch digestion in the intestine, which may be helpful in diabetes treatment (Yuhong *et al.*, 2014). Amylase inhibitors also reduce dietary starches absorption by the body (Jayaraj *et al.*, 2013). Paim, α-amylase inhibitors was extracted from *S. corchorushii* (Song *et al.*, 2014). *Streptomyces toxytricini* produces a pancreatic lipase inhibitor, Lipstatin, is used to treat obesity and diabetes (Rodgers *et al.*, 2012). Some strains of LAB are able to decrease erum cholesterol levels, by digesting bile in the gut, thus preventing its re-absorption (Mekkes *et al.*, 2014).

Actinomycetes are well-known producer of bioactive metabolites including 10,000 antimicrobial products (Demain, 2009). The secondary metabolites synthesized by actinomycetes express various biological activities such as antimicrobial, antiviral, antitumor, anticholesterol, anti-aging and immunosuppressant. Particularly 80 per cent of natural antibiotics have beenributed by different actinomycetes (Khanna *et al.*, 2011). The discovery of antimicrobial products from actinomycetes has significantly contributed to therapeutics and medicine.

Streptomyces species, the most frequent producer contributes 74 per cent of all products (7600 compounds) while remaining 26 per cent is produced by rare species. The most significant antibacterial products of *Streptomyces sp.* are Chloramphenicol (Fernandez-Martínez *et al.*, 2014), Fosfomycin (Woodyer *et al.*, 2006), Lincomycin (El-Sherbini and Khattab, 2018), Tetracycline (Mark Nelson *et al.*, 2001), Tunicamycin (Atta, 2015), Mycangimycin (Atta, 2010) and Novel anti-infective Guadinomine (Holmes *et al.*, 2012). Vancomycin (glycopeptide antibiotic) and Streptomycin (aminoglycoside) were derived from the actinobacteria species *Amycolatopsis orientalis* and *Streptomyces griseus* respectively (Butler *et al.*, 2014; Vakulenko and Mobashery, 2003). Neomycin (aminoglycoside antibiotic) usually present in topical medicines such as ointments and eye drops, is produced naturally by the bacterium *Streptomyces fradiae* (Vastrad and Neelagund, 2014).

Daptomycin, a cyclic lipopeptide antibiotic isolated from *S. roseosporus* is effective against methicillin resistant *S. aureus* and vancomycin resistant enterococci (Baltz, 2009). Lynamicins, produced by *Marinispora* sp. exhibited broad spectrum antimicrobial activities against drug-resistant microbes (McArthur *et al.*, 2008). Marinopyrroles from *Streptomyces* spp, are the first naturally occurring alkaloids which exhibit cytotoxic activity against cancer and antimicrobial effect against resistant pathogenic strains (Hughes *et al.*, 2008). Echinomycin, a quinoxaline compound produced by *S. echinatus* express antimicrobial, antiviral and anticancerous activities (Kong *et al.*, 2005). Diazepinomicin, an alkaloid isolated from *Micromonospora* sp. demonstrated antitumor, antiparasitic, antioxidant and anti-protease activities (McAlpine *et al.*, 2008; Adelmohsen *et al.*, 2012). Spinosyns, polyketide insecticide were produced by *Saccharopolyspora* species (Kirst *et al.*, 2010). Mainly *S. griseus*, *S. griseoviridis*, *S. coelicolor*, *S. echinoruber* and *S. shaanxiensis* were reported to produce pigments (Darshan and Manonmani, 2015; Lin *et al.*, 2012; Gupta *et al.*, 2011). *Streptomyces coelicolor* A3 also produce Actinohodine-related blue pigments (Newman and Cragg, 2007).

Immunosuppressants are used during transplantation to minimize the chance of rejection. A number of microbial compounds with this unique activity have been reported. *Streptomyces tsukubaensis* produces Tacrolimus (FK506), an immno suppressor that is highly active than cyclosporin A. In addition to this, tacrolimus inhibits tumor specific signalling pathway and collagen production in human lung fibroblastic cells. Therefore, tacrolimus may be helpful in pulmonary fibrosis treatment (Nagano *et al.*, 2006). Rapamycin produced by *S. hygroscopicus* is mainly used in kidney transplants due to absence of nephrotoxicity effect (Kim *et al.*, 2014).

Fungi

In contrast to bacteria, fungi are present in all ecological niches and have broad substrate utilization ability. Recently, studies of fungal metabolites have gained an intensive attention due to the requirement for compounds possessing novel biological functions. Fungi have an incredible capacity for biosynthesis of compounds with novel structure and function. Approximately 1.5 million fungal species are present on earth and only a small fraction has been explored (Kim *et al.*, 2004). Many fungi are wonderful sources of secondary metabolites (Wang *et al.*, 2010). The fungal natural products have a diverse structure and function and are synthesized in different enzymatic pathways (Deacon, 1997). Fungal secondary metabolites have significant role in wellbeing of the human (Hoffmeister and Keller, 2007). Some of the most significant examples include antibiotics; immunosuppressive; cholesterol-lowering agents; potent poisons; and some ergot alkaloids. All of these natural products, including fungal metabolites, are grouped together as secondary metabolites.

Several types of metabolites are produced by phylogenetically diverse fungi (Cole and Schweikert, 2003; Larsen *et al.*, 2005). Different fungal species may synthesize same secondary metabolites *e.g. Z. masonii* (a zygomycete), *M. anisopliae* and *E. goetzii* are producer of same product Cytochalasin D (Cole and Schweikert, 2003). Secondry metabolite pathways have been studied only from a few species like *Aspergillus*, *Penicillium* and *Fusarium* that can be easily maintained in the laboratory (Lazarus *et al.*, 2014).

Among fungal derived natural products, main focus is on compound with antimicrobial activity, due to a rapid increase in antibiotic resistant microbes and decrement in effectiveness of existing antibiotics, which is considered as a serious risk to global health (Aiken *et al.*, 2014). Penicillins and cephalosporins, are the most significant antibiotics worldwide: cephalosporins accounts for 28 per cent and penicillins for 19 per cent of the total market of antibiotics (Hamad, 2010). a novel metabolite, Liamocin oil from the fungus *A. pullulans* has antibacterial effect against *Streptococcus* sp. (Price *et al.*, 2017).

Recently, investigation for natural products of symbiotic fungi that grow in association with land plants, lichens, marine organisms and insects have been increased. Clavatol (antimicrobial) and Javanicin (antibiotic) have been isolated from endophytic fungi *Aspergillus clavatonanicus* and *Chloridium* sp. (Leuchtman, 2003; Verma *et al.*, 2007). Antifungal compounds, Jesterone and lactones were isolated from fungus *Pestalotiopsis jester* and *Phomopsis* sp. respectively (Li *et al.*, 2001; Huang *et al.*, 2008). Endophytic fungi also synthesize antiviral agents. *Cytonaema* sp. an endophytic fungus produce two antiviral compounds cytonic acid A and B that are effective against human cytomegalovirus (hCMV) protease (Guo *et al.*, 2000). Nodulisporic acid isolated from endophytes *Nodulispotium* sp. and *Bontia Daphnoides* exhibits strong insecticidal activity against the larvae of the blowfly (Verma *et al.*, 2007).

In addition to antimicrobial property, some fungal metabolites possess cholesterol-lowering property, such as the statins. Lovastatin and mevastatin are produced by *Penicillium citrinum* and *Aspergillus terreus* (Manzoni and Rollini, 2002). These compounds mainly cause selective inhibition of squalene synthase and results in decreased cholesterol biosynthesis (Istvan and Deisenhofer, 2001). Antioxidant are produced by endophytic fungi. Pestacin and Isopestacin were produced by *Pestalotiopsis microspora* from plant *Terminalia morobensis* (Strobel *et al.*, 2002). Gignardic acid was extracted from the culture filterate of an endophytic *Gignardia* sp. isolated from *Spondias mombin* (Kharwar *et al.*, 2009). *Penicillium* and *Aspergillus* produce gallic acid, an ansamycin antibiotic (Kim, 2013).

Some SMs produced from fungi can have immunosuppressant activity. Cyclosporin a non-ribosomal peptide (NRP), isolated from *T. niveum* is widely used to avoid host rejection in transplant surgery (Viaud *et al.*, 2013). Fingolimod, derived from myriocin display activities against multiple sclerosis (Cohen *et al.*, 2010). However, fungi also produce some mycotoxins whose presence as contaminants in food needs to be carefully monitored. Mostly occuring mycotoxins are aflatoxin, produced by *Aspergillus flavus* and the trichothecenes found in *Fusarium* sp. (Olsen *et al.*, 2008; Sudakin, 2003). Patulin (mycotoxin), earlier recognized as antibiotic, has been reported to be acutely toxic (Fung and Clark, 2004).

Sometimes fungal NPs are not used directly in native form but need some kind of derivatization to give semisynthetic derivatives. Alternatively, such natural products may act as a template for the production of synthetic derivatives. Strobilurin derived from plyketide acting as basic framework for the various antifungal componds (Anke *et al.*, 1977). azoxystrobin available in the market is an example of this class of fungicide (Bartlett *et al.*, 2002). Several cell cycle regulators have been identified in fungi (Tsuchiya *et al.*, 2010). Taxol, produced by the endophytic fungi *T. andreanae* and *N. sylviforme*, exhibit toxic effect against mammalian cancer cells by disturbing normal microtubule dissociation during mitosis. Earlier, taxol was approved only for refractory ovarian cancer treatment, however, today it is also used against breast and Kaposi's sarcoma (Newman and Cragg, 2007). Fumagillin, isolated from fungal product induce rounding of endothelial cell and inhibit angiogenesis (Ingber *et al.*, 1990).

In last few decades, natural products have been explored from special environments like cold water (Soldatou and Baker, 2017), Polar Regions (Tian *et al.*, 2017) and deep sea (Wang *et al.*, 2015). Most of the reported products have been isolated from bacteria and actinomycetes, however, very few studies reported the natural products from extremophilic fungi. Sorbicillin type compounds were produced by *Phialocephala* sp. isolated from an underwater sample (Li *et al.*, 2007). Breviones and Cyclopiamides were extracted from *Penicillium* sp. isolated from the deep sea sediment derived (Wang *et al.*, 2016; Zhang *et al.*, 2017). *Aspergilols* were purified from *Aspergillus versicolor* obtained from deep sea (Wu *et al.*, 2016).

REFERENCES

Abdelmohsen, U.R., Szesny, M., Othman, E.M. and Schirmeister, T. 2012. Antioxidant and anti-protease activities of diazepinomicin from the sponge-associated *Micromonospora* strain rv115. *Marine Drugs* 10: 2208-2221.

Adrio, J.L. and Demain, A.L. 2014. Microbial Enzymes: tools for biotechnological processes. *Biomol.* 4(1): 117-139.

Ahmad, W.A., Ahmad, W.Y.M., Zakaria, Z.A., and Yusof, N.Z. 2012. Application of bacterial pigments as colorant. In: Ahmad W. A., Ahmad W. Y. M., Zakaria Z. A., Yusof N. Z. (Eds.). Application of bacterial pigments as colorant: the Malaysian Perspective, Berlin, Springer pp. 57-74.

Aiken, A.M., Allegranzi, B., Scott, J.A., Mehtar, S., Pittet, D. and Grundmann, H. 2014. Antibiotic resistance needs global solutions. *Lancet Infect Dis.* 14: 550-551.

Aksu, Z. and Eren, A.T. 2005. Carotenoids production by the yeast *Rhodotorula mucilaginosa*: use of agricultural wastes as a carbon source. *Process Biochem.* 40: 2985-2991.

Al Atya, A.K., Abriouel, H., Kempf, I., Jouy, E., Auclair, E., Vachee, A. and Drider, D. 2016. Effects of colistin and bacteriocins combinations on the *In vitro* growth of *Escherichia coli* strains from swine origin. *Probiotics Antimicrob Proteins.* 8: 183-190.

Al Atya, A.K., Belguesmia, Y., Chataigne, G., Ravallec, R., Vachee, A., Szunerits, S., Boukherroub, R. and Drider, D. 2016. Anti-MRSA activities of enterocins DD28 and DD93 and evidences on their role in the inhibition of biofilm formation. *Front. Microbiol.* 7: 817.

Anke, T., Oberwinkler, F., Steglich, W. and Schramm, G. 1977. The strobilurins-new antifungal antibiotics from the basidiomycete *Strobilurus tenacellus*. *J. Antibiot.* 30: 806-810.

Asker, D., Beppu, T. and Ueda, K. 2008. *Nubsella zeaxanthinifaciens* gen. nov., sp. nov., a zeaxanthin-producing bacterium of the family *Sphingobacteriaceae* isolated from freshwater. *Int. J. Syst. Evol. Microbiol.* 58: 601-606.

Atta, H.M. 2010. Production, purification, physico-chemical characteristics and biological activities of antifungal antibiotic produced by *Streptomyces antibioticus* AZ-Z71. *American-Eura. J. of Sci. Res.* 5(1): 39-49.

Atta, H.M. 2015. Biochemical studies on antibiotic production from *Streptomyces* sp. Taxonomy, fermentation, isolation and biological properties. *J. Saudi Chem. Soc.* 19(1): 12-22.

Baltz, R.H. 2009. Daptomycin: Mechanisms of action and resistance and biosynthetic engineering. *Current Opinion in Chemical Biology.* 13: 144-151.

Bartlett, D.W., Clough, J.M., Godwin, J.R., Hall, A.A., Hamer, M. and Parr-Dobrzanski, B. 2002. The strobilurin fungicides. *Pest Manag Sci.* 58: 649-662.

Benmechernene, Z., Fernandez-No, I. and Kihal, M. 2013. Recent patents on bacteriocins: food and biomedical applications. *Recent Pat DNA Gene Seq.* 7(1): 66-73.

Bhosale, P. and Bernstein, P.S. 2004. β-Carotene production by *Flavobacterium multivorum* in the presence of inorganic salts and urea. *J. Ind. Microbiol. Biotechnol.* 31: 565-571.

Brogden, K.A. 2005. Antimicrobial peptides: pore formers or metabolic inhibitors in bacteria. *Nat. Rev. Microbiol.* 3: 238-250.

Butler, M.S., Hansford, K.A., Blaskovich, M.A, Halai, R. and Cooper, M.A. 2014. Glycopeptide antibiotics: back to the future. *J Antibiot.* 67: 631-644.

Chen, D., Han, Y. and Gu, Z. 2006. Application of statistical methodology to the optimization of fermentative medium for carotenoids production by *Rhodobacter sphaeroides*. *Process Biochem.* 41: 1773-1778.

Cohen, J.A., Barkhof, F., Comi, G., Hartung, H.P., Khatri, B.O., Montalban, X., Pelletier, J., Capra, R., Gallo, P., Izquierdo, G., Tiel-Wilck, K., de Vera, A., Jin, J., Stites, T., Wu, S., Aradhye, S. and Kappos, L. 2010. Oral fingolimod or intramuscular interferon for relapsing multiple sclerosis. *N. Engl. J. Med.* 362(5): 402-415.

Cole, R.J. and Schweikert, M.A. 2003. Handbook of Secondary Fungal Metabolites, (Vols. 1 and 2) Academic Press: Amsterdam.

Darshan, N. and Manonmani, H.K. 2015. Prodigiosin and its potential applications. *J. Food. Sci. Technol.* 52: 5393-5407.

Deacon, J.W. 1997. Modern Mycology, 3rd edn. Blackwell science, U.K.

Demain, A.L. 2006. From natural products discovery to commercialization: a success story. *J Ind. Microbiol Biotechnol.* 33: 486-495.

Demain, A.L. 2009. Antibiotics: natural products essential to human health. *Medicinal Research Reviews.* 29: 821-842.

Demain, A.L. and Sanchez, S. 2009. Microbial drug discovery: 80 years of progress. *J. Antibiot.* 62: 5-16.

Deng, Z.S., Zhao, L.F., Xu, L., Kong, Z.Y., Zhao, P. and Qin, W. 2011. *Paracoccus sphaerophysae* sp. nov., a siderophore-producing, endophytic bacterium isolated from root nodules of Sphaerophysa salsula. *Int. J. Syst. Evol. Microbiol.* 61: 665-669.

Deshmukh, P.V. and Thorat, P.R. 2014. Detection of antimicrobial efficacy of novel bacteriocin produced from *Lactobacillus similis* RL7. *International Journal of Advanced Research.* 2(1): 987-995.

Dubos, R.J. and Hotchkiss, R.D. 1941. The production of bactericidal substances by aerobic sporulating bacilli. *J. Exp. Med.* 73: 629-640.

Dubos, R.J., Hotchkiss, R.D. 1941.The production of bactericidal substances by aerobic sporulating bacilli. *J. Exp. Med.* 73: 629-640.

Dufosse, L. 2006. Microbial production of food grade pigments. *Food Technol. Biotechnol.* 44: 313-321.

Dufosse, L. 2009. Pigments. *Micro Encyl Microbiol.* 4: 457-471.

Egan, K., Field, D., Rea, M.C., Ross, R.P., Hill, C. and Cotter, P.D. 2016. Bacteriocins: novel solutions to age old spore-related problems. *Front Microbiol.* 7: 461.

El-Sherbini, A. and Khattab, A.A. 2018. Induction of novel mutants of *Streptomyces lincolnensis* with high Lincomycin production. *J. of Applied Pharmaceu. Sci.* 8(02): 128-135.

Epps, V.H.L. and Rene Dubos, R. 2006. Unearthing antibiotics. *J. Exp. Med.* 203: 259.

Fernandez-Martínez, L.T., Borsetto, C., Gomez-Escribano, J.P., Bibb, M. J., Al-Bassam, M.M., Chandra, G. and Bibb, M.J. (2014). New insights into chloramphenicol biosynthesis in *Streptomyces venezuelae* ATCC 10712.*Antimicrobial agents and chemotherapy*,58(12): 7441-50.

Fjell, C.D., Hiss, J.A., Hancock, R.E. and Schneider, G. 2012. Designing antimicrobial peptides: form follows function. *Nat. Rev. Drug Discov.* 11: 37-51.

Franks, A., Haywood, P., Holmstrom, C., Egan, S., Kjelleberg, S. and Kumar, N. 2005. Isolation and structure elucidation of a novel yellow pigment from the marine bacterium *Pseudoalteromonas tunicate. Molecules.* 10: 1286-1291.

Fung, F. and Clark, R.F. 2004. Health effects of mycotoxins: A toxicological overview. *J. Toxicol. Clin. Toxicol.* 42: 217-234.

Gomaa, A.I., Martinent, C., Hammami, R., Fliss, I. and Subirade, M. 2017. Dual coating of liposomes as encapsulating matrix of antimicrobial peptides: development and characterization. *Front Chem.* 5: 103.

Guo, B., Jin-Rui, D., Ng, S., Huang, Y., Leong, C., Ong, W. and B.K. Carte, B.K. 2000. Cytonic acid A and B: Novel tridepside inhibition of Hcmv protease from the endophytic fungus *Cytonaema* species. *Journal of Natural Prod.* 63: 602-604.

Gupta, C., Amar, P.G., Prakash, D., Goyal, S. and Gupta, S. 2011. Microbes as potential source of biocolors. *Pharmacologyonline.* 2: 1309-1318.

Gupta, C., Prakash, D. and Garg, A.P. 2014. Nutraceuticals from microbes. In: Prakash D, Sharma G, editors. *Phytochemicals of nutraceutical importance. UK: CABI International Publishers.* 79-102.

Gupta, C., Prakash, D. and Gupta, S. 2013. Functional foods enhanced with microbial antioxidants. *Acad J Nutr.* 2(2): 10-18.

Hamad, B. 2010. The antibiotics market. *Nat. Rev. Drug Discov.* 9: 675-676.

Hammami, R., Bedard, F., Gomaa, A., Subirade, M., Biron, E. and Fliss, I. 2015. Lassoinspired peptides with distinct antibacterial mechanisms. *Amino Acids.* 47: 417-428.

Hancock, R.E. and Chapple, D.S. 1999. Peptide antibiotics. Antimicrob. *Agents Chemother.* 43: 1317-1323.

Hancock, R.E. and Sahl, H.G. 2006. Antimicrobial and host-defense peptides as new anti-infective therapeutic strategies. *Nat. Biotechnol.* 24: 1551-1557.

Hill, H.Z. 1992. The function of melanin or six blind people examine an elephant. *Bioessays.* 14: 49-56.

Hoffmeister, D. and Keller, N.P. 2007. Natural products of filamentous fungi: enzymes, genes, and their regulation. *Nat Prod Rep.* 24: 393-416.

Holmes, T.C., May, A.E., Zaleta-Rivera, K., Ruby, J.G. and Skewes-Cox, P. 2012. Molecular insights into the biosynthesis of guadinomine: a Type III secretion system Inhibitor. *J. Am. Chem. Soc.* 134(42): 17797-17806.

Hotchkiss, R.D., Dubos, R.J.1940. Fractionation of the bactericidal agent from cultures of a soil *Bacillus. J. Biol. Chem.* 132: 791–792.

Howard, T.D. 1953. The production of Neomycin by *Streptomyces fradiae* in synthetic Media. *Appl. Microbiol.* 1(2): 103-106.

Huang, W., Cai, Y., Hyde, K., Corke, H., Sun, M, 2008. Biodiversity of endophytic fungi associated with 29 traditional Chinese medicinal plants. *Fungal Divers* 33: 61-75.

Hughes, C.C., Prieto-Davo, A., Jensen, P.R. and Fenical, W. 2008. The marinopyrroles, antibiotics of an unprecedented structure class from a marine *Streptomyces* sp. *Organic Letters*. 10: 629-631.

Ingber, D., Fujita, T., Kishimoto, S., Sudo, K., Kanamura, T., Brem, H. and Folkman, J. 1990. Synthetic analogues of Fumagillin that inhibit angiogenesis and suppress tumour growth. *Nature*. 348: 555-557.

Istvan, E.S. and Deisenhofer, J. 2001. Structural mechanism for statin inhibition of HMG-CoA reductase. *Science*. 292: 1160-1164.

Jaswir, I., Noviendri, D., Hasrini, R.F. and Octavianti, F. 2011. Carotenoids: sources, medicinal properties and their application in food and nutraceutical industry. *J. Med. Plants Res*. 5: 7119-7131.

Jayaraj, S., Suresh, S. and Kadeppagari, R.K. 2013. Amylase inhibitors and their biomedical applications. *Starch–Starke*. 65(7-8): 535-542.

Jayaseelan, S., Ramaswamy, D. and Dharmaraj, S. 2014. Pyocyanin: production, applications, challenges and new insights. *World. J. Microbiol. Biotechnol.* 30: 1159-1168.

Jin, L., Quan, C., Hou, X. and Fan, S. 2016. Potential Pharmacological Resources: Natural bioactive Compounds from marine-derived fungi. *Mar. Drugs*. 14(4): 76.

Kamble, K.D. and Hiwarale, V.D. 2012. Prodigiosin production from *Serratia marcescens* strains obtained from farm soil. *Int. J. Environ. Sci.* 3: 631-638.

Khanna, M., Solanki, R. and Lal, R. 2011. Selective isolation of rare actinomycetes producing novel antimicrobial compounds. *Int. J. Adv. Biotech. Res.* 2(3): 357-375.

Kharwar, R.N., Verma, V.C., Kumar, A., Gond, S.K, Harper, J.K., Hess, W.M., Lobkovosky, E., Ma, C., Ren, Y., Strobel, G.A. 2009. Javanicin, an antibacterial naphthaquinone from an endophytic fungus of neem, *Chloridium* sp. *Current Microbiology* 58: 233- 238.

Kim, S.H., Song, Y.S., Kim, S.K., Kim, B.C., Lim, C.J. and Park, E.H. 2004. Antiinflammatory and related pharmacological activities of the nBuOH subfraction of mushroom *Phellinus linteus. Journal of Ethnopharmaco.* 93(1): 141-146.

Kim, S.K. 2013. Marine microbiology: bioactive compounds and biotechnological applications. Wiley-VCH, Weinheim.

Kim, Y.H., Park, B.S., Bhatia, S.K., Seo, H.M., Jeon, J.M., Kim, H.J., Yi, D.H., Lee, J.H., Choi, K.Y., Park, H.Y., Kim, Y.G. and Yang, Y.H. 2014. Production of rapamycin in Streptomyces *hygroscopicus* from glycerol-based media optimized by systemic methodology. *J. Microbiol. Biotechnol.* 24(10): 1319-26.

Kirst, H.A. 2010. The spinosyn family of insecticides: realizing the potential of natural products research. *Journal of Antibiotics.* 63: 101-111.

Kong, D., Park, E.J., Stephen, A.G. and Calvani, M. 2005. Echinomycin, a small-molecule inhibitor of hypoxia-inducible factor-1 DNAbinding activity. *Cancer Res.* 65: 9047-9055.

Larsen, T.O., Smedsgaard, J., Nielsen, K.F., Hansen and M.E., Frisvad, J.C. 2005. Phenotypic taxonomy and metabolite profiling in microbial drug discovery. *Natural Product Reports.* 22: 672-695.

Lazarus, C.M., Williams, K. and Bailey, A.M. 2014. Reconstructing fungal natural product biosynthetic pathways. *Nat. Prod. Rep.* 31: 1339-1347.

Le Lay, C., Dridi, L., Bergeron, M.G., Ouellette, M. and Fliss, I.L. 2016. Nisin is an effective inhibitor of *Clostridium difficile* vegetative cells and spore germination. *J. Med. Microbiol.* 65: 169-175.

Leuchtman, A. 2003. Taxonomy and diversity of *epichloë*endophytes. In: White J.F.J., Bacon C.W., HywelJones N.L., Spatafora J.W. (Eds.). Clavicipitalean fungi: Evolutionary biology, chemistry, biocontrol and cultural impacts. New York, NY, USA: MarcelDekker, pp. 169-175.

Li, D., Wang, F., Cai, S., Zeng, X., Xiao, X., Gu, Q. and Zhu, W. 2007. Two new bisorbicillinoids isolated from a deep-sea fungus, *Phialocephala* sp. FL30r. *J. Antibiot.* 60: 317-320.

Li, J.Y., Harper, J. K. Grant, D.M., Tombe, B.O., Bashyal, B., Hess W.M. and G. A. Strobel, G.A. 2001. Ambuic acid, a highly functionalized cyclohexenone with antifungal activity from *Pestalotiopsis*spp. and *Monochaeti*a sp. *Phytochem.*57: 261-265.

Lin, Y.B., Wang, X.Y., Fang, H., Ma, Y.N., Tang, J. and Tang, M. 2012. *Streptomyces shaanxiensis* sp. nov., a blue pigment-producing streptomycete from sewage irrigation soil. *Int. J. Syst. Evol. Microbiol.* 62: 1725-1730.

Liu, M., Peng, F., Wang, Y., Zhang, K., Chen, G. and Fang, C. 2009. *Kineococcus xinjiangensis* sp. nov., isolated from desert sand. *Int. J. Syst. Evol. Microbiol.* 59: 1090-1093.

Liu, S., Wilkinson, B.J., Bischoff, K.M., Hughes, S.R., Rich, J.O. and Cotta, M.A. 2012. Novel antibacterial polypeptide laparaxin produced by *Lactobacillus paracasei* strain NRRL B-50314 via fermentation. *J. Pet. Environ. Biotechnol.* 3: 3.

Manzoni, M. and Rollini, M. 2002. Biosynthesis and biotechnological production of statins by filamentous fungi and application of these cholesterol-lowering drugs. *Appl. Microbiol. Biotechnol.* 58: 555-564.

Mark Nelson, R.A., Greenwald, W.H. and Mark, L.N. 2001. Tetracyclines in biology, chemistry and medicine. Birkhauser, Switzerland.

McAlpine, J.B., Banskota, A.H., Charan, R.D. and Schlingmann, G. 2008. Biosynthesis of diazepinomicin/eco-4601, a *Micromonospora* secondary metabolite with a novel ring system. *Journal of Natural Products*. 71: 1585-1590.

McArthur, K.A., Mitchell, S.S., Tsueng, G. and Rheingold, A. 2008. Lynamicins a-e, chlorinated bisindole pyrrole antibiotics from a novel marine actinomycete. *Journal of Natural Products*. 71: 1732-1737.

Mekkes, M.C., Weenen, T.C. and Brummer, R.J., 2014. The development of probiotic treatment in obesity: a review. *Benef Microbes*. 5(1): 19-28.

Nagano, J., Iyonaga, K. and Kawamura, K. 2006. Use of tacrolimus, a potent anti-fibrotic agent, in bleomycin-induced lung fibrosis. *Eur Respir. J*. 27(3): 460-469.

Newman, D.J. and Cragg, G.M. 2007. Natural products as sources of new drugs over the last 25years. *J Nat Prod*. 70(3): 461-477.

Newman, D.J. and Cragg, G.M. 2012. Natural products as sources of new drugs over the 30 years from 1981 to 2010. *J Nat Prod*. 75(3): 311-35.

Olsen, M., Johnsson, P., Moller, T., Paladino, R. and Lindblad, M. 2008. *Aspergillus nomius*, an important aflatoxin producer in Brazil nuts? *World mycotoxin Journal*. 1: 123-126.

Panesar, R., Kaur, S. and Panesar, P.S. 2015. Production of microbial pigments utilizing agro-industrial waste: a review. *Curr. Opin. Food. Sci*. 1: 70-76.

Pasupuleti, M., Schmidtchen, A. and Malmsten, M. 2012. Antimicrobial peptides: key components of the innate immune system. *Crit. Rev. Biotechnol*. 32: 143-171.

Peix, A., Berge, O., Rivas, R., Abril, A. and Velazquez, E. 2005. *Pseudomonas argentinensis* sp. nov., a novel yellow pigment-producing bacterial species, isolated from rhizospheric soil in Cordoba, Argentina. *Int. J. Syst. Evol. Microbiol*. 55: 1107-1112.

Peschke, U., Schmidt, H., Zhang, H.Z. and Piepersberg, W. 1995. Molecular characterization of the lincomycin-production gene cluster of *Streptomyces lincolnensis*. *Mol. Microbiol*. 16(6): 1137-1156.

Price, N.P., Bischoff, K.M, Leathers, T.D., Cosse, A.A. and Manitchotpisit, P. 2017. Polyols, not sugars, determine the structural diversity of anti-streptococcal liamocins produced by *Aureobasidium pullulans* strain NRRL 50380. *J. Antibiot*. 70: 136-141.

Raja, **A. and** Prabakarana, **P. 2011.** Actinomycetes and Drug-An Overview. *American Journal of Drug Discovery and Development*. 1(2): 75-84.

Rodgers, R.J., Tschop, M.H. and Wilding, J.P. 2012. Anti-obesity drugs: past, lisent and future. *Dis. Model Mech*. 5(5): 621-626.

Rymbai, H., Sharma, R.R. and Manish, S. 2011. Biocolorants and its implications in Health and Food Industry - a review. *Int. J. Pharm Tech Res*. 3: 2228-2244.

Seyedsayamdost, M.R. and Clardy, J. 2014. Natural products and synthetic biology. *ACS Synth Biol.* 3: 745-747.

Silva, C., Cabral, J.M.S. and Keulen, F.V. 2004. Isolation of a β-total carotenoids over-producing soil bacterium, *Sphingomonas* sp. *Biotechnol. Lett.* 26: 257-262.

Soldatou, S. and Baker, B.J. 2017. Cold-water marine natural products, 2006 to 2016. *Nat. Prod. Rep.* 34: 585-626.

Soliev, A.B., Hosokawa, K. and Enomoto, K. 2011. Bioactive pigments from marine bacteria: applications and physiological roles. *Evid. Based Complement. Alternat. Med.* 670349.

Song, M.K., Bischoff, D.S. and Uyemura, K. 2014. Prevention and treatment of obesity and diabetes and their related complications. *J. Mol. Genet. Med.* S1: 009.

Strobel, G. and Daisy. 2003. Bioprospecting for microbial endophytes and their natural products. *Microbial. Mol. biol. Rev.* 67: 491-502.

Strobel, G.A., Ford, E., Worapong, J., Harper, J.K., Arif, A.M., Grant, D.M., Fung, P.C.W. and Chan, K. 2002. Isopestacin, an isobenzofuranone from *Pestalotiopsis microspora*, possessing antifungal and antioxidant activities. *Phytochemistry.* 60: 179-183.

Sudakin, D.L. 2003. Trichothecenes in the environment: relevance to human health. *Toxicol. Lett.* 143: 97-107.

Surwase, S.N., Jadhav, S.B., Phugare, S.S. and Jadhav, J.P. 2013. Optimization of melanin production by *Brevundimonas* sp. SGJ using response surface methodology. *3 Biotech.* 3: 187-194.

Taylor, P.W. 2013. Alternative natural sources for a new generation of antibacterial agents. *Int J antimicrob. Agents.* 42(3): 195-201.

Tian, Y., Li, Y.L. and Zhao, F.C. 2017. Secondary metabolites from polar organisms. *Mar. Drugs.* 15

Trookman, N.S., Rizer, R.L. and Weber, T. 2011. Treatment of minor wounds from dermatologic procedures: a comparison of three topical wound care ointments using a laser wound model. *Journal of the American Academy of Dermatology.* 64: 224- 239.

Tsuchiya, E., Yukawa, M., Ueno, M., Kimura, K. and Takahashi, H. 2010. A novel method of screening cell-cycle blockers as candidates for anti-tumor reagents using yeast as a screening tool. *Biosci. Biotechnol. Biochem.* 74: 411-414.

Vakulenko, S.B and Mobashery, S. 2003. Versatility of aminoglycosides and prospects for their future. *Clin. Microbiol. Rev.*16: 430-450.

Vastrad, B.M. and Neelagund, S.E. 2014. Optimization of Medium Composition for the Production of Neomycin by *Streptomyces fradiae* NCIM 2418 in solid state fermentation. Biotech. Res. Int. 2014: 1-11.

Venil, C.K., Aruldass, C.A., Dufosse, L., Zakaria, Z.A. and Ahmad, W.A. 2014. Current perspective on bacterial pigments: emerging sustainable compounds with coloring and biological properties for the industry-an incisive evaluation. *RSC Adv.* 4: 39523-39529.

Verma, V.C., Gond, S.K., Kumara, A., Kharwar, R.N. and Strobel, G.A. 2007. The endophytic mycoflora of bark, leaf, and stem tissues of *Azadirachta indica* A. Juss (neem) from Varanasi (India). *Microb. Ecol.* 54: 119-125.

Viaud, W.Y., Saccheri, S.F. and Mignot, G. 2013. The intestinal microbiota modulates the anticancer immune effects of cyclophosphamide. *Science.* 342(6161): 971-976.

Wang J., He, W., Huang, X., Tian, X., Liao, S., Yang, B., Wang, F., Zhou, X. and Liu Y. 2016. Antifungal new oxepine-containing alkaloids and xanthones from the deep-sea-derived fungus *Aspergillus versicolor*SCSIO 05879. *J. Agric. Food Chem.* 64: 2910-2916.

Wang, X.R., Sena-Filho, J.G. and Hoover, A.R. 2010. Chemical epigenetics alters the secondary metabolite composition of guttate excreted by an atlantic-forest-soil-derived *Penicillium citreonigrum*. *Journal of Natural Product.* 73: 942-948.

Wang, Y.T., Xue, Y.R. and Liu, C.H. 2015. A brief review of bioactive metabolites derived from deep-sea fungi. *Mar. Drugs.* 13: 4594-616.

WHO traditional medicine strategy: 2014-2023; 2013. http: //www. who.int/medicines/publications/traditional/trm_strategy14_23/en/.

Woodyer, R.D., Shao, Z., Thomas, P.M., Kelleher, and Neil, L. 2006. Heterologous production of Fosfomycin and identification of the minimal biosynthetic gene cluster. *Chem. Biol.* 13(11): 1171-1182.

Wu, Z., Wang, Y., Liu, D., Proksch, P., Yu, S. and Lin, W. 2016. Antioxidative phenolic compounds from a marine-derived fungus *Aspergillus versicolor*. *Tetrahedron.* 72: 50-57.

Yuhong, H., Wenxu, F. and Yanfen, L. 2014. Comparison of the Effects of Acarbose and TZQ-F, a New Kind of Traditional Chinese Medicine to Treat Diabetes, Chinese Healthy Volunteers. *Evidence-Based Complementary and Alternative Medicine.* 2014: 308126.

Zhang, Z., He, X., Zhang, G., Che, Q., Zhu, T., Gu, Q. and Li, D. 2017. Inducing secondary metabolite production by combined culture of *Talaromyces aculeatus* and *Penicillium variabile*. *J. Nat. Prod.* 80: 3167-3171.

Zhu, H.H., Guo, J., Yao, Q., Yang, S.Z., Deng, M.R. and Phuong le, T.B. 2007. Streptomyces vietnamensis sp. nov., a streptomycete with violet blue diffusible pigment isolated from soil in Vietnam. *Int. J. Syst. Evol. Microbiol.* 57: 1770-1774.

Chapter 5

Impact of different Microorganisms as a Biofertilizers on Yield and Nutrient Content of Common Bean (*Phaseolus vulgaris* L.) under Intercropping System

M.H. Dar[1]*, N. Singh*[2] *and F.A. Khan*[1]

[1]*Division of Basic Sciences and Humanities, Sher-e-kashmir University of Agricultural Sciences and Technology of Kashmir (SKUAST-K) Shalimar – 191 121, Srinagar, J&K*
[2]*Department of Botany, Kurukshetra University, Kurukshetra – 136 119, Haryana*
**E-mail: nsheorankuk@yahoo.com*

ABSTRACT

Use of different microorganisms as a biofertilizer in crop production is most important factor for build up soil biological properties under organic farming besides other organic manure applications. Microorganisms used as bio-fertilizers include selective organisms like bacteria, fungi and algae. These microorganism are capable of fixing atmospheric nitrogen which can't used by plants directly and solubilization of native added nutrients in the soil. In order to study the effect of different microorganisms (*Rhizobium, Azotobacter, Arbuscular mycorrhizae*) as a biofertilizer under different levels of phosphorus (20 and 40 kg/ha) application on yield and nutrient content in seeds **of** common bean (*Phaseolus vulgaris*), an experiment was conducted during kharif seasons of 2012 and 2013 at the Krishi Vigyan Kendra (KVK) of Sher-e-Kashmir University of Agricultural Sciences and Technology, Budgam, Jammu and Kashmir. The research experiment was laid out in complete randomized block design (RBD). The Common bean variety named "Shalimar Rajmash-1" was used for the present research study. The results of the research experiment revealed that the yield parameters of common bean like number of pods, pod length and weight of pods and the nutrient content in seeds of all the inoculated common bean plants was much higher as compared to control plants. The yield and nutrient content was significant in those treatments inoculated with *Rhizobium* + VAM @ 20 kg P/ha as compared to other treatments as well as control. The significant increase in yield and nutrient content of common bean was due to the positive interactions among different microorganisms given to plants.

Keywords: *Common bean, Phosphorus, Intercropping, Biofertilizers, Nutrients.*

Introduction

At present all over the world most growers holding very small land and this small land holding problem is mostly prominent in developing countries like Asia and Africa. So in these developing countries, there is urgent need to enhance production as well as increase the ability of growers to grow different multiple crops in these small areas. Overcoming this problem in these developing countries, intercropping may be the best solution. Intercropping is practised in many parts of the world to produce more food and feed per unit area (Caruthers *et al.*, 2000). Intercropping involves growing of two or more crops on the same area on the same time during same season and also ensures better utility of those sources which are limiting level than the individual crops which are growing separately (Gosh *et al.*, 2007). Cereal intercropping with legumes is very common in humid areas with tropical environments and also in rain fed areas in the most parts of the world. Intercropping system is also used in temperate areas for increasing grain yield (Li *et al.*, 2001) and also decreasing dependency on nitrogen input (Exner *et al.*, 1999).

In modern agriculture sustainable crop production system use of different microorganisms as a biofertilizer have an important role in enhancing various biological properties of soil besides using other organic manures. Different microorganisms used as bio-fertilizers include selective organisms like bacteria, fungi and algae and these microorganisms increase nitrogen content in soil by fixing atmospheric nitrogen and also are capable for solubilisation of native added nutrients in the soil. These microorganisms are particularly ecofriendly, cost-effective and source of various essential plant nutrients like nitrogen, phosphorus and potassium. Use of different microorganisms as a biofertilizer in intercropping system have an important role in keeping the soil healthy and fertile. Use of microorganisms as a biofertilizers are most important to ensure a healthy, sustainable future for the incoming generations. The role of biofertilizers in improving fertility of soil and enhancement in yield attributing characters has been reported by many research workers (Kachroo and Razdan, 2006). Application of different biofertilizers in soil enhances diversity of various microbes and also decreases the use of highly expensive manmade synthetic fertilizers (Subashini *et al.*, 2007).

Use of ecofriendly and low cost effective microorganisms in intercropping system could be a best option for overcoming the various problems which are mostly related to deficiency of nutrients and other different physiological related processes including fixation of nitrogen, photosynthetic process and translocation of various important compounds mostly photsynthesized sugars (Mia and Shamsuddin, 2010). Different research findings showed that legumes form a very important mutalistic association with the bacteria called *Rhizobium* and with the most common fungi named *Arbuscular mycorrhizae* (AM) to increase nitrogen fixation, formation of nodules and solubilisation of fixed phosphorus. (Barea *et al.*, 2009). The macro-symbionts which fixes atmospheric nitrogen by a process commonly called nitrogen fixation happens in nodulated roost of different leguminous plants (Crespi and Galvez, 2000) and *Arbuscular mycorrhizal* fungal type of association mobilize various essential macronutrients mostly phosphate (Miransari *et al.*, 2009). Several research findings of different scientists clearly revealed that combined inoculation of AMF and *Rhizobium* accelerates the different parameters which are related to growth, yield and uptake of nutrients (Abd- Alla *et al.*, 2014). The fungal association (AMF) with the roots of different plants is one the most essential symbiotic association in enhancing uptake of different macro and micronutrients commonly phosphorus (Tajini *et al.*, 2011). It has been worked out that combined application of low phosphorus level with different inoculated microorganisms to the soil importanly increase various plant growth parameters and also enhancement of phosphorus uptake.

Materials and Methods

The Research Site

The present research study was carried out on an experimental field at the research farm of Krishi Vigyan Kendra (KVK) of Sher-e-Kashmir University of Agricultural Sciences and Technology.The temperature of the mostly ranges from 5°C to a maximum of 34°C.

Operation of Field

The experimental field was ploughed smoothly by using a tractor and then smooth levelled beds were prepared to get water and different fertilizers efficiently. The total size of the plot was 16.5 square meters (sqm) with the net size of plot was 9.6 sqm. Complete randomized design (CRD) was used for the present research experiment with three replications.

Treatment Details

T_1: Maize + common bean (control)

T_2: Maize+ common bean + *Rhizobium*

T_3: Maize + common bean + *Azotobacter*

T_4: Maize + common bean + Arbuscular mycorrhizae

T_5: Maize + common bean + 20 kg phosphorus (P)/ha

T_6: Maize + common bean + 40 kg P/ha

T_7: Maize + common bean + *Rhizobium* + Arbuscular mycorrhizae

T_8: Maize + common bean + *Azotobacter* + Arbuscular mycorrhizae

T_9: Maize + common bean + *Rhizobium* + Arbuscular mycorrhizae + 20kg P/ha

T_{10}: Maize + common bean + *Azotobacter* + Arbuscular mycorrhizae + 20 kg P/ha

$T1_1$: Maize + common + *Rhizobium* + Arbuscular mycorrhizae + 40 kg P/ha

T_{12}: Maize + common bean + *Azotobacter* + Arbuscular mycorrhizae + 40 kg P/ha

T_{13}: Maize + common bean + *Rhizobium* + *Azotobacter* + Arbuscular mycorrhizae

Description of Biofertilizers Application

The seeds of the common bean variety used in the present experiment were first sterilized by using 0.1 per cent solution of sodium hypochlorite for almost 2 minutes, then washed properly with distilled water and finally put these seeds in distilled water for almost 6 hours before sowing the seeds. The seeds of the common bean were peated with *Rhizobium leguminosarum* inoculum, and *Azotobacter vinelandi*. Planting holes method was applied for the use of vesicular arbuscular mycorrhizae (*Glomus mosseae*) after seed sowing at the rate of 25 Kg/ha.

The recommended dose of different fertilizers like nitrogen and potassium for common bean variety were 30 Kg/ha N, 30 K_2O_5Kg/ha) were applied on the basis of population of plants present in different research pots of the field. Application of phosphorus was done as per the treatments. At the time of sowing of seeds, basal half dose of nitrogen and potassium were used as basal and the remaining dose of nitrogen was used after 25 days when main true leaves were emerged

Data Analysis

To determine the significant impact of different treatments on various parameters like growth, yield and nutrient content, All the research data was loaded in microsoft excel and then subjected to the analysis of variance by using on line OPSTA, of Hisar Agricultural University, Hisar, Haryana. The critical difference at 5 per cent level was used for significance of research data.

Results

The obtained research data of different yield parameters like pods per plant, pod length and weight of pods revealed that the treatment receiving combined application of *Rhizobium* + VAM + 20 kg phosphorus/ha recorded higher number of pods per plant (6.53) as well as maximum pod length (10.88 cm) compared to other treatments (Table 5.1). The Minimum number of pods, pod length common bean (4.00, 5.40 cm) were recorded in control treatments. In general, all the treatments recorded significantly higher weight of pods compared to control. The weight of pods/plant ranges from 5.54 g to 2.44g. The plants receiving dual inoculation of *Rhizobium* + VAM @ 20 kg P/ha gave higher weight of pods compared to other plants. The lowest weight of pods 2.44 g was recorded with the control. Plants receiving only *Rhizobium* inoculantion alone also gave higher weight of pods (5.40) than the control.

Nitrogen, Phosphorus and Potassium Content

The present research data revealed that highest nitrogen content 4.94 per cent was recorded in treatment receiving dual inoculation of *Rhizobium*+VAM @

Table 5.1: Effect of Biofertilizers Application on Number of Pods, Pod Length and Weight of Pods of Common Bean under Intercropping System

Treatments	*Pods/Plant*	*Pod Length (cm)*	*Pod Weigh (g) per Plant*
T_1	4.00±0.00	5.40±0.13	2.44±0.25
T_2	5.40±0.20	6.74±0.15	3.70±0.16
T_3	4.53±0.06	6.46±0.17	3.15±0.09
T_4	5.40±0.20	6.82±0.08	3.64±0.19
T_5	5.60±0.41	6.72±0.10	3.58±0.09
T_6	5.00±0.20	6.29±0.09	3.15±0.09
T_7	6.00±0.11	7.49±0.15	4.31±0.17
T_8	5.40±0.00	7.36±0.20	4.00±0.05
T_9	6.53±0.06	10.88±0.12	5.54±0.14
T_{10}	6.27±0.13	10.62±0.13	5.12±0.12
T_{11}	5.40±0.00	9.51±0.19	4.48±0.10
T_{12}	5.73±0.06	9.67±0.10	4.77±0.13
T_{13}	6.26±0.13	10.12±0.16	4.87±0.15
C.D. @ 5 per cent	0.454	0.147	0.159

C.D. = Critical difference.

20 kg P/ha than other treatments as well as control. Plants receiving only single inoculation of *Rhizobium* recorded higher nitrogen content 3.92 per cent compared to other single inoculated plants as well as control plants. Plants inoculated with *Azotobacter* + VAM also recorded higher nitrogen content (3.58 per cent) compared to control plants. The lowest nitrogen concentration (3.14 per cent) was recorded in control plants (T_1).

Significantly highest phosphorus concentration (0.73 per cent) was recorded in treatment inoculated with *Rhizobium* + VAM @ 20 Kg P/ha) followed by plants receiving dual inoculation of *Azotobacte*r + VAM @ 20 Kg P/ha T_{10} (0.66 per cent) and the lowest phosphorus concentration (0.47 per cent) was recorded in control treatments (T_1).

Data about concentration of potassium in seeds of common bean showed significant difference among different treatments. The highest potassium concentration (3.67 per cent) was recorded in plants inoculated with *Rhizobium* + VAM + 20 Kg P/ha which was significantly higher compared to other treatments. Plants inoculated with *Azotobacte*r + VAM @ 20 Kg P/ha gave also higher concentration of potassium (3.62 per cent) compared to single inoculated plants as well as control. Lowest potassium concentration (2.75 per cent) was recorded in control plants.

Table 5.2: Effect of different Biofertilizers Application on Nitrogen, Phosphorus and Potassium Content in Seeds of Common Bean under Intercropping with Maize

Treatments	*Seeds*		
	Nitrogen (per cent)	*Phosphorus (per cent)*	*Potassium (per cent)*
T_1	3.14±0.11	0.47±0.01	2.75±0.05
T_2	3.92±0.21	0.51±0.03	3.02±0.02
T_3	3.22±0.19	0.48±0.02	2.88±0.05
T_4	3.82±0.17	0.53±0.05	2.91±0.04
T_5	3.86±0.20	0.58±0.06	3.01±0.04
T_6	3.38±0.16	0.52±0.04	2.92±0.06
T_7	3.51±0.15	0.59±0.03	3.31±0.05
T_8	3.58±0.22	0.54±0.02	3.19±0.05
T_9	4.94±0.16	0.73±0.04	3.67±0.07
T_{10}	4.71±0.14	0.66±0.05	3.62±0.10
T_{11}	3.20±0.18	0.52±0.04	3.47±0.08
T_{12}	3.36±0.20	0.56±0.05	3.56±0.11
T_{13}	3.81±0.17	0.62±0.06	3.59±0.06
C.D.@ 5 per cent	0.02	0.05	0.76

C.D. = Critical difference.

Discussion

In the current research study, the maximum pod number, pod length and pod weight were observed in plants inoculated with *Rhizobium* + VAM with the application of 20 kg P/ha. Similarly, other parameters like seeds per plant and 100-seed weight showed significantly enhancement in plants inoculated with *Rhizobium* + VAM with the phosphorus application of 20 kg/ha applied than other dual inoculated treatments as well as control. Zaffer *et al.* (2011) worked out that combined application of phosphorus along with the special plant growth promoting rhizobacteria (PGPR) increased seed yield which ranges from 27 to 115 per cent when compared to other single inoculated treated plants. Such type of positive yield response in these inoculated treated plants may attribute to the increase in pod number, and seed weight because of more availability of important macronutrient called phosphorus. Phosphorus addition with the plant growth promoting rhizobacteria (PGPR) also had positive impact on various growth and yield components of common bean under intercropping system. Results of Chiezey and Odunze, 2009 also revealed that pod yield and 100-seed weight in soya bean plant significantly increased with the phosphorus application and similarly enhanced economic yield. Zaffer *et al.* (2011) observed higher yield in t plants received integrated application of phosphorus along with different biofertilizers than that of plants receiving sole phosphorus application. The present research findings are in correspondent with results of Zaffer *et al.* (2012) and Abbasi *et al.*

(2011) clearly indicating that phosphorus application is very important in obtaining higher yield in beans.

Results of the present research work revealed that significantly highest nitrogen, phosphorus and potassium content of seeds of common bean was observed when plants were dual inoculated with *Rhizobium* + VAM along with phosphorus application at the rate of 20 Kg/ha. Higher leaf nitrogen uptake increased the synthesis of different amino acids with the result increased nitrogen content in the seeds of common bean by the common process happens in plants called translocation. Further, phosphorus application to the dual inoculated plants might have increased more availability of phosphorus and potassium nutrients and similarly enhancement of these nutrient in the common bean seeds. Different scientists worked out on different plants like chickpea (Aslam *et al.*, 2010), ground nut (Naeem and Khan, 2005) and black gram (Selvakumar *et al.*2012) observed phosphorus application with the combination of different biofertilizers significantly increased nitrogen, phosphorus and potassium content in seeds of these crops which are very much supportive to the present research work. This might be due to the combined impact of rapid release of various essential nutrients by phosphorus mobilization (VAM) and nitrogen fixating bacteria (*Rhizobium*) and increased the availability of nitrogen and phosphorus which are added to the soil through organic and inorganic resources.

Conclusion

It is cocluded tht combined application of phosphorus at different levels with *Rhizobium* and VAM significantly increased the various yield parameters and nutrient content of common bean under intercropping with maize. Highest yield and nutrient content was possible with the dual inoculation of *Rhizobium*, VAM at the application of 20 kg p/ha. Therefore, phosphorus application @ 20 kg/ha with the dual inoculation of *Rhizobium* and VAM can be highly recommended for intercropping of common bean with maize in temperate regions to get significant yield and nutrient content of common bean.

Acknowledgements

I am very thankful to my esteemed supervisor Dr. Narender Singh, Professor and chairman, Department of Botany, Kurukshetra University, Kurukshetra (Haryana) for his valuable guidance, incessant encouragement and promoting care in all the time of research work and authors are also thankful to the Head of KVK Budgam, Sher-e-Kashmir University of Agricultural Sciences and Technology (SKUAST-K), Shalimar, Srinagar for providing necessary research facilities.

REFERENCES

Abbasi, M.K., Sharif, S., Kazmi, M., Sultan, T. and Aslam, M. 2011.Isolation of plant growth promoting rhizobacteria from wheat rhizosphere and their effect on improving growth, yield and nutrient uptake of plants. *Plant Biosyst.*145: 159-168.

Abd-Alla, M. H., El-Enany, A. E., Nafady, N. A., Khalaf, D. M. and Morsy F. M. 2014. Synergistic interaction of Rhizobium leguminosarum bv. viciae and arbuscular mycorrhizal fungi as a plant growth promoting biofertilizers for faba bean (*Vicia faba* L.) in alkaline soil. *Microbiol. Res.* 169 49–58.

Aslam, M., Ahmad, H.K., Himayatullah, A.M., Ahmad, E., Sagoo, A.G., Hussain, I.U.A and Manzoor, M. 2010. Nodulation, grain yield and grain protein contents as affected by rhizobium inoculation and fertilizer placement in chickpea cultivar bittle-98. *Sarhad J. of Agric.* 26: 467-474.

Barea, J.M., Pozo, M.J., Azcón, R. And AzcoÂn-Aguilar, C. 2005.Microbial co-operation in the rhizo-sphere. *J. of Exp. Bot.* 56: 1761–78.

Carruthers, K., Prithiviraj, B.F.Q., Cloutier, D., Martin, R.C.and Smith, D.L 2000. Intercropping maize with soybean lupin and forages: yield component responses.*Eur. J. of Agron.* 12: 103-115.

Chiezey, U.F., Odunze, A.C. 2009.Soybean response to application of poultry manure and phosphorus fertilizer in the Sub-humid Savanna of Nigeria.*J. of Ecol. and Nat. Environ.* 1: 25-31.

Crespi, M. and Galvez, S.2000. Molecular mechanisms in root nodule development. *J.of plant growth regul.*19: 155-166.

Exner, D.N., Davidson, D.G., Ghaffarzadeh, M. and Cruse, R.M. 1999. Yields and returns from strip intercropping on six Iowa farms. *Am. J. of Altern. Agric.***14:** 69-77.

Ghosh, P.K, Bandyopadhyay, K.K., Wanjari, R.H., Manna, M.C., Misra, A.K., Mohanty, M. and Rao, A.S. 2007.Legume effect for enhancing productivity and nutrient use-efficiency in major cropping systems.An Indian perspective: a review. *J.of Sustainable Agric.* 30: 59-86.

Kachroo, D. and Razdan, R. 2006. Growth, nutrient uptake and yield of wheat (*Triticum aestivum*) as influenced by biofertilizers and nitrogen.*Indian J. of Agron.*51: 37-39.

Li, L., Sun, J., Zhang, F., Li, X., Yang, S., and Rengel, Z. 2001. Wheat/maize or wheat/ soybean strip intercrop-ping yield advantage and interspecific interaction on nutrients. *Field Crops Rese.* 71: 123-137.

Mia, M.A.B., Shamsuddin, Z.H., Zakaria, W. and Marziah, M. 2009.The effect of rhizobacterial inoculation on growth and nutrient accumulation of tissue-cultured banana plantlets under low N-fertilizer regime.*Afr. J. of Biotechnol.* 16: 11–15.

Miransari, M., Rejali, F., Bahrami, H.A. and Malakouti, M.J. 2009. Effects of soil comaction and arbuscular mycorrhiza on corn (*Zea mays* L.) nutrient uptake. *Soil Tillage Res.*103: 282–90.

Naeem, M. and Khan, M.M.A. 2005.Growth, physiology and seed yield of Cassia tora (syn. *Cassia obtusifolia*) as affected by phosphorus fertilization. *J. of Med Aromat. Plant Sci.* 27: 4–6.

Selvakumar, G., Reetha, S. and Thamizhiniyan, P. 2012. Response of Biofertilizers on growth, yield attributes and associated protein profiling changes of black gram (*Vigna mungo* L.). *World Appl. Sci. J.* 16: 1368-1374.

Subashini, H.D., Malarvannan, S. and Kumar, P. 2007.Effect of biofertilizers on yield of rice cultivars in Pondicherry, India.*Asian J. of Agric. Res.* 1: 146-150.

Tajini, F. Trabelsi, M. and Drevon J. 2011. Co-inoculation with *Glomus intraradices* and *Rhizobium tropici* CIAT899 increases P use efficiency for N2 fixation in the common bean (*Phaseolus vulgaris* L.) under P deficiency in hydroaeroponic culture. *Symb.*53 123–129.

Zafar, M., Abbasi, M. K., Khan, M. A., Khaliq, A., Sultan, T. and Aslam M. 2012. Effect of plant growth-promoting rhizobacteria on growth, nodulation and nutrient accumulation of Lentil under controlled conditions. *Pedosph.*22: 848–859

Zafar, M., Abbasi, M.K., Khaliq, A. and Rehman, Z. 2011.Effect of combining organic materials with inorgan-ic phosphorus sources on growth, yield, energy content and phosphorus uptake in maize at Rawalakot Azad Jammu and Kashmir, Pakistan.*Arch. of Appl. Sci. Res.* 3: 199– 212.

Chapter 6

Foodborne Pathogens and their Biological Control: A Review

Ram Kumar Pundir[1], Pranay Jain[2], Vikas Sharma[1] and Viraj K. Mishra[1]*

[1]*Faculty of Department of Biotechnology, Ambala College of Engineering and Applied Research, Devsthali, PO Sambhalkha, Ambala Cantt, Haryana*
[2]*Faculty of Department of Biotechnology, University Institute of Engineering and Technology, Kurukshetra University, Kurukshetra, Haryana*
**E-mail: drramkpundir@gmail.com*

ABSTRACT

Microbiological safety of fresh vegetables, fruits and other food products is of great concern which is still growing under present circumstance. Origin of foodborne diseases through contamination of these products with plant and human pathogens can cause considerable economic losses to the industry. Some of the human pathogens like *Escherichia coli* O157:H7, *Listeria monocytogenes, Salmonella* sp., *Staphylococcus aureus* and *Pseudomonas aeruginosa* and Fungi, like *Penicillium expansum, Aspergillus flavus, Rhizopus* and *Mucor sp.* can cause post-harvest diseases of the food products, are most prevalent contaminants of fresh food products. Several factors can increase the risk of foodborne infections. Many control measures are provided in the food industry to prevent or minimize microbial contamination. Biological approach to control or eradicate these foodborne pathogens is the key research area all over the world. Several natural antimicrobial products *i.e.* biocontrol agents, have been produced, including Lactic Acid Bacteria (LAB) for controlling these pathogens. Over several years, these bacteria have been used in food preservation industry and ensure food safety. These biological agents imparts some extrinsic defensive properties to the food, without changing the intrinsic defensive factors, moreover, they are inexpensive and traditional in nature. All these factors have individual as well as cumulative antagonistic effects on pathogenic microorganisms which causes food spoilage. Biological control plays an important role for controlling food spoilage and foodborne pathogens. Bacteriocins produced by these Lactic Acid Bacteria play a key role against foodborne pathogens because of their antimicrobial nature and thus, are the subject of intense research. Lactic acid bacteria apart from antimicrobial properties also have health-promoting attributes, including anti-mutagenic, anti-carcinogenic activities and therefore can also be used as probiotics. Therefore, further studies should be focused on the mechanism of action of Lactic Acid Bacteria within the gastrointestinal tract in order to effectively control the invasion of foodborne pathogens and to avoid the risk of foodborne diseases.

Keywords: *Foodborne pathogens, Biological control, Lactic acid bacteria, Bacteriocins antimicrobials.*

Introduction

The single most important development of civilizations was the ability to produce and store large quantities of food. When it became possible for people to produce more food then they needed, time from gathering food could be apportioned to culture and community knowledge. Food storage has always been at odds with food spoilage. Some of the earliest evidence of food preservation comes from the post-glacial era, from 15,000 to 10,000 BC. The first use of biological methods was from 6000 to 1000 BC when fermentation was used to produce beer, bread, wine, vinegar, yoghurt, cheese and butter. Much later, in 1864, Louis Pasteur proved that microorganisms in foods were the cause of food spoilage, that heat treatment of food killed these microbes and that sealed containers helped to preserve food by preventing recontamination from atmospheric air. A major development in the distribution and storage of foods came in 1940 with the availability of low cost home refrigerators and freezers. Other developments included the artificial drying, vacuum packaging, ionizing radiations and chemical preservation (Soomro *et al.*, 2002).

Now-a-days as consumers are getting aware of side effects of synthetic chemicals used as preservatives in food, resulting in more drift towards less processed food, however, these untreated or less processed foods can harbour dangerous pathogens which can multiply even under refrigeration and anoxygenic conditions. Use of anti-microbial metabolites of fermentative microorganisms is one of the possible solutions to this dilemma.Many antimicrobial chemicals and metabolites produced by the bacteria associated with fermented food are organic in nature hence doesn't have known side effects. These organic compounds have been in use for long time. Fermentation is a metabolite process that in which the raw substrate is converted into a product having acceptable food qualities with the help of microorganism such as bacteria, yeast and molds. In common fermented products such as yogurt, sauerkraut, kefir, kimchi, pickles *etc.*, growth of undesirable microbes doesn't occur attributed to lactic acid produced by the starter culture bacteria. (Ray and Daeschel, 1992). Food fermentation has been accepted as human health boosters, hence, is of great economic value. Food fermentation by Lactic Acid Bacteria (LAB) has contributed to enhanced worldwide consumption of fermented food already containing other probiotic or health promoting bacteria. Micro-organisms involved in these fermentation processes are of genera such as *Lactococcus, Lactobacillus, Leuconostoc, Streptococcus* and *Pediococcus*. In addition, non-LAB bacteria such as species of *Bifidobacterium* are however, part of normal human intestinal microflora exerting a positive effect on human health (Daly and Davis, 1998; Nuraida, 2015; Thakkar *et al.*, 2015; Evivie *et al.*, 2018).

Microbial Contaminants in Food Products

Diseases and infections caused by food contaminants dictates us the complexicity of food web where we are linked with animals, plants and microbial population in an ecosystem. Some of the contaminating pathogens such as norovirus and *Salmonella* serotype *typhi* remain or sustain in the human reservoir and contaminate the food supply *via* the excreta of infected humans. Furthermore, others sustain in animal reservoirs and contaminate food supply through flesh, milk, or eggs, or because of their presence in the excreta of infected animals, subsequently contaminating the food. Some pathogens exist in multiple hosts in the environment, and thus, can contaminate food supply *via* various pathways reflecting or representing variety of ecosystems that they thrive in. Hence, food safety depends upon the understanding of these pathways well enough to prevent them.

Pathogenic organisms can pass through food supply as they reside in the animals (Host) without showing any symptoms of infection. Hence, their presence doesn't come as apparent to the farmer or breeder thus escaping through the health inspection at slaughter. These then cause infection when consumed by humans. Therefore, prevention paradigm for these microbes requires different approach, reducing the microbial contamination load throughout the food chain. One of the approach could start from the farm or ranch itself where animals are raised, checking their fodder, water and biosecurity. Apart from animals plants also contribute to inclusion of human pathogens in food chain by getting contaminated through manures, water we use and incursion of wild or farm animals (Lynch *et al.*, 2009; Karp *et al.*, 2015).

Emergence of new foodborne pathogens even before the documentation of well-known microbes as foodborne pathogen or identification of newly designated food pathogens is a matter of serious concern. Table 6.1 lists the agents involved in food pathogenesis in past three decades which includes bacteria, viruses, parasites, biotoxins, and a prion. Seventy per cent of them emerge from animal reservoirs; 70 per cent are sustained in animal populations and affecting humans only incidentally.

Contamination Source: Dairy and Dairy Product

Dairy and dairy products such as milk are also potential source of human pathogens. These pathogens can be transferred to milk through farm animals and since milk provide potential growth medium for their growth and development (Farrokh *et al.*, 2013). The predominant human bacterial pathogens that can potentially be transferred to milk include *Staphylococcus aureus, Listeria monocytogenes, Salmonella* spp. and pathogenic *Escherichia coli*.

Sterilization processes such as pasteurization destroys potential pathogenic microorganisms, however, post-pasteurization processing can lead to the recontamination of dairy products. Post pasteurization contaminating bacteria such as *L. monocytogenes* can cause illnesses extending from those with mild flu-like symptoms or gastroenteritis to more serious, such as meningitis and bacteremia

Table 6.1: Major Microbial Pathogens identified as Foodborne Microorganisms

Bacterial Pathogens	*Parasitic Pathogens*
Arcobacterbutzleri	*Cryptosporidium*
Campylobacter jejuni	*Cyclosporacayetenensis*
Campylobacter fetus	*Sarcocystis*
Cronobactersakazaki	*Trypanosoma cruzi*
Escherichia coli 0157: H7	*Viral Pathogens*
Escherichia coli non 0157 STEC	Astrovirus
E. coli enteroaggregative/STEC	Members of Calciviridae
E. coli other diarrhoeagenic	Hepatitis E
Listeria monocytogenes	Rotavirus
Vibrio cholerae 0139 toxigenic	Nipah Virus
Vibrio vulnificus	Prion agent
Vibrio parahaemolyticus	new Variant Creutzfeld Jacob Disease prion
Yersinia enterocolitica	*Fungal Pathogens*
Yersinia pseudotuberculosis	*Aspergillus flavus*

which are potentially fatal in nature. During pregnancy and child birth the infection can cause preterm delivery to neonatal infection and even infant death (Cartwright *et al.,* 2013). *Listeria* can enter human food supply chain through contaminated dairy environment causedby manure or improperly fermented silage. The control of this pathogen in the food industry remains a challenge because of its ubiquitous character such as its ability to grow at low temperatures and to survive and persist even in hostile environments. One of the post pasteurization contaminating process is making of soft cheese as it can support the growth of *Listeria.* Recalls of dairy products, mainly soft cheeses contaminated with the pathogen are relatively frequent. Listeriosis after salmonellosis is known to be the second most frequent foodborne disease causing agent and leads to infection-related deaths in Europe because of its high case-fatality rate (Allerberger and Wagner, 2010). *Staphylococcus aureus,* common environmental microorganism which is normally found in raw milk is capable of producing thermo-stable enterotoxins which causes bovine mastitis disease in animals (Jackson *et al.,* 2012). In humans enterotoxigenic *S. aureus* can cause abdominal cramps, nausea, vomiting, and diarrhea and are found in low levels in dairy products (Benett, 2012). However, poor starter culture activity during fermentation and temperature abuse above 10 °C are factors involved in dairy related outbreaks of staphylococcal intoxication (Cretenet, 2011).

Other toxin producing microbes such as *Escherichia coli* O157:H7, a shiga toxin-producing strain (STEC) serotype of high virulence can also cause severe infection; however, the number of cases in dairy products has remained quite low, probably due to the compliance with good hygienic practices at the farm level as shown by Farrokh *et al.* (2013). The main reservoirs of STEC are ruminants, contaminating milk through subclinical mastitis or faecal routes, and the bacteria that can persist in milking equipment. Vulnerable groups such as young children and elderly people though get affected however, severe cases of bloody diarrhoea or haemolytic uremic syndrome caused by STEC are uncommon,

Contamination Source: Meat and Poultry

Animal faeces is the major sources of microbial contamination in slaughter houses, their raising and roaming places. Cleaning of hides, removing them with care, and preventing the contamination of meat with intestinal contents are few of the efforts slaughter house are focused on, since animal hides and intestinal contents may have pathogens. Apart from animal farms, poultry or bird farms with large populations of birds are one of the setting where infectious agents can spread rapidly which is attributed to hot water dips of slaughtered birds which help remove feathers but can spread intestinal contents to subsequent carcasses. At least 40 per cent of chicken breasts get contaminated by *Campylobacter jejuni,* resulting in major source of *Campylobacter* infections in humans (FDA, 2011; Thomson *et al.,* 2017). People become infected by consuming inadequately cooked poultry or other foods that become cross-contaminated via contact with infectious poultry products. Moreover, infants riding in shopping carts containing raw poultry are at increased risk (Patrick *et al.,* 2010).

Salmonella has long been recognized as a cause of severe human health problem of economic significance in animal and humans. Intestinal tract of humans, wild birds, reptiles, and occasionally of insects and farm animals are the main niche of *Salmonella* serovars. It can also be present in the feedstuff, soil bedding, litter, and fecal matter are commonly identified as sources of *Salmonella* contamination in farms (Le Minor, 1999; Sanchez *et al.,* 2002; Rodriguez *et al.,* 2006; Hoelzer *et al.,* 2011). *Salmonella entering dairy products apart from meat and eggs from the above mentioned route* is the main cause of salmonellosis. Dairy products get contaminated either through inadequate pasteurization or post process handling. Furthermore, most cheeses from raw or pasteurized milk, properly manufactured and aged, appear to pose no significant health risk of *Salmonella* infection. Several factors can increase the risk of foodborne infections and the severity of the diseases, resulting in a reduction in the number of pathogens needed to cause the disease and increased severity. The occurrence of infection in groups of people, with a high risk, accounts for 15–20 per cent of the general population in developed countries as was reviewed by Lund and O'Brien (2011) which includes people mainly with immunodeficiency, pregnancy, children, and the elderly The risk of foodborne disease should be minimized for these vulnerable groups. Several control measures such as good manufacturing practices, sanitation and hygiene measurements for raw material are some of the few steps taken in the food industry to prevent or minimize bacterial foodborne contamination. thus, can prevent or avoid the occurrence of microbial food associated outbreaks.

Biological Control of Foodborne Pathogens

In order to achieve improved food safety against such pathogens, chemical preservatives or physical treatments (*e.g.* high pressure and temperatures) are routinely used in food industry. However, these preservation techniques used for decades have many drawbacks such as toxicity by chemicals used as preservatives (*e.g.* nitrites), alteration in its nutritional and organoleptic properties due to high temperature and pressure. Furthermore, recent increase in consumer's demand for safe but minimally processed products without additives has forced food industry to look for an alternative. Hence, to cater for consumer's demand with the necessary safety standards, biological means of controlling microbial spoilage came into picture. Though these biological means were used traditionally for long time they were replaced by combinations of innovative technologies that include lactic acid bacteria and/or their metabolites as biological antimicrobial systems which don't injure the host or the environment. Biotechnological tools are used to genetically improve useful microorganisms to increase the product of interest or to increase the production of desired product as well as their technical application in food quality. This improvement in microbiological safety and extension of the shelf life of food using non-pathogenic microorganisms and/or their metabolites is defined as biopreservation (De Martinis *et al.*, 2001; Delesa, 2017).

To prevent the outbreaks of foodborne diseases there has been focus on the biological approach through development of natural antimicrobial products for the biocontrol of pathogens by exploiting plant (plant extracts) and microbes (especially lactic acid bacteria) for controlling and eradication of foodborne pathogens (Grasson, 2002). Lactic acid bacteria have a long history in preventing food from spoilage microorganisms and are therefore commonly used in food fermentation. Moreover, they are generally recognized as safe (GRAS) as they produce several metabolites with beneficial health effects. These bacteria have also been exploited for delivery of vaccines, competitive exclusion of pathogens, and synthesis of bioactive compounds (Rawat, 2015).

Population boom has led the usage of high doses of chemicals in the food preservation industry to prevent foodborne diseases. This high usage of chemical preservatives led to increase in resistant food spoilage microorganisms, however, consumer's high demand for safe, minimally processed food and the hazards associated with the use of high doses of chemical preservatives has led to the need for finding safer alternatives in food preservation.

The application of lactic acid bacteria having broad spectrum activity such as fungi-staticability can help minimize food spoilage. Therefore selection and addition of novel isolates apart from improved strains of lactic acid bacteria may be the key to reduce the use of chemicals, extension the shelf life of food products and enhancing their nutritional properties. In food industry starter and protective cultures are currently used in a number of products to safeguard the microbial and sensory quality. Thus, lactic acid bacteria are responsible for natural transformation of agricultural industry through transforming its primary products into safe, shelf stable and delicious and food for human consumption (Lucera *et al.*, 2012).

Lactic Acid Bacteria as Antimicrobial and Bio-preservative Agents

Lactic Acid Bacteria are gram-positive, catalase negative, non-sporulating rod or coccoid in shaped. They are strictly fermentative in nature, producing either two or more different products (mixture of lactic acid, carbon dioxide, acetic acid and/or ethanol (heterofermentation)) or almost one kind of product (entirely lactic acid (homofermentation)) as the major metabolic end-product (Rogosa, 1974; Priest and Campbell, 1996). All members of the genera *Pediococcus, Streptococcus, Lactococcus*and *Vagococcus* are homofermentors, along with some lactobacilli sp., while all *Leuconostoc* sp. as well as some lactobacilli are heterofermentors in nature.

Lactic acid bacteria are ubiquitous in nature, found naturally in variety of food materials rich in nutrients such as dairy, vegetables and meat products, for the fastidious metabolism of lactic acid bacteria (Carr *et al.*, 2002; Bjorkroth and Holzapfel, 2003; Hammes and Hertel, 2003). Some lactic acid bacteria are associated with the mouth, intestine and vagina of mammals (Whittenbury, 1964), while others such as *Lactobacillus plantarum* and *L. reuteri*, and are present in fermented seafood, such as pla-som, afermented Thai fish (Saithong *et al.*, 2010). Lactic acid bacteria are also used in the fermentation industry of dairy products, such as cheese, sour milk, yogurt, and butter, and in combination with yeast they ferment cereal products such as dough (Lavermicocca *et al.*, 2000; Ryan *et al.*, 2008; Muhialdin *et al.*, 2011).

LAB in Human

Humans and animals gastrointestinal tract has complex ecosystem of microbes called gut microflora consisting of both good and bad bacteria in a balance that favours good ones. Among good bacteria, strains of lactic acid bacteria have a history of use as probiotics, in case of human health and controlling human pathogens in farm animals. Controlling of range of human pathogens including *E. coli, Campylobacter jejuni* and *Clostridium perfringens* have been demonstrated by researchers using *Lactobacillus* species (Grasson, 2002). The most common and promising probiotic candidate within the human digestive tract is the *Lactobacillus* spp., furthermore, some of them have natural antibiotic producing ability and cancer fighting properties (Rettger *et al.*, 1935; Gilliland and Speck, 1977; Gill and Guarner, 2004; Zielińska and Krajewska, 2018).

Some strains of *Lactobacillus acidophilus* are particularly beneficial against infectious bacteria such as *Streptococcus, Staphylococcus, E. coli, Salmonella* and *Clostridium botulinum* (Liao *et al.*, 1994). Some strains of *L. acidophilus* have shown to be very effective against viral infections such as polio, HIV, and herpes (Cherpes *et al.*, 2003; Siddiqee *et al.*, 2012), and can also produce hydrogen peroxide thus, showing biocidal effects against undesirable *Candida* yeast and prevent its overgrowth (Mallett *et al.*, 1989). Apart from antibiotic production and cancer fighting ability, lactic acid bacteria are known to produce various enzymes and vitamins into the intestinal lumen (Gorbach, 2000), which exert synergistic effects on digestion, alleviating symptoms of intestinal malabsorption (Gill and Guarner, 2004). Bioavailability of proteins and fats may be increased by enzymatic hydrolysis due to bacteria present in the gut, furthermore, increase in production of short chain fatty acids (SCFA), free amino acids, lactic acid, propionic acid and butyric acid also takes place which when absorbed, contribute to the available energy pool of the host andmay protect against pathological changes in the colonic mucosa (Leavitt *et al.*, 1978; Fernandes *et al.*, 1987; Rombaeu *et al.*, 1990; Young, 1996; Leopold and Eileler, 2000; Rolfe, 2000; Siddiqee *et al.*, 2012).

Botanicals as Antimicrobials

Plant extracts, juices, oils, paste and powders were used to assess their bioactivities including antioxidant and antimicrobial. Most of the work on effectiveness of natural antimicrobial products was directed towards foodborne pathogen (Dhanze *et al.*, 2013).

Food spoilage and their causative agents can be prevented by using chemical preservatives, however, these chemicals despite having high efficacy in controlling the outbreak of food poisoning, their repeated applications has resulted in the accumulation of chemical residues in food and feed chain leading to side effects in humans. Apart from accumulation, extensive use of these chemicals also results in acquisition of microbial resistance (Akinyemi *et al.*, 2006, Bialonska *et al.*, 2010). There have been considerable efforts on developing healthy, sustainable, potentially effective, eco-friendly and safer natural food preservatives as there are considerable side effects of chemical preservatives. One of the natural preservatives having the above mentioned paradigm is plant extract as it has antimicrobial properties for food preservation since they are natural antimicrobial agents, which are nutritionally safe and easily degradable (Cowan, 1999, Duffy and Power, 2001, Nasar-Abbas and Kadir, 2004, Hara Kudo *et al.*, 2004, Mathabe *et al.*, 2005; Berahou *et al.*, 2007, Chika *et al.*, 2007, Akinpelu *et al.*, 2015, Singh, 2018).

Plant Antimicrobial Peptides (pAMPs)

Microbes, animals and plants produce antimicrobial peptides as natural defense compounds against pathogens. They are small molecules which are able to pass through the membrane; however, their size is still debatable among researchers (Rai *et al.*, 2016). Joerger (2003), have estimated their molecular mass between 1-5kDa, however, according to Choon Koo *et al.* (2002), it's between 2-9kDa. Furthermore, they are amphiphilic in nature and are predominantly positively charged (Jenssen *et al.*, 2006). Although antimicrobial peptides are produced by many organisms, this section specifically deals with Plant derive AMPs and their mode of action (Boto *et al.*, 2017).

Plant derived AMPs play a key role in plant's defense against its pathogens both by getting triggered upon infection as well as being a part of pre-existing defense barrier, *i.e.*, either overexpression of the peptide due to triggering by pathogens results in enhanced tolerance of the plant to the pathogen or peptide-sensitive mutants of pathogens show decreased virulence in plant tissues due to preexisting specific peptide (Jenssen *et al.*, 2006). These peptides are broad spectrum acting against both Gram-negative and Gram-positive bacteria; however, some are narrow spectrum (Pelegrini *et al.*, 2011). Recent review by Salas *et al.* (2015) has listed more than twenty different pAMPs isolated from different plant species and from different tissues further more these peptides are predominantly cysteine-rich in nature. These pAMPs can be divided into distinct protein families based on their structure and amino acid sequence, although there is debate amongst researchers characteristics

(Choon Koo *et al.*, 2002). These plant antimicrobial peptide families are (i)Thionins (ii) Plant Defensins and four defensin groups (iii) Lipid Transfer Proteins (LTPs) (iv) Myrosinase-Binding Proteins (MBPs), (v) Hevein- and Knottin-Like Peptides (vi) Snakins (vii) Cyclotides (viii) Peptides from Hydrolysates (Nawrot *et al.*, 2014).

Phytochemicals

Phytochemicals are simple organic compounds which are confer organoleptic properties and therefore are nonnutritive in nature; furthermore they serve as serve as antimicrobial agents. Chemically Phytochemicals are simple phenolic compounds (Simple bioactive phytochemicals with single substituted phenolic ring), flavonols (phenolic compounds with a carbonyl and a 3-hydroxyl group), Flavones (Phenolic compounds with one carbonyl group), flavonoids (hydroxylated phenolic structures with a C3–C6 aromatic ring linkage), Essential oils, or terpenes (secondary metabolites based on an isoprene structure), Quinones (aromatic compounds with two carbonyls, providing a stable source of free radicals), Alkaloids (heterocyclic nitrogen compounds having limited microbicidal activity as well as possessing an antidiarrheal activity) (Hintz *et al.*, 2015).

Antimicrobial properties of these phytochemicals is determined by composition, concentration, structure and functional groups, furthermore, phenolic phytochemicals among all the phytochemicals are the most effective, (Holley and Patel, 2005). Most widely used phytochemicals is essential oil which has been in use as food preservatives across continents since ancient times. Other phytochemicals such as alkaloids from plants have also been used as antimicrobials in food. Recently, Negi (2012) has listed many different phytochemicals and their antibacterial activities. Various plant extracts include phenolic compounds and essential oil having antifungal activities have also been recently reviewed by Cabral *et al.* (2014). Polyphenolic compounds from fruits such as cranberry, pomegranate, blueberry, raspberry, and grape were also summarized for their antiviral activities against human enteric viruses (D'Souza, 2014; Hintz *et al.*, 2015).

Conclusion

The biological control using generally recognized as safe (GRAS) organisms which include bacteriocin-producing strains of Lactic Acid Bacteria along with antimicrobial products such as Bacteriocins are of great interest in food industry. These products are called bio-preservatives. However, effect of environmental factors, on these bacteriocinogenic strains and efficacy of their products both qualitative and quantitative have to be established before recognizing them as bio-preservatives in the food industry.

Plant extracts, can also be used effectively both as stand-alone and as concoction or in combinations with other extracts, apart from useful microbes, to protect the consumer from foodborne pathogens and by preventing early spoilage of the foods. However, these natural products such as plant-derived antimicrobial compounds and alternative preservatives should be fully assessed for their feasibility for food application. These new fields of research though have great possibility of employing antimicrobials from the extensive source of plants and lactic acid bacteria worldwide; however, evaluation to meet regulatory requirements is need of the hour.

Acknowledgement

The authors are grateful to Management of Ambala College of Engineering and Applied Research, Devsthali, Near Mithapur, Sambhalkha, Ambala (Haryana, India) and also thankful to Hon'ble Vice-Chancellor, Kurukshetra University, Kurukshetra and Director UIET, KUK, Haryana, India for encouraging us to write this manuscript.

REFERENCES

Akinpelu, D.A., Aiyegoro, O.A., Akinpelu, O.F., Okah, A.I. 2015. Stem bark extract and fraction of *Perseaamericana* (Mill) exhibits bactericidal activities against strains of *Bacillus cereus* associated with food poisoning. *Molecul.* 20: 416-429.

Allerberger, F. and Wagner, M. 2010.Listeriosis: a resurgent foodborne infection. *Clin.Microbiol. Infect.* 16(1): 16–23.

Bialonska, D., Ramnani, P., Kasimsetty, S.G., Muntha, K.R., Gibson, G.R., Ferreira, D. 2010. The influence of pomegranate by-product and punicalagins on selected groups of human intestinal microbiota. *Int. J. Food Microbiol.*140: 175-182.

Bjorkroth, J. and Holzapfel, W. 2003. Genera *Leuconostoc, Oenococcus*and*Weissella,* In M. Dworkin *et al.* (Eds).The Prokaryotes: An Evolving Electronic Resource for the MicrobiologicalCommunity,3rd edn, Springer-Verlag, New York.

Boto, A., Pérez de la, J. M., Lastra and González, C. 2017.The Road from Host-Defense Peptides to a New Generation of Antimicrobial Drugs.Molecules.23: 311.

Cabral, L. D. C. and Pinto, V. F. and Patriarca, A. 2013. Application of plant derived compounds to control fungal spoilage and mycotoxin production in foods. *Int. J. Food Microbiol.* 166(1): 1–14.

Carr, F. J., Chill, D. and Maida, N. 2002. The lactic acid bacteria: a literature survey. *Crit. Rev. Microbiol.* 28: 281-370.

Cartwright, E. J.,Jackson, K. A., Johnson, S. D., Graves, L. M., Silk, B. J. and Mahon, B. E. 2013. Listeriosis outbreaks and associated food vehicles, United States, 1998–2008. *Emerg. Infect. Dis.* 19(1): 1–9.

Cherpes, T. L., Meyn, L. A., Krohn, M. A. and Hillier, S. L. 2003. Risk factors for infection with herpes simplex virus type 2: role of smoking, douching, uncircumcised males, and vaginal flora. *Sex Transm.Dis.* 30: 405- 410.

Chika, C.O, Jude, N.O., Ifeanyi, C.O., Anyanwu, N.B. 2007. Antibacterial activities and toxicological potentials of crude ethanolic extracts of *Euphorbia hirta.J. Am. Sci.,* 3 (3): 11-16

Choon Koo, J., Jin Chun, H., Cheol Park, H.et al. 2002. Over-expression of a seed specific hevein-like antimicrobial peptide from Pharbitis nil enhances resistance to a fungal pathogen in transgenic tobacco plants *Plant Mol. Biol.* 50(3): 441–452.

Cowan, M.M. (2000). Plant products as antimicrobial agents. *Clin.Microbiol. Rev.,* 12: 564-582

Cretenet, M., Even, S. and Le Loir, Y. 2011. Unveiling *Staphylococcus aureus* enterotoxin production in dairy products: a review of recent advances to face new challenges. *Dairy Sci. Technol.* 91(2): 127–150.

Daly, C. and Davis, R. 1998. The biotechnology of lactic acid bacteria with emphasis on applications in food safety and human health. Agri. Food Sci Finland, 7: 2, 251- 264.defense peptides, *Biopolym.* 47 (6): 479–491.

Delesa, D. A. 2017. Bacteriocin as an advanced technology in food industry.*Int. J. Adv. Res. Biol. Sci.* 4(12): 178-190

De Martinis, Elaine, C. P. and Bernadette, D. G. M. Franco, B.D.G.M. 2001. Inhibition of *Listeria monocytogenes* in a pork product by a *Lactobacillus sake* strain, *Intern. J. Food Microbiol.* 42(1-2): 119-126.

Dhanze, H., Khurana, S.K. and Mane, B.G. 2013.Effect of seabuckthorn leaf extract on microbiological quality of raw chicken during extended periods of storage.*J. Food Quality*, 36(1): 59-66.

D'Souza, D. H. 2014. "Phytocompounds for the control of human enteric viruses," *Cur. Opinion in Virol.* 4: 44–49.

Duffy, C.F., Power, R.F. 2001.Antioxidant and antimicrobial properties of some Chinese plant extracts. *Int. J. Antimicrob. Agent.* 17: 527-529

Evivie, S. E., Huo, G. C. and Bian, X. 2017. Some current applications, limitations and future perspectives of lactic acid bacteria as probiotics. *F. Nutri. Res.* 61: 1318034.

Farrokh, C., Jordan, K. and Auvray, F. 2013. Review of Shiga-toxinproducing *Escherichia coli* (STEC) and their significance in dairy production. *Int. J. Food Microbiol.*162(2): 190–212.

FDA (Food and Drug Administration). 2011. 2010 Retail meat report–National Antimicrobial Resistance Monitoring System. Rockville, MD: Food and Drug Administration.

Fernandes, C. F., Shahani, K. M. and Amer, M. A. 1987. Therapeutic role of dietary lactobacilli and lactobacillic fermented dairy products. *FEMS Microbiol.Rev.* 46: 343–356.

Gill, H. S. and Guarner, F. 2004. Probiotics and human health: a clinical perspective. *Post Grad. Med. J.*80: 516-26.

Gilliland, S. E. and Speck, M. L.1977.Deconjugation of bile acids by intestinal lactobacilli. *Appl Environ Microbiol.* 33: 5-18.

Gorbach, S. L. 2000. Probiotics and gastrointestinal health. *Am. J. Gastroenterol.* 95: S2–S4.

Hammes, W. P. and Hertel, C. 2003.The Genera *Lactobacillus* and *Carnobacterium*. In *The Prokaryotes: An Evolving Electronic Resource for the Microbiological Community.* M. Dworkin, M. (Ed.). Springer-Verlag, New York.

Hara, Y., Kudo, Kobayashi, A., Sugita-Konishi, Y., Kondo, K. 2004.Antibacterial activity of plants used in cooking for aroma and taste. *J. Food Prot.* 67: pp. 2820-2824.

Hintz, T., Karl, K., Matthews and Rong, Di. 2015.The Use of Plant Antimicrobial Compounds for Food Preservation. *BioMed Res. Int.* Pp 1-12.

Hoelzer K., Switt A. I. M., Wiedmann M. Animal contact as a source of human non-typhoidal salmonellosis. *Veterinary Research*.2011;42(1, article 34) doi: 10.1186/1297-9716-42-34.

Holley, R. A. and Patel, D. 2005. Improvement in shelf-life and safety of perishable foods by plant essential oils and smoke antimicrobials, *Food Microbiol.*, 22(4): 273–292.

Jenssen, H., Hamill, P. and Hancock, R. E. W. 2006. Peptide antimicrobial agents, *Clinical Microbiol. Rev.*, 19(3): 491–511.

Joerger, M. C. and Klaenhammer, T. R. 1986.Characterization and purification of helveticin J and evidence for a chromosomally determined bacteriocin produced by Lactobacillus helveticus 481.*J. Bacteriol.* 167: 439-446.

Joerger, R. D. 2003. Alternatives to antibiotics: bacteriocins, antimicrobial peptides and bacteriophages, *Poultry Sci.* 82(4): 640–647.

Johnson, T. J., Shank, J. M. and Johnson, J. G. 2017.Current and Potential Treatments for Reducing *Campylobacter* Colonization in Animal Hosts and Disease in Humans. Front. Microbiol.8: 487.

Karp, D. S. Baur, P., Atwill, E. R., De Master, K., Gennet, S., Iles, A., Nelson, J. L., Sciligo, A. R. and Kremen, C. 2015. The Unintended Ecological and Social Impacts of Food Safety Regulations in California's Central Coast Region. *BioScience,* 65: 1173–1183.

Lavermicocca, P., Valerio, F. and Visconti, A. 2003. Antifungal activity of phenyllactic acid against molds isolated from bakery products. *Appl. Environ. Microbiol.*69: 634-640.

Le Minor, L. 1999. *The Prokaryotes: A Handbook on the Biology of Bacteria: Ecophysiology, Isolation, Identification, Applications.* 2nd. Ed, New York, NY, USA, The genus *Salmonella* 1: pp. 2760–2774.

Leavitt, J., J. C. Barrett, B. D. Crawford, and P. O. Tso.1978. Butyric acid suppression on the in vitro neoplastic state of Syrian hamster cells. *Nature*271 (5642): 262-265.

Lund, B. M. and O'Brien, S. J. 2011.The occurrence and prevention of foodborne disease in vulnerable people.*Foodborne Pathog. Dis.* 8(9): 961–973.

Lucera, A., Costa, C., Conte, A. and Del Nobile, M. A. 2012. Food applications of natural antimicrobial compounds.*Front. In Microbiol.*3: 1-13.

Lynch, M. F., Tauxe, R. V. and Hedberg, C. W. 2009. The growing burden of foodborne outbreaks due to contaminated fresh produce: Risks and opportunities. *Epidemiol. Infect.*137: 307-315.

Mallett, A. K., C. A. Bearne, and I. R. Rowland. (1989). The influence of incubation pH on the activity of rat and human gut flora enzymes. J. Appl. Bacteriol. 66: 433–437.

Mathabe, M.C., Nikolova, R.V., Lall, N., Nyazema, N.Z. 2005. Antibacterial activities of medicinal plants used for the treatment of diarrhea in Limpopo Province, *South Africa. J. Ethnopharmocol.* 105: 286-293

Nasar-Abbas, S.M. and Kadir, H. A. 2004.Antimicrobial effect of water extract of sumac (*Rhuscoriaria* L.) on the growth of some foodborne bacteria including pathogens.*Int. J. Food Microbiol.* 97: 63-69.

Nawrot, R., Barylski, J. Nowicki, G., Broniarczyk, J., Buchwald, W. Gozdzicka-Jozefiak, A. 2015. Plant antimicrobial peptides.*Folia Microbiol.* 59: 181–196

Negi, P. S. 2012. "Plant extracts for the control of bacterial growth: efficacy, stability and safety issues for food application," *Int. J. Food Microbiol.* 156 (1): 7–17.

Nuraida, L. 2015. A review: Health promoting lactic acid bacteria in traditional Indonesian fermented foods. *Food Sc. Hum.Welln.*4(2): 47-55.

Patrick, M. E., Mahon, B. E., Zansky, S. M., Hurd, S. andScallan, E. 2010. Riding in shopping carts and exposure to raw meat and poultry products: Prevalence of, and factors associated with, this risk factor for *Salmonella* and *Campylobacter* infection in children younger than 3 years.*J. Food Protect.*73(6): 1097-1100.

Pelegrini, P. B., Del Sarto, R. P., Silva, O. N. Franco, O. L. and Grossi-de-Sa, M. F. 2011. Antibacterial peptides from plants: what they are and how they probably work,*Biochem. Res. Int.* 9.

Priest, F. G. and Campbell, I. 1996. Brewing Microbiology.2nd Edn.International Centre for Brewing and Distilling.Chapman and Hall, UK. pp 134-156.

Rawat, S. 2015. Food Spoilage: Microorganisms and their prevention. *Asian J. Plant Sci. Res.* 2015, 5(4): 47-56.

Ray, B. and Daeschel, M. 1992. Food biopreservatives of microbial.origin. CRC Press, In. Boca Raton, Florida: 3-11

Rai, M., Pandit, R., Gaikwad, S. and Kövics, G. 2016.Antimicrobial peptides as natural bio-preservative to enhance the shelf-life of food.*J. Food SciTechnol* 53(9): 3381–3394

Rodriguez A., Pangloli P., Richards H. A., Mount J. R., Draughon F. A. 2006. Prevalence of *Salmonella* in diverse environmental farm samples. *J. Food Prot.*69(11): 2576–2580.

Rogosa, M. 1974. Gram Positive Asporogenous, Rod-Shaped Bacteria. In: Bergey's Manual of Determinative Bacteriology. Buchanan, R.E. and Gibbons, N.E. (Eds.) 8th Edn. Williams and Wilkins, Baltimore. Pp 576-593.

Rolfe, R. D. 2000. The role of probiotic cultures in the control of gastrointestinal health. *J. Nutr.*130: 396S–402S.

Ryan, L. A. M., Dal Bello, F. and Arendt, E. K. 2008.The use of sourdough fermented by antifungal LAB to reduce the amount of calcium propionate in bread. *Int.J. Food Microbiol.* 125: 274-278.

Saithong, P., Panthavee, W., Boonyaratanakornkit, M. and Sikkhamondhol, C. 2010. Use of a starter culture of lactic acid bacteria in plaa-som, a Thai fermented fish. *J. Biosci. Bioengg.*110: 553-557.

Salas, C. E, Badillo-Corona, J. A., Ramírez-Sotelo, G. and Oliver-Salvador, C. 2015."Biologically active and antimicrobial peptides from plants," *BioMed Res. Internation.* 11.

Sanchez S., Hofacre C. L., Lee M. D., Maurer J. J., Doyle M. P. 2002.Animal sources of salmonellosis in humans.*J. Amer. Veter. Med. Assoc.* 221(4): 492–497. doi: 10.2460/javma.2002.221.492.

Shan, B., Cai, Y., Brooks, J.D., Corke, H. (2007). The *in vitro* antibacterial activity of dietary spice and medicinal herb extracts. *Int. J. Food Microbiol.* 117: 112-119

Siddiqee, M. H., Sarker, H. and Shurov, K. M. 2012. Assessment of probiotic application of lactic acid bacteria (LAB) isolated from different food items *Stamford J. Microbiol.* 2(1): 10-14.

Singh, V. P. 2018. Recent approaches in food bio-preservation-a review. *Open Veter. J.* 8(1): 104-111.

Soomro, A.H., Masud, T. and Anwaar, K. 2002.Role of Lactic Acid Bacteria (LAB) in Food Preservation and Human Health – A Review.*Pak. J. Nut.*1(1): 20-24.

Thakkar, P., Modi, H. A. and Prajapat, J. B. 2015. Isolation, characterization and safety assessment of lactic acid bacterial isolates from fermented food products. *Int. J. Curr. Microbiol. App. Sci.* 4(4): 713-725.

Whittenbury, R. 1964. Hydrogen peroxide formation and catalase activity in the lactic acid bacteria. *J. Gen. Microbiol.*35: 13-26.

Yamamura, A., Murai, A., Takamatsu, H., Watabe, K. 2000. Antimicrobial effect of chemical preservatives on enterohemorrhagic*Escherichia coli* O157: H7. *J. Health Sci.*, 46: 204-208

Young, G. 1996. Prevention of colon cancer: role of short chain fatty acids produced by intestinal flora. *Asia Pacific J. Clin.Nutr.*5: 44-47.

Zielińska, D. and Krajewska, D.K. 2018. Food-Origin Lactic Acid Bacteria May Exhibit Probiotic Properties: Review. *BioMed Res. Internation.* Article ID 5063185; 15.

Chapter 7

Optimized Production of Crude Cellulase and Xylanase Enzymes from *Bacillus subtilis* through Response Surface Methodology

*Chakarvati Sango and Jitender Sharma**

Department of Biotechnology, Kurukshetra University, Kurukshetra – 136 119, Haryana
**E-mail: jsharma@kuk.ac.in*

ABSTRACT

Response surface methodology and central composite design were used for optimized production of cellulase and xylanase enzymes from *Bacillus subtilis* in submerged state fermentation using rice bran as carbon substrate at pH 9.0. The Variables used in this research were inoculum size, inoculum age, fermentation production time, peptone, yeast extract, xylose, pH, temperature, KNO_3 and shaking. Statistical analysis in Plackett Burman design and central composite design showed inoculum age (6h), production time (42h), peptone (1.2g/50ml), yeast extract (1.1g/50ml), xylose (0.3g/50ml), temperature (37°C) and shaking (180 rpm) had a significant effect on cellulase and xylanase production. The resultant improvement in enzyme activity of cellulase and xylanase were 15.67IU/ml and 326.76 IU/ml respectively while it predicted 15.29 IU/ml for cellulase and 317.53 IU/ml for xylanase.

Introduction

Screening of micro-organisms on the basis of their ability to produce various enzymes and to analyze their actions on various environmental substrates is the key factor to develop any economical feasible technology (Aristidou and Penttilä, 2000). Lignocellulose is one of the most abundent sources of energy on earth which is composed of Cellulose, hemicellulose and lignin monomer units (Sanchez, 2009). Cellulose is an unbranched glucose polymer in which anhydro-D-glucose units linked by 1, 4-β-D-glucoside bonds, while hemicellulose is a polymer of xylan subunits attached by β-1, 4-D-xylopyranose residues, present between lignin and cellulose residues (Robson *et al.*, 1989; Lama *et al.*, 2004; Holker *et al.*, 2005; Collins *et al.*, 2005). Cellulases and xylanases are the key enzymes for hydrolysis of such lignocellulosic waste (Esterbauer *et al.*, 1991; Nieves *et al.*, 1998). Both these enzymes have gained importance in microbiology due to their applications in various industries such as food, paper and pulp, fermentation, textiles, manufacturing as well as biofuel industries due to their specific properties such as thermostability and thermotolerence (Chakdar *et al.*, 2016; Polizeli *et al.*, 2005).

Various Bacterial and fungal strains have been used for production of cellulase and xylanase enzymes in which bacterial strain *Bacillus subtilis* showed significant production of both enzymes after optimization of all factors. Optimization involves improvement of the biochemical processes in order to obtain a valid product in more ecofriendly and less time consuming approch. Designing a proper production media and conditions (temperature, inoculum size and age, pH and incubation time and carbon substrate) is important to improve the productivity of microbial metabolites via fermentation processes (Hegde *et al.*, 2013). Response surface methodology involves statistical and mathematical approches which depends upon the fit of polynomial equation to experimental data in order to make statistical predictions of an optimized response (Bezerra *et al.*, 2008). RSM is powerful approch for optimizing effect of indipandant variable in production processes for enzymes and chemicals (Khurana *et al.*, 2007; Vimalashanmugam *et al.*, 2012; Rajeshkannan *et al.*, 2009). Greater yield of cellulase with cost effective method can be achieved by optimization of process parameters. The objective of present study is to statistically optimize the various production parameters for submerged fermentation through central composite design (CCD) in response surface methodology using rice bran as carbon substrate.

Experimental

Isolation of Bacterial Strain

Soil samples were collected from paper mill campus, yamuna nagar. Screening studies showed the presence of clear zones in cellulose and wheat bran enriched petri plates. Characterization studies showed the specific strain to be the alkalophilic *Bacillus subtilis* sp. The following was stored in nutrient agar slant cultures at 4°C.

Enzyme Production

Crude cellulase enzyme was produced in Erlenmeyer flask (250 ml) containing 50 ml of basal medium having (g/L): yeast extract, 5.0; $MgSO_4.7H_2O$, 0.2; KH_2PO_4, 1.0 supplemented with rice bran, 5.0 and peptone, 5.0; pH 9.0. The flasks were inoculated with 2 per cent of the 24 h grown inoculum and incubated at 37°C under shaking conditions (180 rpm) for incubation period of 48 h. The enzyme was harvested by centrifuging at 10,000×g for 15 min.

Enzyme Assay

Both cellulase and xylanase activities were determined by measuring the amount of reducing sugars using Dinitrosalicylic acid (DNS) or Bailey's method (1992) and Miller's method (1959) respectively. For CMCase activity, the reaction mixture of 500µl of 2 per cent w/v Carboxy methyl cellulose, as a substrate prepared in 0.2 M glycine NaOH buffer (pH 9.0), was incubated at 50°C in water bath for 30 min which includes 10µl of crude enzyme filterate incubated in same water bath with proper dilution. Control for both, heat killed substrate and enzyme were included simultaneously which were treated for 20 min at 100°C. After the proper incubation, 1.5ml of DNS reagent was added to the reaction mixture, and was kept in boiling water for 15 min for proper color development indicating the reaction between substrate and enzyme. After cooling down, the reaction mixture was mixed and optical density was measured at 540 nm at UV-VIS spectrophotometer. The values were compared with the values of standered curve for D-glucose.

For xylanase activity, 10 µl of suitable enzyme dilution was added to the 490 µl of 1 per cent birchwood xylan which was prepared in 0.2M glycine NaOH buffer (pH 9.0) at 50°C. Controls were also included as described above. The reaction mixture was incubated at 50°C for 15 min. After that, 1.5 ml of DNS reagent was added to the reaction mixture and boiled for 15 min. Optical density for xylanase was determined under UV-VIS spectrophotometer and reducing sugar was determined by comparing the values using standered curve prepared for D-xylose.

Production of Cellulo-Xylanolytic Enzymes under Submerged Fermentation using Plackett-Burman Design (PB) and Central Composite Design Methods (CCD)

Plackett Burman is two level factorial design used to screen a large number of factors which significantly affects the process. A total of 12 experiments were generated in Placket-Burman design for cellulase and xylanase production simultaneously in the same fermentation medium. Ten variables, *i.e.* Inoculum size and age, Production Time, Rice bran, Temperature, pH, Peptone, Yeast extract, xylose and shaking speed were selected to study their effect on cellulase and xylanase production. Each variable was studied at two extreme levels, *i.e.* higher and lower for their screening (Table 7.1). Central composite design (CCD) was used for the further evaluation of significant factors obtained through Plackett Burman design. CCD investigates each variable at minimum and maximum level for obtaining best condition for the enhanced production of cellulase and xylanase under SmF. Range and levels of significant variables in terms of actual and coded factors for central composite design for the simultaneous production of cellulase and xylanase are given in Table 7.2. Each factor was studied at five different levels, that is, lowest (-3.36), low (-1), medium (0), high (+1), and highest (+3.36) to estimate the experimental error and providing extra information region (Haaland, 1990). Total of (152) experiments were generated in CCD.

Table 7.1: Experimental Variables at Various Levels Used in the Production of Cellulase and Xylanase Enzymes using Plackett-Burman Design

Variables	*Units*	*Symbol Code*	*Experimental Values*	
			Lower	*Higher*
Inoculum size	per cent	A	0.2	1
Inoculum age	h	B	2	10
Production time	h	C	24	60
pH	-	D	7	11
Temperature	°C	E	30	45
Peptone	g/50ml	F	0.2	2
Yeast extract	g/50ml	G	0.2	2
KNO_3	g/50ml	H	0.1	0.5
Xylose	g/50ml	J	0.1	0.5
Shaking	rpm	K	120	240

Table 7.2: Central Composite Design for the Simultaneous Production of Cellulase and Xylanase are given in Terms of Actual and Coded Factors

Variables		*Levels*				
		-3.36	*-1*	*0*	*+1*	*+3.36*
A	Inoculum age (h)	-7.45	2	6	10	19.45
B	Production time (h)	-18.54	24	42	60	102.54
C	Temperature (°C)	12.27	30	37.5	45	62.72
D	Peptone (g/50ml)	-1.92	0.2	1.1	2	4.12
E	Yeast extract (g/50ml)	-1.92	0.2	1.1	2	4.12
F	Xylose (g/50ml)	-0.37	0.1	0.3	0.5	0.97
G	Shaking (rpm)	-21.81	120	180	240	381.81

Statistical Analysis

The statistical software design expert 6.0.8 Stat-Ease, Inc. Minneapolis, USA was used for regression analysis of the experimental data and for plotting the response surface. Variance analysis (ANOVA) was used to estimate the statistical parameters.

Results

Screening of Significant Variables using Plackett-Burman Design

A total of 12 variables were analyzed for co-production of cellulo-xylanolytic enzymes at 37°C using Plackett Burman design. The average cellulase and xylanase activity of observed values and predicted values by the equation were taken as response Y (Table 7.3).

Table 7.3: Plackett–Burman Design with Experimental and Predicted Values of Cellulase and Xylanase Production

Sl.No.	*Variables*										*Cellulase (IU/mL)*		*Xylanase (IU/mL)*	
	A	*B*	*C*	*D*	*E*	*F*	*G*	*H*	*J*	*K*	*Experimental Value*	*Predicted Value*	*Experimental Value*	*Predicted Value*
1	1	2	60	7	30	0.2	2	0.5	0.5	120	3.75±0.45	3.73	23.73±1.19	24.35
2	1	10	24	11	30	0.2	0.2	0.5	0.5	120	1.26±0.18	1.28	8.36±0.85	7.74
3	0.2	10	60	7	45	0.2	0.2	0.1	0.5	240	1.28±0.19	1.26	12.57±0.91	13.19
4	1	2	60	11	30	2	0.2	0.1	0.1	240	10.42±1.32	10.40	121.18±3.02	121.80
5	1	10	24	11	45	0.2	2	0.1	0.1	120	5.69±0.57	5.67	66.27±1.37	66.89
6	1	10	60	7	45	2	0.2	0.5	0.1	120	11.57±1.78	11.59	147.63±3.14	147.01
7	0.2	10	60	11	30	2	2	0.1	0.5	240	**13.89±1.97**	**13.91**	**163.89±3.48**	**163.27**
8	0.2	2	60	11	45	0.2	2	0.5	0.1	240	4.57±0.78	4.59	34.59±1.16	33.97
9	0.2	2	24	11	45	2	0.2	0.5	0.5	120	6.18±0.82	6.16	77.95±2.39	78.57
10	1	2	24	7	45	2	2	0.1	0.5	240	7.53±0.94	7.55	88.39±2.47	87.77
11	0.2	10	24	7	30	2	2	0.5	0.1	240	13.37±1.74	13.35	146.33±3.25	146.95
12	0.2	2	24	7	30	0.2	0.2	0.1	0.1	120	1.63±0.29	1.65	12.61±0.62	11.99

A: Inoculum size (per cent), B: Inoculum age (h), C: Production Time (h), D: Ph, E: Temperature (°C), F: Peptone (g/50ml), G: Yeast extract (g/50ml), H: KNO_3 (g/50ml), J: Xylose (g/50ml), K: Shaking (rpm), L: Dummy.

The 2nd order polynomial equation that represented a suitable model for cellulase production under Plackett-Burman design is given below:

Cellulase activity (IU/ml) = 6.77-0.066A-1.08 B+0.82C+
0.24D--0.62E+3.73 F+1.37 G+0.022 H-1.11J-0.36K *eq (1)*

Analysis of variance (ANOVA) of Plackett–Burman design for cellulase production showed thatthe Model F- value of 4866.32 make themodel significant (Table 4). There is only a 1.12 per cent chance that a "Model F-Value" this large could occur due to noise. In case of cellulase production, P value (0.0112) less than 0.0500 indicates that the model terms are significant. In this case B, C, E, F, G, J, K are significant model terms. Values of "Prob>F" greater than 0.1000 indicate that the model terms are not significant. The "Pred R-Squared" of 0.9970 is in reasonable agreement with the "Adj R-Squared" of 0.9998. "Adeq Precision" measures the signal to noise ratio. A ratio greater than 4 is desirable. This value of 190.706 indicates an adequate signal.

The 2nd order polynomial equation that represented a suitable model for xylanase production under Plackett-Burman design is given below:

Xylanase activity (IU/ml) = 75.21+0.71A+15.55 B+8.64C+
3.42D--4.06E+48.94 F+11.91G-2.19H-12.81J-6.72K *eq (2)*

Where A, B, C, D, E, F, G, H J, K, L represents inoculum size, inoculum age,production time, pH, temperature, peptone, yeast extract, KNO_3, xylose, shaking and dummy.

Analysis of variance (ANOVA) of Plackett–Burman design for xylanase production showed thatthe Model F- value of 805.33 make themodel significant (Table 7.4). There is only a 2.74 per cent chance that a "Model F-Value" this large could occur due to noise. In case of xylanase production, P value (0.0274) less than 0.0500 indicates that the model terms are significant. In this case B, C, F, G and J are significant model terms. Values of "Prob>F" greater than 0.1000 indicate that the model terms are not significant. The "Pred R-Squared" of 0.9821 is in reasonable agreement with the "Adj R-Squared" of 0.9998. "Adeq Precision" measures the signal

Table 7.4: Results of Analysis of Variance (ANOVA) of Plackett–Burman Design for Cellulase and Xylanase

Terms	*Cellulase Yield*	*Xylanase Yield*
F-value	4866.32	805.33
P-value (Prob >*F*)	0.0112	0.0274
Mean	6.76	75.29
R-Squared	1.0000	0.9999
Adj R-Squared	0.9998	0.9986
Pred R-Squared	0.9970	0.9821
C.V.	1.02	2.85
Adeq Precision	127.010	75.636
PRESS	190.706	664.24

to noise ratio. A ratio greater than 4 is desirable. This value of 75.636 indicates an adequate signal. The cellulase activity varied between 1.26 and 13.89IU/ml, while the xylanase activity varied between 8.36 and 163.89 IU/ml.

Optimization of Significant Variables through CCD

Central composite design (CCD) was used for the further evaluation of significant factors obtained through Plackett Burman design. Range and levels of seven significant variables (±A- Inoculum age (h), B- Production time (g/50ml), C- Temperature (°C), D- Peptone (per cent),E- Yeast extract (g/50ml), F- Xylose (g/50ml), G- Shaking (rpm), in terms of actual and coded factors for central composite design for the simultaneous production of cellulase and xylanase are given in Table 7.4. Each factor was studied at five different levels, that is, lowest (-3.36), low (-1), medium (0), high (+1), and highest (+3.36). Total of (152) experiments were generated in CCD. The full experimental design with experimental and predicted response is shown in Table 7.5.

The 2nd order polynomial equation that represented a suitable model for cellulase production under CCD is given below:

Cellulase=11.54+1.09A+0.14B-0.12C+1.41D+1.33E+0.26F+ 0.22G-1.04A2-1.14B2-0.29C2-0.76D2-0.82E2-0.56F2-1.07G2+ 0.10AB-0.36AC+0.61AD+0.37AE-0.12AF-0.17AG+0.023BC+ 0.049BD-0.076BE-0.061BF+0.075BG-0.25CD-0.34CE+ 0.19CF-0.036CG+0.19DE+0.12DF+0.038DG-0.28EF+0.20EG-0.37FG *eq (3)*

Results of analysis of variance (ANOVA) of central composite design for cellulase showed that Model F-value of 48.91, which implies that the model is significant (Table 7.6). There is only a 0.01 per cent chance that a "Model F-Value" this large could occur due to noise. P value less than 0.0500 indicates that the model terms are significant. In this case A, D, E, F, G, A2, B2, C2, D2, E2, F2, G2, AC, AD, AE, AG, CD, CE, CF, DE, EF, EG, FG are significant model terms.The "Pred R-Squared" of 0.8312 is in reasonable agreement with the "Adj R-Squared" of 0.9174. "Adeq Precision" measures the signal to noise ratio. A ratio greater than 4 is desirable. The ratio of 36.662 indicates an adequate signal.

The 2nd order polynomial equation that represented a suitable model for xylanase production under CCD is given below:

Xylanase=211.95+23.41A+1.91B-1.66C+29.34D+26.65E+3.91F+5.56G-15.60A2-22.17B2+7.16C2-15.02D2-16.70E2-13.05F2-20.57G2+2.33AB-4.50AC+15.11AD+8.68AE-3.42AF-3.42AG-0.19BC-1.48BD+0.22BE-1.41BF+2.11BG-1.08CD-9.34CE+4.20CF-2.27CG+7.37DE+2.61DF+1.49DG-6.95EF+4.40EG-7.22FG *eq (4)*

Results of analysis of variance (ANOVA) of central composite design for xylanase showed that Model F-value of 32.71, which implies that the model is significant (Table 7.6). There is only a 0.01 per cent chance that a "Model F-Value" this large could occur due to noise. P value less than 0.0500 indicates that the model terms are significant. In this case A, D, E, F, G, A2, B2, C2, D2, E2, F2, G2, AC, AD, AE, CE, CF, DE, EF, EG, FG are significant model terms.The "Pred R-Squared"

Table 7.5: Central Composite Design with Experimental and Predicted Value of Cellulase and Xylanase Production

Sl.No.	Parameters							Cellulase Activity (IU/ml)		Xylanase Activity (IU/ml)	
	A	B	C	D	E	F	G	Experimental Value	Predicted Value	Experimental Value	Predicted Value
1	-1	-1	-1	-1	-1	-1	-1	1.35±0.45	1.33	11.42±2.05	19.81
2	1	-1	-1	-1	-1	-1	-1	2.58±0.81	2.75	27.58±2.18	37.07
3	-1	1	-1	-1	-1	-1	-1	1.25±0.43	1.49	12.76±1.89	20.48
4	1	1	-1	-1	-1	-1	-1	3.14±0.94	3.21	37.65±3.85	47.05
5	-1	-1	1	-1	-1	-1	-1	2.58±0.81	2.75	32.17±3.19	42.84
6	1	-1	1	-1	-1	-1	-1	2.47±0.64	2.62	33.84±3.22	42.12
7	-1	1	1	-1	-1	-1	-1	2.41±0.58	2.89	31.77±2.92	42.75
8	1	1	1	-1	-1	-1	-1	3.51±0.75	3.17	42.49±3.88	51.33
9	-1	-1	-1	1	-1	-1	-1	2.39±0.36	2.74	21.78±2.26	30.46
10	1	-1	-1	1	-1	-1	-1	6.27±1.13	6.49	97.85±4.77	108.17
11	-1	1	-1	1	-1	-1	-1	2.11±0.55	2.98	20.79±1.24	25.20
12	1	1	-1	1	-1	-1	-1	7.11±1.35	7.14	121.26±5.28	112.22
13	-1	-1	1	1	-1	-1	-1	2.37±0.57	3.03	37.93±2.73	49.17
14	1	-1	1	1	-1	-1	-1	5.12±0.83	5.35	96.47±4.36	108.89
15	-1	1	1	1	-1	-1	-1	2.99±0.71	3.37	31.64±2.18	43.15
16	1	1	1	1	-1	-1	-1	6.18±1.35	6.09	106.53±4.54	112.17
17	-1	-1	-1	-1	1	-1	-1	3.73±0.79	3.96	56.96±3.26	64.35
18	1	-1	-1	-1	1	-1	-1	6.58±1.59	6.75	104.68±4.73	116.33
19	-1	1	-1	-1	1	-1	-1	3.66±0.73	3.70	52.48±3.12	65.92
20	1	1	-1	-1	1	-1	-1	6.89±1.37	6.90	120.15±4.95	127.20
21	-1	-1	1	-1	1	-1	-1	2.87±0.68	3.90	47.62±2.12	50.01
22	1	-1	1	-1	1	-1	-1	5.43±1.15	5.26	80.26±3.82	83.99
23	-1	1	1	-1	1	-1	-1	3.09±0.73	3.74	43.15±2.62	50.81

Sl.No.	Parameters							Cellulase Activity (IU/ml)		Xylanase Activity (IU/ml)	
	A	B	C	D	E	F	G	Experimental Value	Predicted Value	Experimental Value	Predicted Value
24	1	1	1	-1	1	-1	-1	4.63±0.81	5.50	88.53±3.51	94.10
25	-1	-1	-1	1	1	-1	-1	5.89±1.13	6.02	99.13±4.42	104.47
26	1	-1	-1	1	1	-1	-1	11.34±2.38	11.25	212.18±6.84	216.88
27	-1	1	-1	1	1	-1	-1	5.94±1.34	5.96	95.37±3.88	100.11
28	1	1	-1	1	1	-1	-1	11.75±2.63	11.60	226.93±6.96	221.83
29	-1	-1	1	1	1	-1	-1	4.81±0.98	4.94	84.38±3.76	85.80
30	1	-1	1	1	1	-1	-1	9.48±1.85	8.76	166.49±4.61	180.23
31	-1	1	1	1	1	-1	-1	4.13±0.91	4.99	74.55±2.81	80.68
32	1	1	1	1	1	-1	-1	9.84±1.93	9.20	174.71±5.25	184.41
33	-1	-1	-1	-1	-1	1	-1	2.74±0.38	3.00	36.86±1.82	51.99
34	1	-1	-1	-1	-1	1	-1	3.07±0.42	3.84	57.84±2.64	55.58
35	-1	1	-1	-1	-1	1	-1	2.13±0. 25	2.80	34.58±1.72	47.04
36	1	1	-1	-1	-1	1	-1	4.33±0.51	4.05	62.57±3.13	59.93
37	-1	-1	1	-1	-1	1	-1	4.11±0.57	5.06	72.88±3.29	91.84
38	1	-1	1	-1	-1	1	-1	4.33±0.67	4.47	66.18±3.26	77.44
39	-1	1	1	-1	-1	1	-1	4.14±0.53	4.96	73.74±3.38	86.12
40	1	1	1	-1	-1	1	-1	4.28±0.55	4.77	76.25±3.65	81.02
41	-1	-1	-1	1	-1	1	-1	4.41±0.59	4.78	63.93±3.42	73.08
42	1	-1	-1	1	-1	1	-1	7.39±1. 25	8.07	137.26±4.27	137.11
43	-1	1	-1	1	-1	1	-1	4.52±0.58	4.78	63.48±2.62	62.20
44	1	1	-1	1	-1	1	-1	8.18±1.32	8.47	62.97±2.15	137.54
45	-1	-1	1	1	-1	1	-1	5.75±0.91	5.84	139.76±4.29	108.60
46	1	-1	1	1	-1	1	-1	7.69±1. 22	7.70	137.26±4.27	154.64
47	-1	1	1	1	-1	1	-1	5.85±0.92	5.93	89.62±3.35	96.95

Sl.No.	Parameters							Cellulase Activity (IU/ml)		Xylanase Activity (IU/ml)	
	A	B	C	D	E	F	G	Experimental Value	Predicted Value	Experimental Value	Predicted Value
48	1	1	1	1	-1	1	-1	8.49±1.33	8.19	157.68±5.28	152.30
49	-1	-1	-1	-1	1	1	-1	4.95±0.53	4.40	56.64±2.11	68.74
50	1	-1	-1	-1	1	1	-1	6.16±0.84	6.72	99.28±4.29	107.03
51	-1	1	-1	-1	1	1	-1	3.13±0.58	3.90	59.73±2.62	64.68
52	1	1	-1	-1	1	1	-1	6.79±0.71	6.63	121.59±4.86	112.28
53	-1	-1	1	-1	1	1	-1	4.69±0.64	5.11	66.49±2.74	71.20
54	1	-1	1	-1	1	1	-1	5.78±0.75	5.99	81.75±3.49	91.51
55	-1	1	1	-1	1	1	-1	4.73±0.49	4.70	56.39±2.18	66.38
56	1	1	1	-1	1	1	-1	6.61±0.93	6.00	87.93±3.45	95.99
57	-1	-1	-1	1	1	1	-1	6.37±0.87	6.97	96.88±4.12	119.29
58	1	-1	-1	1	1	1	-1	11.58±1.65	11.72	221.46±6.28	218.03
59	-1	1	-1	1	1	1	-1	6.33±0.77	6.65	115.92±4.86	109.31
60	1	1	-1	1	1	1	-1	11.37±1. 61	11.82	223.42±6.34	217.35
61	-1	-1	1	1	1	1	-1	6.36±0.77	6.66	111.63±4.69	117.44
62	1	-1	1	1	1	1	-1	9.39±0.89	9.99	193.86±5.62	198.18
63	-1	1	1	1	1	1	-1	6.26±0.65	6.45	96.79±4.18	106.69
64	1	1	1	1	1	1	-1	9.68±1.39	10.18	196.58±6.21	196.74
65	-1	-1	-1	-1	-1	-1	1	2.37±0.41	2.40	25.83±2.13	40.75
66	1	-1	-1	-1	-1	-1	1	2.88±0.66	3.04	31.49±2.38	44.34
67	-1	1	-1	-1	-1	-1	1	1.82±0.32	2.75	40.84±2.65	49.85
68	1	1	-1	-1	-1	-1	1	3.47±0.46	3.79	47.59±2.81	62.74
69	-1	-1	1	-1	-1	-1	1	3.66±0.45	3.56	44.98±2.72	54.72
70	1	-1	1	-1	-1	-1	1	2.79±0.38	2.76	28.95±2.05	40.32
71	-1	1	1	-1	-1	-1	1	3.73±0.54	4.00	58.32±2.97	63.05

Sl.No.	*Parameters*							*Cellulase Activity (IU/ml)*		*Xylanase Activity (IU/ml)*	
	A	*B*	*C*	*D*	*E*	*F*	*G*	*Experimental Value*	*Predicted Value*	*Experimental Value*	*Predicted Value*
72	1	1	1	-1	-1	-1	1	3.99±0.58	3.61	48.63±2.79	57.95
73	-1	-1	-1	1	-1	-1	1	2.95±0. 25	3.84	44.72±2.71	57.36
74	1	-1	-1	1	-1	-1	1	6.41±0.95	6.92	116.79±4.68	121.39
75	-1	1	-1	1	-1	-1	1	4.27±0.41	4.38	53.57±3.13	60.53
76	1	1	-1	1	-1	-1	1	8.37±1. 22	7.87	124.7±5.15	133.86
77	-1	-1	1	1	-1	-1	1	3.81±0.45	4.00	58.74±3.28	67.00
78	1	-1	1	1	-1	-1	1	5.24±0.62	5.64	112.81±4.74	113.04
79	-1	1	1	1	-1	-1	1	4.49±0.68	4.63	62.63±3.18	69.40
80	1	1	1	1	-1	-1	1	5.79±0.84	6.68	115.34±4.78	124.75
81	-1	-1	-1	-1	1	-1	1	5.21±0.75	5.72	94.38±4.26	102.90
82	1	-1	-1	-1	1	-1	1	7.31±0.95	7.84	135.94±4.98	141.20
83	-1	1	-1	-1	1	-1	1	5.35±0.78	5.77	112.99±4.52	112.89
84	1	1	-1	-1	1	-1	1	8.71±1.15	8.29	148.93±4.95	160.49
85	-1	-1	1	-1	1	-1	1	5.76±0.85	5.52	77.31±3.32	79.49
86	1	-1	1	-1	1	-1	1	5.48±0.79	6.21	96.71±4.28	99.80
87	-1	1	1	-1	1	-1	1	5.38±0.75	5.66	85.28±3.95	88.72
88	1	1	1	-1	1	-1	1	6.81±0.68	6.75	114.68±4.65	118.33
89	-1	-1	-1	1	1	-1	1	7.16±0.84	7.93	148.22±4.98	148.97
90	1	-1	-1	1	1	-1	1	12.93±1.15	12.49	243.67±6.52	247.71
91	-1	1	-1	1	1	-1	1	7.88±0.95	8.17	146.47±4.92	153.04
92	1	1	-1	1	1	-1	1	13.61±1.57	13.14	251.58±6.68	261.08
93	-1	-1	1	1	1	-1	1	6.49±0.58	6.73	113.46±4.58	121.24
94	1	-1	1	1	1	-1	1	9.97±0.35	9.86	199.87±5.46	201.98
95	-1	1	1	1	1	-1	1	7.78±0.92	7.06	113.86±4.59	124.54

Sl.No.	Parameters							Cellulase Activity (IU/ml)		Xylanase Activity (IU/ml)	
	A	B	C	D	E	F	G	Experimental Value	Predicted Value	Experimental Value	Predicted Value
96	1	1	1	1	1	-1	1	9.71±1.12	10.60	212.48±5.61	214.59
97	-1	-1	-1	-1	-1	1	1	2.05±0.24	2.47	36.95±2.38	44.06
98	1	-1	-1	-1	-1	1	1	2.42±0.29	2.64	25.38±2.02	33.97
99	-1	1	-1	-1	-1	1	1	2.05±0.24	2.57	44.75±2.73	47.53
100	1	1	-1	-1	-1	1	1	2.98±0.31	3.15	42.55±2.62	46.74
101	-1	-1	1	-1	-1	1	1	3.43±0.39	4.39	64.5±3.38	74.84
102	1	-1	1	-1	-1	1	1	2.72±0.28	3.13	35.77±2.94	46.76
103	-1	1	1	-1	-1	1	1	4.17±0.46	4.59	62.85±3.51	77.54
104	1	1	1	-1	-1	1	1	3.16±0.31	3.73	52.07±2.86	58.77
105	-1	-1	-1	1	-1	1	1	4.41±0.48	4.40	57.69±2.93	71.10
106	1	-1	-1	1	-1	1	1	7.26±0.80	7.02	119.26±4.27	121.45
107	-1	1	-1	1	-1	1	1	4.17±0.42	4.70	62.33±3.13	68.64
108	1	1	-1	1	-1	1	1	7.86±0.85	7.72	141.57±4.91	128.31
109	-1	-1	1	1	-1	1	1	4.37±0.43	5.32	78.63±3.74	97.56
110	1	-1	1	1	-1	1	1	6.83±0.72	6.50	116.72±4.34	129.92
111	-1	1	1	1	-1	1	1	5.41±0.61	5.71	83.58±3.68	94.34
112	1	1	1	1	-1	1	1	6.93±0.75	7.30	132.81±4.53	136.00
113	-1	-1	-1	-1	1	1	1	3.88±0.42	4.68	89.66±3.87	78.41
114	1	-1	-1	-1	1	1	1	6.79±0.73	6.33	90.85±3.94	103.02
115	-1	1	-1	-1	1	1	1	4.14±0.42	4.48	69.37±3.19	82.78
116	1	1	-1	-1	1	1	1	5.81±0.65	6.54	91.36±3.97	116.70
117	-1	-1	1	-1	1	1	1	4.35±0.41	5.25	62.47±3.05	71.81
118	1	-1	1	-1	1	1	1	5.86±0.67	5.46	72.61±3.37	78.44
119	-1	1	1	-1	1	1	1	5.25±0.59	5.14	76.39±3.51	75.41

Sl.No.	Parameters							Cellulase Activity (IU/ml)		Xylanase Activity (IU/ml)	
	A	B	C	D	E	F	G	Experimental Value	Predicted Value	Experimental Value	Predicted Value
120	1	1	1	-1	1	1	1	5.94±0.62	5.76	86.77±3.83	91.35
121	-1	-1	-1	1	1	1	1	7.03±0.78	7.39	128.43±4.29	134.92
122	1	-1	-1	1	1	1	1	11.34±1.16	11.48	215.84±5.44	219.98
123	-1	1	-1	1	1	1	1	7.13±0.78	7.38	129.47±4.31	133.36
124	1	1	-1	1	1	1	1	11.67±1.34	11.88	223.47±5.51	227.73
125	-1	-1	1	1	1	1	1	6.25±0.64	6.95	117.73±4.13	123.99
126	1	-1	1	1	1	1	1	9.81±1.15	9.61	187.76±5.19	191.06
127	-1	1	1	1	1	1	1	6.56±0.73	7.03	119.86±4.16	121.67
128	1	1	1	1	1	1	1	9.83±1.18	10.10	195.88±5.24	198.04
129	-3.36	0	0	0	0	0	0	0	-3.91	0	-43.33
130	3.36	0	0	0	0	0	0	2.19±0.24	3.44	148.47±4.94	114.15
131	0	-3.36	0	0	0	0	0	0	-1.79	0	-45.26
132	0	3.36	0	0	0	0	0	0	-0.87	0	-32.39
133	0	0	-3.36	0	0	0	0	9.93±1.26	8.60	176.94±5.18	136.58
134	0	0	3.36	0	0	0	0	9.13±1.08	7.80	162.69±5.08	125.40
135	0	0	0	-3.36	0	0	0	0	-1.77	0	-56.64
136	0	0	0	3.36	0	0	0	8.58±0.94	7.69	161.74±5.06	140.73
137	0	0	0	0	3.36	0	0	0	-2.20	0	-66.59
138	0	0	0	0	3.36	0	0	7.19±0.73	6.73	123.72±4.22	112.66
139	0	0	0	0	0	-3.36	0	4.74±0.53	4.28	94.57±3.97	51.15
140	0	0	0	0	0	3.36	0	8.25±0.84	6.05	111.68±4.17	77.45
141	0	0	0	0	0	0	-3.36	0	-1.35	0	-39.48
142	0	0	0	0	0	0	3.36	1.43±0.17	0.15	36.11±2.37	-2.08
143	0	0	0	0	0	0	0	11.69±1.68	11.54	218.62±4.94	211.95

Sl.No.	Parameters							Cellulase Activity (IU/ml)		Xylanase Activity (IU/ml)	
	A	B	C	D	E	F	G	Experimental Value	Predicted Value	Experimental Value	Predicted Value
144	0	0	0	0	0	0	0	12.99±1.76	11.54	238.71±5.16	211.95
145	0	0	0	0	0	0	0	14.62±1.92	11.54	283.68±5.41	211.95
146	0	0	0	0	0	0	0	10.69±1.28	11.54	195.91±4.46	211.95
147	0	0	0	0	0	0	0	14.81±1.96	11.54	297.83±5.52	211.95
148	0	0	0	0	0	0	0	11.76±1.51	11.54	245.64±5.15	211.95
149	0	0	0	0	0	0	0	12.43±1.63	11.54	240.81±4.94	211.95
150	0	0	0	0	0	0	0	12.53±1.68	11.54	246.28±5.14	211.95
151	0	0	0	0	0	0	0	10.69±1.28	11.54	191.28±4.76	211.95
152	0	0	0	0	0	0	0	11.72±1.48	11.54	295.74±5.72	211.95

of 0.7653 is in reasonable agreement with the "Adj R-Squared" of 0.8803. "Adeq Precision" measures the signal to noise ratio. A ratio greater than 4 is desirable. The ratio of 28.598 indicates an adequate signal.

Table 7.6: Results of Analysis of Variance (ANOVA) of Central Composite Design for Cellulase and Xylanase Production

Terms	*Cellulase Yield*	*Xylanase Yield*
F-value	48.91	32.71
P-value (Prob >*F*)	< 0.0001	< 0.0001
Mean	5.90	102.69
R-Squared	0.9365	0.9080
Adj R-Squared	0.9174	0.8803
Pred R-Squared	0.8312	0.7653
C.V.	16.19	22.93
Adeq Precision	36.662	28.598
PRESS	281.80	1.641E+005

Validation for Model for Cellulase and Xylanase Production

To validate and confirm the predictions obtained after CCD, 25 experiments were designed with different levels of significant factors according to predicted response generated by the model, taking the optimum production conditions after CCD. The response surface showing the effects of yeast extact and peptone as well as temperature and production time on cellulase and xylanase production is shown in Figures 7.1–7.4 respectively.

Discussion

Because of commercial significance of alkaline cellulase and xylanase enzymes, we studied the effects of different variables in submerged fermentation process using response surface methodology. Plackett burman showed range and levels of seven significant variables (A- Inoculum age (h), B- Production time (g/50ml), C- Temperature (°C), D- Peptone (per cent), E- Yeast extract (g/50ml), F- Xylose (g/50ml), G- Shaking (rpm), in terms of actual and coded factors for central composite design for the simultaneous production of cellulase and xylanase The improvement in successive activities of cellulase and xylanase at alkaline pH were descibed in various previous studies (Sharma *et al.*, 2011). We need to have optimum enzyme production parameters within limited period of time. In recent years, *Streptomyces* sp. showed maximum xylanase production at pH 9 (Techapun *et al.*, 2001). The optimum inoculum size and age reported were 0.5 per cent and 6 h respectively. This might be due to decrease in enzyme titre with increase in inoculum size (Battan *et al.*, 2007). The expeditious growth of micro-organisms and enzyme production process showed mutual association explaining less inoculum time (Kumar *et al.*, 2012; Nagar *et al.*, 2010). Rice bran showed maximum cellulo-xylanolytic enzyme activity as it contained significant amount of nutrients and remained loose for moist conditions rather than other cheap organic sources (Dhillon and Khanna, 2000; Santos *et*

Table 7.7: Validation for Model for Cellulase and Xylanase Production

Sl. No.	Variables							Cellulase Activity (IU/ml)		Xylanase Activity (IU/ml)	
	A	B	C	D	E	F	G	Experimental Value	Predicted Value	Experimental Value	Predicted Value
1	5.5	44	38	0.8	1.2	0.4	160	11.35±1.58	11.79	217.46±3.44	205.18
2	6	43	36	0.9	1.3	0.2	150	13.68±1.64	13.80	262.76±3.87	266.96
3	**6**	**42**	**37**	**1.2**	**1.1**	**0.3**	**180**	**15.67±1.93**	**15.29**	**326.76±6.98**	**317.53**
4	6.25	43	36	0.8	1	0.2	150	10.83±0.82	10.18	197.25±4.75	194.51
5	6.75	44	39	1.2	0.8	0.3	170	13.43±1.75	13.22	267.71±4.19	266.26
6	5.75	41	36	1.2	0.9	0.2	160	10.86±0.87	11.01	192.78±3.37	190.26
7	5.5	43	37	1.2	1	0.3	170	12.79±1.83	12.31	259.38±5.79	247.10
8	6.25	43	38	1.3	1	0.2	160	12.47±1.69	12.16	267.81±4.58	264.75
9	6.5	44	37	1	0.9	0.3	170	12.35±1.71	11.09	253.97±5.19	242.83
10	6.25	43	38	0.9	1	0.3	160	11.82±1.58	10.90	241.85±4.78	229.04
11	5.5	43	37	1	1.2	0.15	190	11.97±1.62	10.94	245.78±4.92	230.12
12	6	44	36	0.9	1	0.3	180	12.43±1.85	1.03	254.76±5.18	256.56
13	6.25	42	37	1	0.9	0.2	160	13.16±1.42	12.57	279.76±4.12	272.53
14	6	43	37	0.9	1	0.3	180	14.06±1.73	14.04	303.78±5.19	231.84
15	5.75	42	36	1	0.8	0.2	170	11.87±1.78	10.38	203.41±3.49	196.37
16	6.75	43	38	0.8	1	0.3	170	12.84±1.92	10.90	214.73±3.86	209.08
17	6.25	43	38	1.2	1.10	0.2	160	12.42±1.81	11.21	204.42±3.71	205.73
18	5.50	41	36	1	1	0.2	160	11.71±1.73	10.52	191.73±3.42	196.19
19	6.25	42	37	1.10	1.10	0.3	180	13.14±1.91	13.61	266.85±5.24	263.45
20	6.75	43	37	1.30	0.90	0.2	170	12.78±1.84	12.27	251.93±3.49	247.07
21	6.25	44	36	1.00	1.30	0.2	180	13.87±1.97	13.50	263.79±5.22	261.62
22	5.50	41	36	0.90	1.20	0.4	170	12.97±1.84	11.08	238.43±4.73	221.10

Sl. No.	*Variables*							*Cellulase Activity (IU/ml)*		*Xylanase Activity (IU/ml)*	
	A	*B*	*C*	*D*	*E*	*F*	*G*	*Experimental Value*	*Predicted Value*	*Experimental Value*	*Predicted Value*
23	6.75	42	37	1.00	0.90	0.4	190	13.89±1.85	13.14	229.18±3.92	223.43
24	5.50	42	37	1.00	0.90	0.25	160	12.17±1.77	10.57	210.72±3.61	205.42
25	6	42	38	1.00	0.90	0.4	150	14.95±1.73	14.80	289.58±3.45	295.93

Inoculum age (h), B: Production Time (h), C: Temperature (°C), D: Peptone (g/50ml), E: Yeast extract (g/50ml), F: Xylose (g/50ml), G: Shaking (rpm).

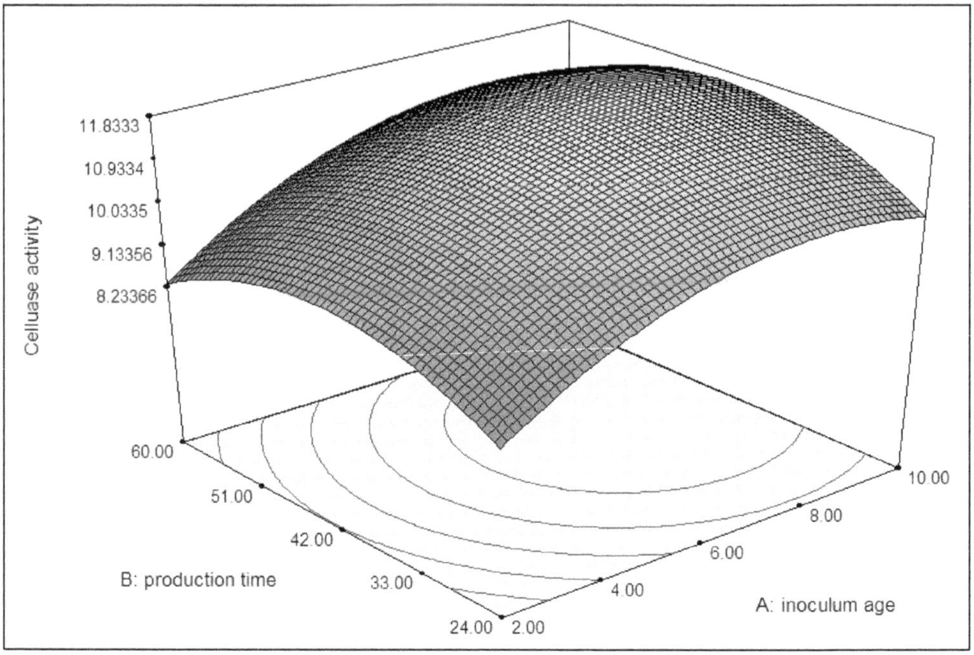

Figure 7.1: 3D Response Surface Diagram Showing Interactions of Production Time and Inoculum Age to Enhance the Cellulase Activity (IU/ml).

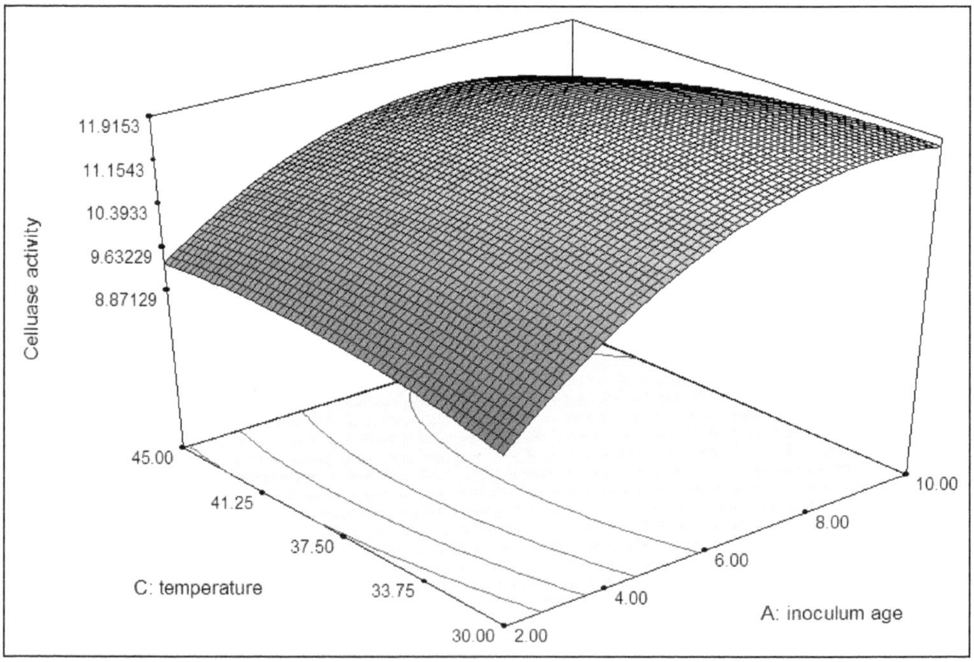

Figure 7.2: 3D Response Surface Diagram Showing Interactions of Temperature and Inoculum Age to Enhance the Cellulase Activity (IU/ml).

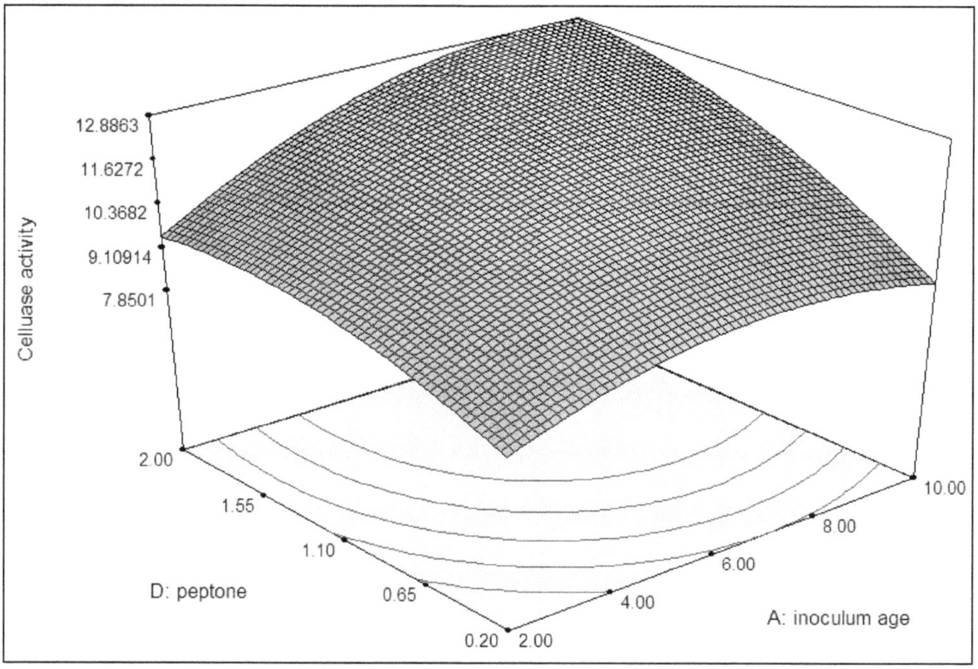

Figure 7.3: 3D Response Surface Diagram Showing Interactions of Peptone and Inoculum Age to Enhance the Cellulase Activity (IU/ml).

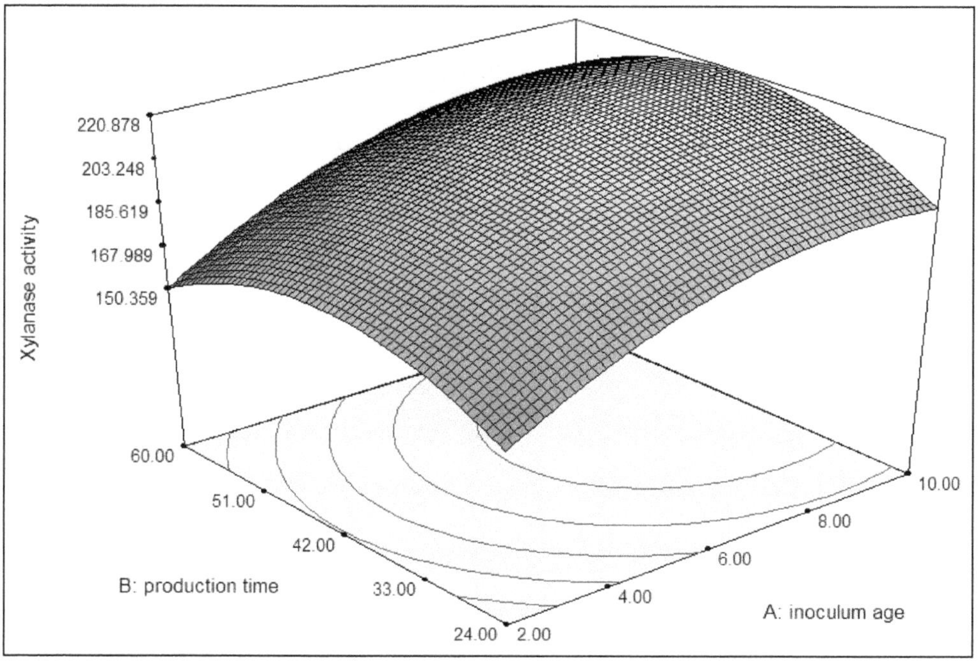

Figure 7.4: 3D Response Surface Diagram Showing Interactions of Production Time and Inoculum Age to Enhance the Xylanase Activity (IU/ml).

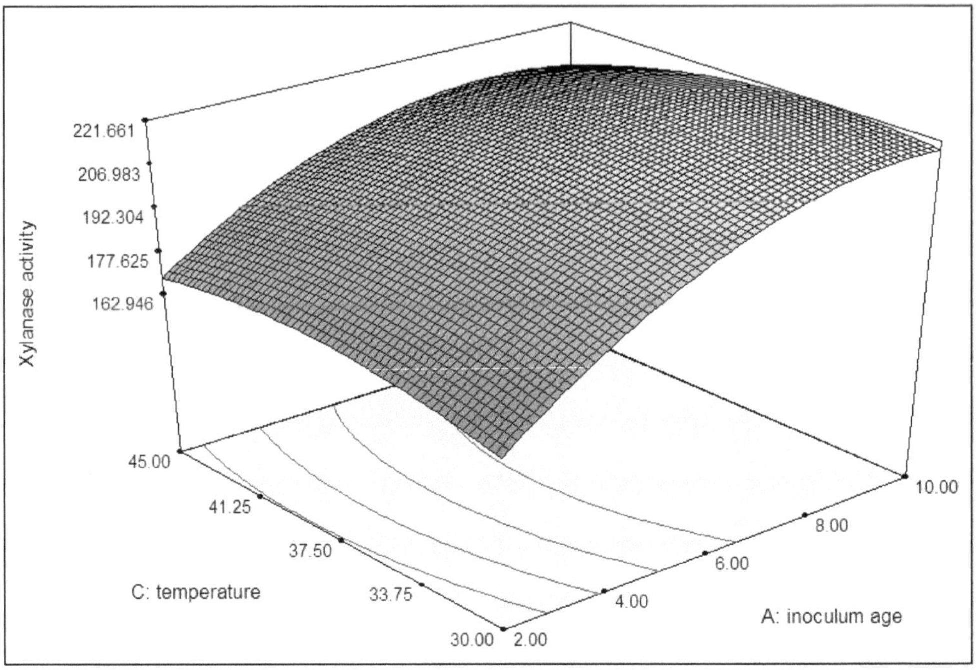

Figure 7.5: 3D Response Surface Diagram Showing Interactions of Temperature and Inoculum Age to Enhance the Xylanase Activity (IU/ml).

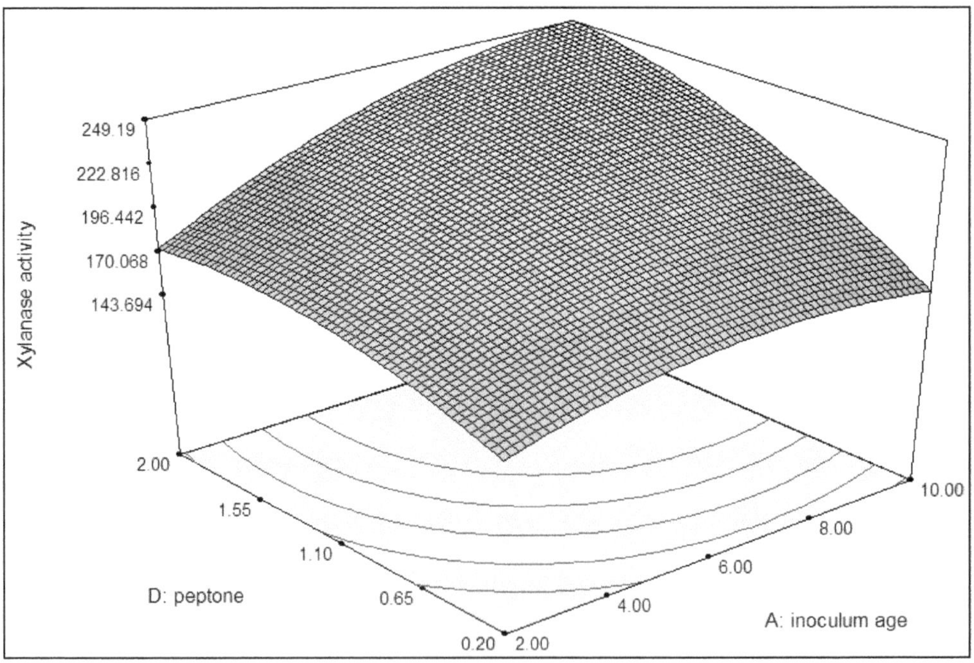

Figure 7.6: 3D Response Surface Diagram Showing Interactions of Peptone and Inoculum Age to Enhance the Xylanase Activity (IU/ml).

al., 2002; Techapun *et al.*, 2001). However, previous studies showed commercial substrates and wheat bran as the cheap carbon source (Battan *et al.*, 2007; Van *et al.*, 2011). Lu *et al.* (2003) reported that incubation temperature between 30-45°C was the optimum temperature range for Cellulo-xylanolytic enzyme production. Previously, incubation temperature of 40°C was reported for enhanced cellulase and xylanase production from *Bacillus subtilis* 115 and *Bacillus subtilis* (Vyas *et al.*, 2005; Jansova *et al.*, 1993). There was significant enhancement in enzymes activity with different concentrations of peptone, yeast extract and xylose concentrations (Haltrich *et al.*, 1997; Hoq *et al.*, 1994). Increase in xylanase activity showed similar results with temperature 37°C and shaking speed of 200 rpm using wheat bran as carbon substrare (Khurana *et al.*, 2007). Statistical optimization of enzyme production by plackett burman and central composite design showed a specific production time of 42°C which was similar to the production time for maximum xylanase production (Sharma *et al.*, 2007). Other production time discussed recently was 48 h on different *Bacillus megaterium* MTTC 2444 and *Bacillus mojavensis* AG137 for xylanase production (Simphiwe *et al.*, 2011; Mrudula at al., 2012)

Conclusion

In our study, the optimum modified horikoshi media formulation and growth conditions for production of cellulase and xylanase enzymes from *Bacillus subtilis* under submerged fermentation was achieved by growing bacteria at 37°C and pH 9 under statistical design methods such as placket burman as well as central composite design. Production rate of cellulase was found higher than that of xylanase. Production of cellulo-xylanolytic enzymes has been directly proportional to biomass concentration indicating that specific medium composition and optimum cultural conditions are the key parameters for higher biomass production. Growth of microorganisms requires primary metabolites *i.e.* Cellulo-xylanolytic enzymes, that convert complex substrates into reducing sugars that is absorbed into the cells. Rice bran found out to be the optimum carbon source stimulating initial growth of micro-organisms in submerged fermentations. After validation of CCD results various parameters such as inoculum size, inoculum age, pH, production time, carbon substrate, and nitrogen source, shaking enhanced the enzyme production as much as 15.67 IU/ml of cellulase activity and 326.76 IU/ml of xylanase.

Acknowledgement

The author greatfully acknowledges Department of Biotechnology, Kuruksketra University, Kurukshetra for providing facilities while carrying out experimental work.

REFERENCES

Aristidou, A., and Penttilä, M. (2000). *Current Opinion in Biotechnology*, 1, 187–198

Bailey MJ, Biely P and Poutanen K (1992) Interlaboratory testing of methods for assay of xylanase activity. *J Biotechnol* **23**(3): 257

Battan B, Sharma J, Dhiman SS, Kuhad RC (2007) Enhanced production of cellulase-free thermostable xylanase by Bacillus pumilus ASH and its potential application in paper industry. *Enzym Microb Technol* **41**(6–7): 733–739. doi: 10.1016/j.enzmictec.2007.06.006

Bezerra MA, Santelli RE, Oliveira EP, Villar LS and Escaleira LA (2008) Response surface methodology (RSM) as a tool for optimization in analytical chemistry. *Talanta* **76**, 965-977

Chakdar H, Kumar M, Pandiyan K, Singh A, Nanjappan K, Kashyap PL (2016) Bacterial xylanases: biology to biotechnology. *3 Biotech* **6**, 1-15

Collins T, Gerday C, and Feller G (2005) Xylanases: xylanase families and extremophilic xylanases. *Microbiology Reviews*, **29**, 3–23.

Dhillon A, Khanna S (2000) Production of a thermostable alkalitolerent xylanase from Bacillus circulans AB 16 grown on wheat straw. *World J Microbiol Biotechnol*, **16**, 325–327

Esterbauer H, Steiner J, Labudova I, Hermann A, Hayn M (1991) Production of Trichoderma cellulase in laboratory and piolet scale. *Bioresource Technology* **36**, 21-35

Haaland PD (1990) Experimental design in biotechnology. Elsevier, New York

Haltrich D, Nidetzky B, Kulke KD, Steiner W and Zupanie S (1997) Production of fungal xylanases. *Bioresource Technology*, **58**, 137 -161.

Hegde S, Bhadri G, Narsapur K, Koppal S, Oswal P, Turmuri N, Jumnal V and Hungund B (2013) Statistical optimization of medium components by response surface methodology for enhanced production of bacterial cellulase by *Gluconacetobacter persimmonis*, *J Bioproces Biotechniq* **4**(1): 1-5

Holker U, Hofer M, Lenz J (2004) Biotechnological advantages of laboratory scale solid state fermentation with fungi. *Appl Microbiol Biotechnol*, **64**, 175-86. http: //dx.doi.org/10.1007/s00253-003-1504-

Hoq, M., Hempel, C., and Deckwer, W. D. 1994. Cellulasefree xylanase by Thermomyces lanuginosus RT9: effect of agitation, aeration, and medium components on production. *Journal of Biotechnology*, **37**, 49-58.

Jansova E, Schwarzova Z, Chaloupka J (1993) Sporulation and synthesis of extracellular proteinases in Bacillus subtilis are more temperature-sensitive than growth. *Folia Microbiol*. 38, 22–24

Khurana S, Kapoor M, Gupta S, and Kuhad RC (2007) Statistical optimization of alkaline xylanase production from *Streptomyces violaceoruber* under submerged fermentation using response surface methodology, *Indian Journal of Microbiology* **47**, 2: 144-152

Khurana S, Kapoor M, Gupta S, Kuhad RC (2007) Statistical optimization of alkaline xylanase production from *Streptomyces violaceoruber* under submerged fermentation using response surface methodology. *Indian J Microbiol* 47: 144–152

Kumar A, Gupta R, Shrivastava B, Khasa YP, Kuhad RC (2012) Xylanase production from an alkaliphilic actinomycete isolate Streptomyces sp. RCK2010, its characterization and application in saccharification of second generation biomass. *J. Mol. Catal. B: Enzym.* **74**, 170–177.

Lama L, Calandrelli V, Gambacorta A, Nicolaus B (2004) Purification and characterization of thermostable xylanase and β-xylosidase by thermophilic bacterium *Bacillus thermantarcticus*. *Res Microbiol*, **155**: 283-89.http: //dx.doi.org/10.1016/j.resmic.2004.02.001

Lu W, Li D and Wu Y (2003) Influence of Water Activity and Temperature on Xylanase Biosynthesis in Pilot-Scale Solid-State Fermentation by *Aspergillus sulphureus*. *Enzyme and Microbial Technology*, **32**, 305-311.http: //dx.doi.org/10.1016/S0141-0229(02)00292-2

Mrudula S and Shyam N (2012) Immobilization of Bacillus megaterium MTCC 2444 by Ca-alginate entrapment method for enhanced alkaline protease production. *Braz. Arch. Biol. Technol.* **55**, 135–144

Nagar S, Gupta VK, Kumar L, Kuhad RC (2010) Production and optimization of cellulase-free, alkali-stable xylanase by Bacillus pumilus SV-85S in submerged fermentation. *J. Ind. Microbiol. Biotechnol.* **37**, 71–83.

Nieves RA, Ehrman CI, Adney WS, Elander RT, Himmel ME (1998) Survey and analysis of commercial cellulase preparations. *World Journal of Microbiology and Biotechnology* **14**(2): 301–304

Polizeli MLTM, Rizzatti ACS, Monti R, Terenzi HF, Jorge JA, Amorim DS (2005) Xylanases from fungi: properties and industrial applications. Appl Microbiol Biotechnol **67**, 577-591

Rajeshkannan R, Rajamohan NR, Rajasimman M (2009) Removal of malachite green from aqueous solution by sorption on hydrilla verticillata biomass using response surface methodology, *Front Chem. Eng. China* **3**(2): 146-154

Robson LM, Chambliss GH (1989) Cellulases of bacterial origin.*Enzyme Microb Technol*, **11**, 626-44.

Sanchez C. (2009) Lignocellulosic residues: Biodegradation and bioconversion by fungi.

Santos ESD, Guirardello R, Franco TT (2002) Preparative chromatography of xylanase using expanded bed adsorption. *J Chromatogr* **944**, 217–224

Sharma A, Adhikari S and Satyanarayana T (2007) Alkali-Thermostable and cellulase-free xylanase production by an extreme thermophile Geobacillus thermoleovorans. *World J. Microbiol. Biotechnol.* **23**, 483–490.

Sharma SCD, Shovon MS, Asaduzzaman AKM, Sarowar Jahan MG, Yeasmin T, and Roy N, *J. Biosci.*, 2011 19(2): 7–14

Simphiwe P, Ademola B, Olaniran O and Pillay B (2011) Sawdust and digestive bran as cheap alternate substrates for xylanase production. *Afr. J. Microbiol. Res.* **5**, 742–752.

Techapun C, Sinsuwongwat S, Poosaran N, Watanabe M, Sasaki K (2001) Production of a cellulase-free xylanase from agricultural waste materials by a thermotolerant *Streptomyces* sp. *Biotechnol Lett* **23**, 1685–1689

Van HV, Tuan AP and Keun K (2011) Improvement of Fungal Cellulase Production by Mutation and Optimization of Solid State Fermentation, *Mycobiology*, **39**(1): 20-25.

Vimalashanmugam K, Viruthagiri T (2012) Statistical optimization of media components for xylanase production by Aspergillus fumigatus using SSF, *IJERA* **2**(5): 1320-1329

Vyas A, Vyas D and Vyas KM (2005) Production and Optimization of Cellulases on Pretreated Groundnut Shell by *Aspergillus terreus* AV49. *Journal of Scientific and Industrial Research*, **64**, 281-286.

Chapter 8

Plant Growth Promoting Rhizobacteria (PGPR): A Sustainable Approach for Managing Soil Fertility and Crop Productivity

Satyavir Singh Sindhu* and Ruchi Sharma

Department of Microbiology, CCS Haryana Agricultural University, Hisar – 125 001, Haryana
****E-mail: sindhuss58@gmail.com***

ABSTRACT

Soil fertility and health has deteriorated recently due to current intensive agricultural practices and cultivation of exhaustive crops by the farmers. The most important constraint for obtaining enhanced crop yield among resource-poor farmers is soil infertility and especially in developing nations. Therefore, maintenance of soil fertility and quality is needed to stabilize or improve the rapidly declining crop production levels. Microorganisms present in the soil or rhizosphere cause mineralization/solubilization of bound/insoluble nutrient minerals through biogeochemical cycling and affect the chemical and biological properties of the soil system leading to enhanced soil fertility. Thus, rhizospheric microorganisms provide an attractive option to promote plant growth either as free-living cells, endophytically or through symbiosis using different mechanisms. These plant growth promoting rhizosphere bacteria (PGPR) may act directly by increasing the acquisition of plant nutrients or by enhancing hormone levels on the plant surface, or may act indirectly by suppressing the diseases caused by the bacterial/fungal plant pathogens. Different beneficial microorganisms could be used as biofertilizers to improve the nitrogen, phosphorus and potassium levels in the soil and they become important components of integrated nutrients management to improve soil fertility. Some antagonistic microorganisms also suppress the diseases caused by phytopathogens. Moreover, some PGPR strains alleviate abiotic stress conditions and also contribute in bioremediation of pollutants, heavy metals and toxic pesticide chemicals in the soil. Thus, these potential biological fertilizers/biopesticides/bioremediating agents act as eco-friendly cost effective input and play key roles in productivity and sustainability of soil and protection of the environment as well. With the use of these biological fertilizers/ biopesticides, a low input system can be optimized for achieving sustainability of agriculture.

Keywords: *Plant growth promting rhizobacteria, Biofertilizers, Bioremediation, Sustainable agriculture.*

Introduction

In modern cultivation practices, indiscriminate use of nitrogenous and phosphate fertilizers to enhance crop productivity has caused substantial pollution in soil, air and water leading to human and environmental hazards. Moreover, farmers are applying synthetic pesticides for control of pests and phytopathogens in agriculture. Excessive use of these agrochemicals adversely affects the population of soil microorganisms (Walia *et al.*, 2018) and affects the soil health and fertility. In addition, non-renewable resources such as oil and natural gas also gets depleted during the production of chemical fertilizers. These problems have necessitated the development of eco-friendly microbe-based technologies for sustainable production system (Dadarwal *et al.*, 1997; Sindhu *et al.*, 2014a; 2017; Sharma *et al.*, 2018a). Many microorganisms have recently been characterized for their possible use in nutrient and pest management programmes to improve the crop productivity.

Microbial population constitutes most extraordinary and ubiquitous life on the earth but the prevalence of microorganisms is higher in nutrient-rich niches like the rhizosphere. Therefore, rhizosphere is considered as the 'hotspot' for microbial colonization and activity (Mohanram and Kumar, 2019). Thus, screening of microorganisms from the rhizosphere to identify beneficial microorganisms for use as biofertilizer and biopesticides is one of the most lucrative options. Members of the rhizosphere microbiome contribute towards nutrient acquisition (Pii *et al.*, 2015; Parmar and Sindhu, 2018; Sindhu *et al.*, 2019a, b), abiotic stress tolerance (Yang *et al.*, 2008; Chaudhary and Sindhu, 2015; Sindhu *et al.*, 2016), protection against soil borne pathogens and insects (Sindhu *et al.*, 2002; 2016) and host immune regulations (van Hulten *et al.*, 2006; Sindhu *et al.*, 2017). Moreover, these PGPR strains actively colonize the rhizosphere of various plants and production of various growth promoting substances and hormones, quorum sensing, *etc.* also contribute towards enhanced plant growth (Bhattacharya and Jha, 2012; Sindhu *et al.*, 2014b).

Based on the experiments with radish plants, rhizobacteria were identified as a community (Kloepper *et al.*, 1980), that competitively colonized plant roots and enhanced the plant growth by providing solubilized nutrients alongwith biocontrol ability for reduction in plant diseases (Weller *et al.*, 2002; Vessey, 2003). Depending upon their intimacy and interactions with different crop plants, PGPR were further classified as extracellular plant growth rhizobacteria (ePGPR) or intracellular plant growth rhizobacteria (iPGPR) (Viveros *et al.*, 2010). The ePGPR group includes bacteria from genera like *Azotobacter, Bacillus, Chromobacterium, Agrobacterium, Caulobacter, Pseudomonas, Serratia, etc.* and is mainly localized in the rhizosphere, rhizoplane or between cells of root cortex (Gray and Smith, 2005). Another class iPGPR includes various endophytic and symbiotic rhizobia (*Allorhizobium, Azorhizobium, Bradyrhizobium, Mesorhizobium, etc.*) and *Frankia* species (Wang and Martinez-Romero, 2000; Verma *et al.*, 2011; Laranjo *et al.*, 2014).

Seed bacterization or soil inoculation of PGPR strains has been found to increase nodulation and nitrogen fixation in many plants including soybean (*Glycine max* (L.) Merr.) (Zhang *et al.*, 1996) and chickpea (Sindhu *et al.*, 2002). Beneficial effects of PGPR inoculation has also been widely reported on many other plant species including field crops, vegetables, ornamentals, horticultural crops and forest trees (Kloepper *et al.*, 2004; Sindhu *et al.*, 2010). Combined inoculation of *Bacillus amyloliquefaciens* strain IN937a and *Bacillus pumilus* strain T4 into soil alonwith application of 75 per cent fertilizer rate, gave the tomato yield as obtained with the application of 100 per cent fertilizer dose (Adesemoye *et al.*, 2009). The application of biofertilizers was also found to significantly increase plant growth and uptake of nitrogen (N) and phosphorus (P) in maize and also improved soil organic matter and total N (Wu *et al.*, 2005). Seed bacterization with phosphate-solubilizing bacteria (PSB) gave significantly higher dry weight and total N and P uptake of tomato shoots as compared with uninoculated control treatment (Kim *et al.*, 1998; Hariprasad and Niranjana, 2009).

Plant-Microbe Interactions Involved in Plant Growth Promotion

Crop productivity in agriculture is the outcome of complex network interactions among soil, plant and microbes. There is an urgent need for developing an ecologically compatible, environment-friendly technology for agriculture production system that may provide adequate supply of essential nutrients in utilizable form for enhancing food production to feed the ever-increasing human population. Conventional agriculture plays a crucial role to fulfill the increasing food demands, but it has led to enhanced use of pesticides and chemical fertilizers. Therefore, an understanding of soil microbiological processes and microbial diversity is needed to improve soil quality, soil health and agricultural productivity. The root surfaces are nutrient-rich niches and root exudates provide a large number of nutrients for the growth and energy of soil microbes, and therefore great diversity of microorganisms, including phytopathogens are chemo-attracted towards the plant roots (Degenhardt *et al.*, 2003). Thus, released root exudates provide substrate and energy sources for the soil microbial community and the quantity and type of compounds released in the rhizosphere influence the shape and size of microbiome community (Kandeler *et al.*, 2002; Marschner and Baumann, 2003). Plant genotypes and species are considered as one of the most important factors in shaping rhizobacterial communities (Prashar *et al.*, 2014; Wen *et al.*, 2017). More studies are still required to fully understand the beneficial and harmful plant-microbe interactions in the rhizosphere (Micallef *et al.*, 2009; Ryan *et al.*, 2009).

Rhizosphere microorganisms can be classified either as pathogenic, saprophytic and beneficial depending upon their interactions on plants (Lugtenberg *et al.*, 2002; Sindhu *et al.*, 2018). Many microorganisms *viz. Azospirillum, Azotobacter, Klebsiella, Enterobacter, Alcaligenes, Arthrobacter, Burkholderia, Bacillus, Pseudomonas, Rhizobium* and *Serratia* have been reported to enhance the growth and yield of different crops under pot house or field conditions (Safdar *et al.*, 2012; Joseph *et al.*, 2012; Gupta *et al.*, 2015). Direct stimulatory effect on plant growth is caused by providing fixed nitrogen and solublization of phosphate and potassium or by production of phytohormones, whereas indirect stimulation of plant growth is caused by suppression of phytopathogens' growth (Juanda, 2005; Sindhu *et al.*, 2014c). Some rhizosphere bacteria enhance plant growth by producing growth-promoting substances in the culture media, rhizosphere and the rhizoplane of forage grasses and cereals (Whipps, 2001).

Various strains of rhizobia, *Azotobacter* and *Azospirillum* have been applied commercially as biofertilizers on different cereal, legume, vegetable and horticulture crops. Other category of microbial inoculants includes auxin-producing and root-elongating phytostimulators viz. *Azospirillum* and *Azotobacter*, and pollutant degrading rhizobacteria and biopesticides (Glick and Bashan, 1997; Kumari *et al.*, 2014; Sindhu *et al.*, 2016). Siderophores-producing and phosphate-solubilizing bacteria provide solubilized inorganic phosphate, iron and other essential nutrients for stimulation of plant growth (Idris *et al.*, 2007; Chabot *etal.*, 1996; Rajkumar *et al.*, 2006; Saharan and Nehra, 2011). Other mechanisms of plant growth improvement by PGPR include N_2 fixation by associative or endophytic bacteria, promotion of mycorrhizal symbiosis, regulation of stress hormone ethylene by ACC deaminase containing bacteria and imparting tolerance to heavy metal toxicity (Saharan and Nehra, 2011). Suppression/killing of plant pathogens and reduction in disease occurrence is often indirectly connected with improvement in plant growth (Freitas and Germida, 1990; Barazani and Friedman, 1999; Nelson, 2004, Sharma *et al.*, 2018a). Thus, microbial inoculants having beneficial plant-microbe interactions have been reported to improve crop productivity (Dadarwal *et al.*, 1997; Berg, 2009; Sangwan *et al.*, 2012; Sindhu *et al.*, 2019b).

Mechanism of Action for Stimulation of Plant Growth

Plant growth stimulation by inoculation of PGPR strains has been attributed to the net outcome of multiple mechanisms that may be activated simultaneously (Hafeez *et al.*, 2004; Shaharoona *et al.*, 2008; Viveros *et al.*, 2010; Sindhu *et al.*, 2018) (Figure 8.1). On the basis of the mechanism of action, PGPR are categorised into three broad forms *i.e.*, biofertilizers, phytostimulators and biopesticides (Choudhary and Sindhu, 2015; Sharma *et al.*, 2018a). Different plant growth-promoting microorganisms (PGPMs) possess the ability to either fix the atmospheric inert nitrogen to plant utilizable ammonia or solubilize the bound form of nutrients (phosphorus and potassium) in soil and facilitate their absorption by plants, and consequently enhance the plant growth and yield (Kumar *et al.*, 2018; Sindhu *et al.*, 2019b). Phytohormones production, induction of pathogen resistance and promotion of mycorrhizal functioning also contribute towards enhanced crop productivity (Glick *et al.*, 1999; Zahir *et al.*, 2004; Spaepen *et al.*, 2007; Castro *et al.*, 2009). Some rhizosphere bacteria contribute to plant growth improvement by reducing the level of abiotic and biotic stress, and contribute towards scavenging of heavy metal toxicity (Lugtenberg and Kamilova, 2009; Yang *et al.*, 2009; Singh *et al.*, 2011; Jain and Bhatt, 2014; Sindhu and Sharma, 2019). Recent application of biochemical, molecular and computational approaches are providing new information on the genetic basis

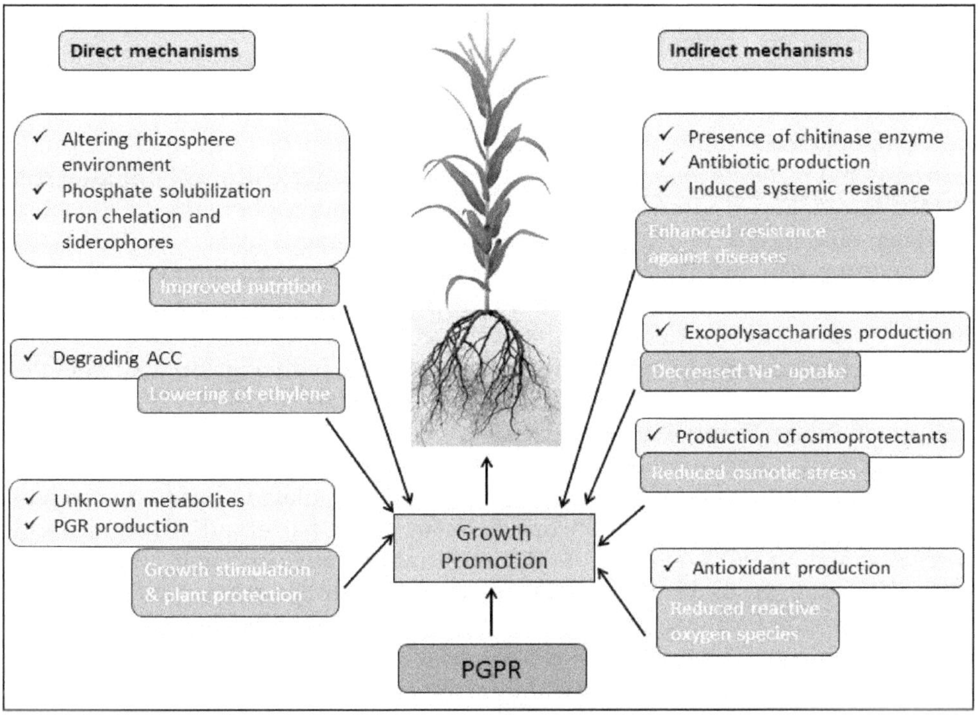

Figure 8.1: Mechanism of Action of PGPR Involved in Growth Enhancement of Crops.

of various biosynthetic pathways and their regulation for further improvement in production of growth-promoting substances by the PGPR (Joshi and Bhatt, 2011). However, the establishment and survival of the introduced PGPR to maintain a critical population for effective functioning is needed in the the rhizosphere.

PGPR Role as Biofertilizer

Biofertilizers colonize the rhizosphere after their application on to the seeds, seedlings or on the plant surfaces (McCully, 2001) and thereby promote root and shoot growth by providing nutrients. In endophytic relationship, PGPR reside within the intercellular apoplastic spaces of parenchyma tissue inside the host plants (Dong *et al.*, 1997) and within xylem vessels (James *et al.*, 2002). Various rhizobial strains form nitrogen-fixing root nodules on the leguminous plants and act as as biofertilizers (Vessey, 2003; Sindhu *et al.*, 2019b). Besides nitrogen-fixing bacteria, phosphorus and potassium solubilizing bacteria are involved in releasing P and K from insoluble/bound minerals. Cyanobacteria such as *Anabaena* and *Nostoc* provide fixed nitrogen to the rice plants. Some bacteria and fungi produce cellulases, hemicellulases and lignolytic enzymes to degrade the lignocellulosic material of plant residues leading to formation of compost (Sindhu *et al.*, 2013; Nisha *et al.*, 2014). Different physicochemical properties of rhizospheric soil such as soil pH, water potential, partial pressure of O_2 and release of root exudates were found to affect the survival, persistence and establishment of the inoculated PGPR strains in the rhizosphere of plants (Griffiths *et al.*, 1999).

Maintenance of Soil Fertility and Nutrient Availability

Rhizobacteria help in solubilization of the bound or unavailable forms of nutrients and facilitate their transport and uptake in plants (Glick, 1995; Sindhu *et al.*, 2014b; Parmar and Sindhu, 2018). The production of siderophores by rhizobacteria has been observed to increase uptake of nutrient elements like Ca, K, Fe, Cu, Mn and Zn (Mantelin and Touraine, 2004). *Bacillus* and *Microbacterium* were found to improve uptake of mineral elements by crop plants (Karlidag *et al.*, 2007). Thus, improvement of the nutrient use efficiency (NUE) to increase crop yield is related with economic saving of natural resources besides, reducing input of chemical fertilizers as nutrients into the environment. The inoculation of PGPR in soil is one of the methods used to enhance NUE (Kloepper *et al.*, 1989; Adesemoye *et al.*, 2010). A significantly higher amount of N and P is taken up by plants from the soil after inoculation with with PGPR as compared to the uninoculated plots (Adesemoye *et al.*, 2010). Single inoculation of phosphate solubilizing bacteria (PSB) significantly increased soil $NaHCO_3$-extractable P concentration compared with uninoculated plots (Kim *et al.*, 1998; Wu *et al.*, 2012). The use of controlled release fertilizer (CRF) and slow release fertilizer (SRF) is another approach to improve N use efficiency by various plants. The gradual and slow release of nitrogen in the CRF and SRF processes extend the N availability in the entire growing season and it reduces the impact of added N fertilizer on the environment (Shoji and Kanno, 1994; Guertal, 2009). Moreover, significantly less N leaching was observed using SRF approach than application of ammonia and nitrate in the soil (Xiaolin and Zongwen, 2009).

Nitrogen Fixation by PGPR

Nitrogen is the most vital nutrient for plant growth and productivity. The atmospheric inert nitrogen is converted into plant utilizable form by the process termed as biological nitrogen fixation (BNF) using a enzyme nitrogenase (Gaby and Buckley, 2012). These nitrogen-fixing PGPR strains are categorized into symbiotic, associative or non-symbiotic. Symbiotic nitrogen fixation is carried out by rhizobia, which forms nodules on the roots of leguminous plants (Figure 8.2), whereas, *Frankia* from symbiotic association with non-leguminous trees and shrubs (Zahran, 2001). The differentiated bacteroids inside the nodules fix the nitrogen to ammonia and provide it to the plant (Sindhu *et al.*, 1990; Ahemad and Kibret, 2014). On the other hand, non-symbiotic nitrogen fixation is carried out by free-living diazotrophs and their inoculation has been reported to promote growth of non-legume plants such as radish, wheat, maize, mustard and rice (Lakshminarayana *et al.*, 2000; Chaudhary *et al.*, 2013). These diazotrophic bacteria belong to different genera including *Azoarcus*, *Azotobacter*, *Acetobacter*, *Azospirillum*, *Burkholderia*, *Enterobacter*, *Gluconacetobacter*, *Pseudomonas* and cyanobacteria (Bhattacharya and Jha, 2012; Vessey, 2003). Non-symbiotic N_2 fixation by rhizobia is another attractive characteristic in which some of the rhizobia has been found to fix N in association with non-legume plants. For example, acetylene reduction acivity has been detected in wild rice (*Oryza breviligulata*) plants inoculated with photosynthetic bradyrhizobia (Chaintreuil *et al.*, 2000). The genes involved in synthesis of nitrogenase enzyme are termed as *nif* genes and are found in both symbiotic and free-living bacterial systems (Sindhu and Dadarwal, 1985; Sindhu *et al.*, 2003; Reed *et al.*, 2011).

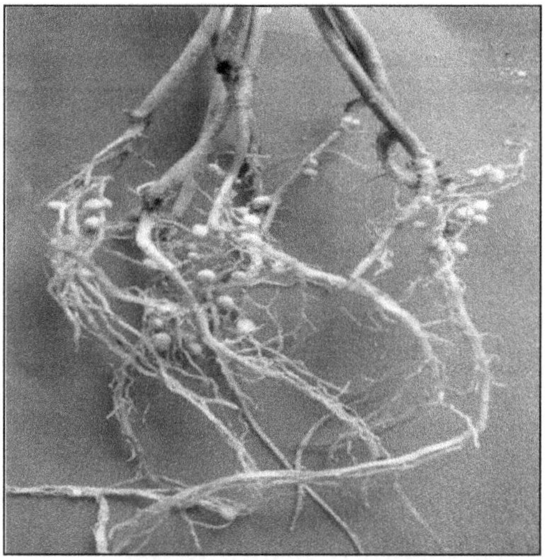

Figure 8.2: Nodulation in Pigeon Pea by Coinoculation of PGPR and *Bradyrhizobium* Strain.

The nitrogen fixed by symbiotic rhizobia has been found to vary in the range from 57-600 Kg N ha^{-1} annually (Elkan, 1992). This legume-*Rhizobium* symbiotic association reduces about 70-80 per cent of the total 17.2 x 10^7 tonnes biologically fixed nitrogen per year on the global basis (Ishizuka, 1992). Inoculation of effective symbiotic rhizobia has been reported to improve the plant biomass and grain yield in various legumes (Sindhu and Dadarwal, 1992; Goel *et al.*, 2001). Moreover, the left-over nitrogen in the field after harvesting of legumes is often equivalent to the application of 30-80 kg of fertilizer-N ha^{-1} (Sindhu *et al.*, 1992). Rhizobial strains have also been found to release secondary metabolites, which minimize the harmful effects of phytopathogens (Datta and Chakrabartty, 2014; Sindhu *et al.*, 2014b; 2017). The combined inoculation of rhizobia with PGPR strains has been found to enhance nodulation, plant biomass and yield in different legumes as observed by mixed inoculation of *B. japonicum* with *P. fluorescens* in soybean (Nishijima *et al.*, 1988), *R. meliloti* with *Pseudomonas* in alfalfa (Li and Alexander, 1988), *Mesorhizobium* or *Bradyrhizobium* strains with *Pseudomonas* sp. in green gram and chickpea (Sindhu *et al.*, 1999; Goel *et al.*, 2002; Chaudhary and Sindhu, 2017).

PGPR as Phosphorous Solubilizers

Phosphorus (P) is another essential nutrient for growth and development of plants that has profound impact on global agriculture and food production (Song *et al.*, 2014). Because the vast majority of soil P is found in insoluble forms, therefore the amounts available to plants are usually low. The plants can only absorb phosphorus in two soluble forms, the monobasic ($H_2PO_4^-$) and the diabasic (HPO_4^{2-})

ions (Kim *et al.*, 2007). Several phosphate-solubilizing microorganisms (PSMs) have been characterized, which solubilize the insoluble form of phosphorus either by acidification (Abd-Alla, 1994; Rajkumar and Freitas, 2008), secretion of organic acids (Richardson *et al.*, 2009) and by chelation and exchange reactions (Hameeda *et al.*, 2008). Organic acids released in the plant root exudates also contribute towards phosphorus solubilization and improve the concentration of available P in rhizosphere soil (Hinsinger, 2001).

The inoculation of phosphate-solubilising microorganisms was found promote growth as well as P uptake of maize plants and improved tolerance of plants under water-deficit stress conditions (Ehteshami *et al.*, 2007). The inoculation of these PSMs into a microcosm containing barren lakeside area soil enhanced the plant growth significantly (Jeon *et al.*, 2003; Paul and Sarma, 2006). Six native peanut bacteria were found to solubilize phosphate from FeO_4 and $AlPO_4$ by producing gluconic acid and cofactor PQQ into the medium (Anzuay *et al.*, 2017), and their inoculation on peanut and maize was found to increase seed germination, plant growth and phosphorus content in the root as well shoot. Similarly, significant growth promotion of poplar was observed by inoculation of phosphate solubilizing bacterium *Pseudomonas frederiksbergensis* JW-SD2 in both sterilized and non-sterilized soils (Zeng *et al.*, 2017).

Production of Plant Growth Regulators

The soil and the entire subterranean root system establish an intimate association with microbes (bacteria, fungi) and other small organisms (nematodes, earthworms, *etc.*) (Schmülling, 2002; Stirk and van Staden, 2010; Kim*et al.*, 2018). Production of different phytohormones like IAA, gibberellic acid and cytokinins by the PGPR strains have been reported to alter root architecture, leading to more adsorption of nutrients and promotion of plant growth (Cheryl and Glick, 1996; Han *et al.*, 2005; Boiero *et al.*, 2007; Malik and Sindhu, 2011; Park *et al*,. 2015). Indole acetic acid (IAA) is the most common natural auxin found in plants (Miransari and Smith, 2014). Moreover, up to 80 per cent of soil- and plant associated bacteria have been found to synthesize indole acetic acid (IAA) (Spaepen and Vanderleyden, 2011).

The inoculation of indole acetic acid and amino levulinic acid (ALA) producing strains of *Rhodopseudomonas* sp. KL9 and BL6 was found to cause growth enhancement of tomato seedlings under axenic conditions (Hyun and Song, 2007). On the other hand, retardation effects were observed on growth of *Phalaris minor* weed by inoculation of *Bacillus subtilis* strain SYB101 and *Providentia* strain CPS67 having different levels of IAA and ALA production (Phour and Sindhu, 2018). Similarly, the inoculation on the roots of cherry cuttings with a recombinant strain of *P. fluorescens* that produced increased amount of IAA inhibited root growth at high density, whereas lower densities on black-currant promoted grow of the cuttings of sour cherry (*Prunus cerasus*) and black-currant (*Ribes nigrum*) (Dubeikovsky *et al.*, 1993). Inhibitory effects on the growth of *Convolvulus arvensis* were observed by inoculation of *Enterobacter taylorae* isolate having high auxin-producing potential (72 µg ml^{-1}) (Sarwar and Kremer, 1995). Differential effects of IAA production ability of *Pseudomonas trivialis* strain X33d were observed on the growth of weed and wheat (Mejri *et al.*, 2010), and its inoculation caused growth suppression of great brome weed and promoted the growth of durum wheat. Similar differential effects of inoculation with *Bacillus* strains having variation in IAA production ability (ranging from 4.0-24.3 µg/mL) were observed on the growth and yield of chilli pepper (*Capsicum annuum* L. cv Jalapeno) (Pena-Yam *et al.*, 2016). Inoculation with *B. subtilis* strain ITC-N67 caused significant gain in stem diameter and root volume, whereas flower buds, biomass of roots and total biomass were increased by inoculation with *B. cereus* strain ITC-BL18. On the other hand, inoculation with *B. flexus* strain ITC-P4 and *B. flexus* strain ITC-P22 adversely affected the root volume and total biomass of chilli pepper.

Other plant hormones involved in plant growth promotion such as cytokinins or gibberellins, or both were produced by several PGPR strains (Kang *et al.*, 2010; Shi *et al.*, 2017). Cytokinins are also produced by some strains of phytopathogens. However, it appears that PGPR produce lower cytokinin levels in comparison to the phytopathogens and therefore, the effect of the PGPR on plant growth is stimulatory, whereas higher levels of cytokinin production from pathogens is inhibitory. Arkhipova *et al.* (2005) reported that inoculation of lettuce plants with different *B. subtilis* strains having the ability to produce zeatin riboside-type cytokinin, dihydro-zeatin riboside and iso-pentyl adenosine, caused 30 per cent increase in plant weight.

Contribution towards Abiotic Stress Tolerance

Drought, ssalinity, flooding, low temperature and heavy metals are key sources of abiotic stress. Exposure of crop plants to various abiotic stress factors may cause losses in yield ranging from 50 to 82 per cent (Barrios *et al.*, 2008; Kang *et al.*, 2014). Crop yield is limited in semiarid and arid regions of the world by increased salinisation of irrigation water as well as soil (Han and Lee, 2005; Krasensky and Jonak, 2012). Some soil bacteria has been reported to provide strong growth adaptation potential under stressful conditions (Grover *et al.*, 2011). Tripathi *et al.* (2002) reported that *Azospirillum*-inoculated sorghum plants were less drought-stressed over uninoculated plants and inoculated plants showed more water content, higher water potential and lower canopy temperature in their foliage, and thus these plants. Besides salinity, drought is another environmental factor that affects cultivation of crops in semi-arid regions. Drought stress resulted in seed yield reduction of different chickpea genotypes varying from 34.3 to 78.1 per cent (Talebi *et al.*, 2013). However, seed inoculations with PGPR are able to reduce water deficit damage in asparagus (*Asparagus officinalis* L.) and enhanced plant growth under drought conditions (Liddycoat *et al.*, 2009). Exopolysaccharides (EPS) secreted by PGPR enabled the bacterium to survive under prevailing drought stress and improved drought tolerance of plants through osmotic and intracellular adjustment (Mayak *et al.*, 2004; Sandhya *et al.*, 2009). The inoculation with EPS-producing PGPR strains extended the tolerance of drought-exposed barley plants for 2 weeks than uninoculated control plants (Timmusk, 2003).

Similarly, inoculation with a consortium of three ACC-deaminase producing rhizobacterial strains significantly enhanced the seed germination, length of root and shoot, and dry weight of *Vigna mungo* and *Pisum sativum* plants under drought stress in agricultural fields (Saikia *et al.*, 2018). Enhanced plant tolerance against a wide range of abiotic stresses was observed by inoculation of ACC deaminase-containing PGPR strains (Stepien and Klobus, 2005; Shahzad *et al.*, 2017; Chaudhary and Sindhu, 2017). Peanut plants inoculated with *Bradyrhizobium* strains SEMIA 6144 and ESA 123 reduced the water stress effects on some genotypes, indicating that *Bradyrhizobium* strains could be used as inoculants under water stress condition (Barbosa *et al.*, 2018). The role of poly β-hydroxyl butyrate (PHB) synthesis and its accumulation was found to tolerate various stresses in *Azospirillum* (Kadouri*et al.*, 2003). The plant/PGPR interactions were also found to mitigate the effects of stress hormone ethylene and significantly increased plant biomass and it also caused the decontamination of persistent petroleum and organic contaminants in soil (Glick and Stearns, 2011; Gamalero and Glick, 2012).

Production of Volatile Organic Compounds

Production of volatile organic compounds (VOCs) by rhizobacteria constitutes an important mechanism for the enhancement of plant growth. Some PGPR strains *i.e., Bacillus subtilis* GB03, *B. amyloliquefaciens* IN937a and *Enterobacter cloacae* JM22 were found to release a range of volatile compounds, particularly, 2,3-butanediol and acetoin, which promoted the growth of *Arabidopsis thaliana* (Ryu *et al.*, 2003). Recently, VOCs produced by the rhizobacterial strains have been reported to act as signaling molecule to trigger the plant responses in various plant–microbe interactions (Farmer, 2001; Ryu *et al.*, 2003). However, more studies are required to understand the role of the volatile components in signaling and plant defense in the plant–rhizobacterial system.

Application of Microbes as Biopesticides

Certain microbes from the rhizosphere facilitate plant protection from various pathogens including fungi and bacteria, thus act as biocontrol agents (Mishra *et al.*, 2015; Jha and Saraf, 2015; Sindhu *et al.*, 2016). Various bacterial and fungal genera (Table 8.1) have been characterized for their role in the suppression of plant pathogens along with stimulation of plant growth (Sindhu *et al.*, 2010; Bonaldi *et al.*, 2015; Bhattacharyya *et al.*, 2016; Sharma *et al.*, 2018), and many microbe-based products are available in the market. However, intensive screening and selection of microorganisms is required before their use in disease control and plant growth promotion in various cropping systems. Furthermore, knowledge of marketing of biocontrol agents cost of mass-scale production and registration procedures also to be revised to popularize the market status of these biocontrol and biofertilizer strains.

The plant beneficial rhizospheric microorganisms protect plants from the attack of pathogens either by direct killing of pathogens through the production of antibiotics (Shaharoona *et al.*, 2006; Avis *et al.*, 2008; Meena *et al.*, 2017). The strains belonging to *Bacillus* and *Pseudomonas* genera have been studied extensively for antibiosis mechanisms in the disease management practices (Kinsella *et al.*, 2009; Dominguez-Nunez *et al.*, 2016). Various antibiotics are produced by pseudomonads such as 2,4-diacetylphloroglucinol (DAPG), oomycin A, phenazines, pyoluteorin, pyrrolnitrin, phenazine carboxylic acid (PCA), tropolone and cyclic lipopeptides

Table 8.1: Biocontrol of Pathogens by Rhizosphere Bacteria

Biocontrol Organism	*Suppressed Pathogen*	*Disease on Crops*	*Reference*
P. fluorescens strain strain CHA0	Thievaloviopsis basicola Pythium ultimum	Tobacco Cucumber	Stutz *et al.* (1986) Maurhofer *et al.* (1992)
P. fluorescens strain strain MDU2	Rhizoctonia solani	Rice	Nagrajkumar *et al.* (2004)
P. putida strain WCS358	Ralstonia solanacearum	*Eucalyptus urophylla*	Ran *et al.* (2005)
P. aureofaciens strain 30-84	*G. graminis* var. *tritici*	Wheat	Harrison *et al.* (1993)
Pseudomonas aeruginosa strain 7NSK2	*Pythium ultimum*	Tomato	Buysens *et al.* (1996)
B. amyloliquefaciens	*Fusarium oxysporum*	Tomato	Koumoutsi *et. al.* (2004)
B. subtilis strain QST713	*R. solani*	Tomato	Kloepper *et. al.* (2004)
B. subtilis strain CA32, *Trichoderma harzianum* strain R40I	*R. solani*	*Solanum melongena,* *C. annuum*	Abeysinghe (2009)
Bacillus sp. strain BPR7	*Macrophomina phaseolina, Fusarium oxysporum, R. solani*	*Phaseolus vulgaris*	Kumar *et al.* (2012)
P. fluorescens, B. subtilis	*F. oxysporum* f. sp. *lycopersici*	Tomato	Sundaramoorthy and Balabasar (2013)
B. subtilis, Paenibacillus polymyxa, P. fluorescens, P. putida, Sinorhizobium meliloti	*F. oxysporum, P. solani, R. solani*	Alfalfa	Sarhan and Shehata (2014)
Bacillus subtilis B006	*F. oxysporum* f. sp. *cucumerinum*	Cucumber	Yang *et al.* (2014)
Bacillus sp. HCS43, *Pseudomonas* sp. HCS36	*R. solani*	Clusterbean	Choudhary and Sindhu (2015)
Chryseobacterium sp. strain ISE14	*Phytophthora capsici*	Pepper	Sang *et al.* (2018)
Pseudomonas chlororaphis strain 30-84	*Gaeumannomyces graminis* var. *tritici*	Wheat	Yu *et al.* (2018)
Pseudomonas chlororaphis strain O6	*Meloidogyne hapla* (nematode)	Tomato	Kang *et al.* (2018)
Stenotrophomonas sp. strain Rz291Ce, *Curtobacterium* sp. strain R208Ag, R209Ag, *Leucobacter* sp. strain Rz274Cm	*Xanthomonas citri* pathotype A (Xac306)	*Citrus sinensis*	Felestrino *et al.* (2018)
Bacillus amyloliquefaciens strain BS_Abi	*Ralstonia solanacearum*	Brinjal (egg plant)	Sakthivel *et al.* (2019)

(Loper and Gross, 2007). Antibiotics such as oligomycin A, kanosamine, zwittermicin A and xanthobaccin are produced by *Bacillus, Streptomyces* and *Stenotrophomonas* sp. to prevent the growth of plant pathogens (Compant *et al.*, 2005). These secondary metabolites produced by biological control agents have the advantage due to their different mode of action over the chemical fungicides (Setiawati and Handayanto, 2010; Wackett, 2013).

Pseudomonas sp. was found to suppress take-all disease of wheat caused by the fungus *Gaeumanomyces graminis* var. *tritici* by producing the antibiotic 2, 4- DAPG under field conditions (de Souza *et al.*, 2003). Similarly, *P. fluorescens* strains which produced the antibiotic phenazine-1- carboxylic acid (PCA) resulted in significant suppression of take-all disease in 60 per cent of field trials (Weller, 2007). Ongena and Jacques (2008) reported biological control activity against soil borne pathogens and plant growth promotion activity by inoculation of lipopeptide and polyketide producing *Bacillus amyloliquefaciens* strain. On the other hand, the biocontrol of black root rot disease of tobacco caused by *Thielaviopsis basicola* was carried out by hydrogen cyanide (HCN) producing rhizobacteria (Sacherer *et al.*, 1994). Lanteigne *et al.* (2012) reported that the production of both 2,4-diacetylphloroglucinol antibiotic and HCN by *Pseudomonas* strains resulted in the biological control of bacterial canker in tomato.

Some PGPR strains produced hydrolytic enzymes such as chitinases, β-glucanase and proteases, which exhibited hyperparasitic activity (Joshi *et al.*, 2012; Hayat *et al*, 2010). The hydrolytic enzymes produced by antagonistic rhizobacteria were found to suppress the growth of different pathogenic fungi including *Botrytis cinerea, Sclerotium rolfsii, Fusarium oxysporum, Phytophthora* sp. and *Pythium ultimum* and their inoculation resulted in significant gains in plant growth (Upadyay *et al.*, 2012; Nadeem *et al.*, 2013). The inoculation of different PGPR strains was found to suppress the diseases in various crops. For example, *Pseudomonas putida* suppressed the disease caused by *Macrophomina phaseolina* in chickpea and *Azotobacter chroococcum* reduced the disease incidence caused by *Fusarium oxysporum* in *Sesamum indicum* under field conditions (Maheshwari *et al.*, 2012). *Pseudomonas fluorescens* has been reported to protect plants against a wide range of important fungal diseases such as root-rot of mustard and damping-off disease of sugar beet under field conditions (Fenton *et al.*, 1992; Kumar *et al.*, 2002; Arora *et al.*, 2008).

The use of multistrain inoculants has been found to exhibit synergistic effects and improved the biocontrol activity in comparison to treatment with single antagonistic strains. For example, the coinoculation of antagonistic *Trichoderma* fungi with *Pseudomonas* and rhizobial strains protected chickpea plants from infection by the collar rot pathogen *Sclerotium rolfsii* (Singh *et al.*, 2013). Similarly, seed bacterization with some PGPR strains conferred systemic resistance against a broad spectrum of plant pathogens ranging from fungal, bacterial, viral, insects and nematodes (Naznin *et al.*, 2012). Moreover, jasmonate and ethylene signaling within the plant was found to stimulate the host plant's defense responses against a variety of plant pathogens through induced systemic resistance (Glick, 2012). Bacterial components such as lipopolysaccharides, flagella, siderophores, cyclic lipopeptides, 2, 4-diacetylphloroglucinol, homoserine lactones and volatiles like acetoin and 2, 3-butanediol were reported to induce systemic resistance (Press *et al.*, 2001; Doornbos *et al.*, 2012).

Contribution in Bioremediation

Pesticides and heavy metals are well known hazardous environmental toxicants and a potential threat to the life of microbes, insects and humans. Excessive accumulation of heavy metals in the human body may cause DNA breakage, failure of vital organs like kidney and damage of brain tissues (Alvarez-Ayuso, 2008; Kristanti *et al.*, 2011; Singh *et al.*, 2011). Heavy metals also influence germination, growth and development of the annual as well as perennial crop plants (Sethy and Ghosh, 2013). To scavenge heavy metals from the environments, many microbes including bacteria, yeast, fungi and algae have been characterized, which could be used in developing eco-friendly, cost effective and economically sustainable bioremediation technology (Aken *et al.*, 2010; Gupta *et al.*, 2016). The deleterious effects of heavy metals on plants were reduced with the application of rhizobacteria (Belimov *et al.*, 2005; Rajkumar and Freitas, 2008; Glick, 2010; Ahemad and Kibret, 2014). On the other hand, P-solubilising bacteria, mycorrhizal-helping bacteria (MHB) and arbuscular mycorrhizal fungi (AMF) were found to play an important role in phytoremediation in metal-contaminated soils (Khan, 2005; Gerhardt *et al.*, 2009). These soil microbes mitigated the toxic effects of heavy metals on the plants through secretion of acids, proteins, phyto-antibiotics and other chemicals (Denton, 2007).

The inoculation of ACC deaminase containing PGPR isolates was found to protect plants against the inhibitory effects of cadmium (Dell'Amico *et al.*, 2008; Govindasamy *et al.*, 2009). The inoculation of beneficial bacteria stimulated plant growth of *Brassica juncea* and protected the plant from metal toxicity under greenhouse conditions (Wu *et al.*, 2006). The inoculated plants produced much larger aboveground biomass and altered metal bioavailability in the soil. *Pseudomonas putida* has been found to tolerate a number of heavy metals at higher concentrations and therefore, it is used as an excellent candidate for field application in contaminated soil (Chacko *et al.*, 2009). The hydroxamate siderophores produced in the culture filtrates of *Streptomyces acidiscabies* strain E13 were found to promote cowpea growth under nickel contamination. The siderophores bound the iron and nickel, and thus played a dual role of supplying iron for plant use and protecting the plants against nickel toxicity (Dimkpa *et al.*, 2008; Badri *et al.*, 2009). Similarly, a wide range of soil bacteria and fungi in agricultural soils may utilize residual toxic pesticides as energy and nutrient source (Parte *et al.*, 2017; Walia *et al.*, 2018). In addition, the enzymatic transformation of the pesticides makes these xenobiotics to nontoxic and many bacteria, fungi and actinomycetes have been characterized as the pesticide degraders in different laboratories around the world (Kumar *et al.*, 2018; Huang *et al.*, 2018).

Crop Responses to Inoculation with PGPR

Inoculation of different beneficial rhizosphere bacteria and arbuscular mycorrhizal (AM) fungi has been found to increase the root and shoot biomass and yield of various cereal, legume and oilseed crops (Tambekar *et al.*, 2009; Sindhu *et al.*, 2014b; 2019b) (Table 8.2). Seed bacterization of maize (*Zea mays*) with *Azospirillum brasilense* was found to significantly increase the root surface area as well as root and

Table 8.2: Growth Promotion Effects of PGPR in different Crops and Mechanisms Involved

PGPR Isolates	*Host Plant*	*Mechanisms Involved*	*References*
Paenibacillus strain K165	Eggplant, potato	Induced systemic resistance	Tjamos *et al.* (2004, 2005)
Serratia plymuthica strain HRO-C48	Oilseed rape	Quorum sensing	Müller *et al.* (2009)
Serratia marcescens strain UPM39B3	Banana	Growth promotion	Ting *et al.* (2008)
B. amyloliquefaciens strain Bg-C31	Capsicum	Production of antimicrobial proteins	Hu *et al.* (2010)
Pseudomonas *fluorescens* strain EB69	*Solanum melongena*	Production of siderophores and inhibitory compounds	Ramesh and Phadke (2012)
Streptomyces virginiae strain Y30 and E36	Tomato	Production of siderophores and ACC deaminase activity	Tan *et al.* (2011)
B. amyloliquefaciens strain BZ6-1	Peanut	Production of antimicrobial compounds	Wang and Liang (2014)
Species of *Bacillus*, *Alcaligenes* and *Proteus*	Chili	Increased root and shoot length	Patel *et al.* (2017)
Pseudomonas frederiksbergensis strain JW-SD2	Poplar	Promoted growth of poplar in both sterilized and non-sterilized soils by phosphate solubilization	Zeng *et al.* (2017)
Enterobacter sp.	Rice	Reduced ethylene production and stimulated the growth of rice seedlings	Sarkar *et al.* (2018)

foliage dry weight of the maize seedling (Zaddy *et al.*, 1993), whereas *Azospirillum* inoculation of maize enhanced dry weight of seed by 59 per cent, which was similar to application of 60 kg urea N ha^{-1} (Fulchieri and Frioni, 1994). The co-inoculation of *Azospirillum* and *Pseudomonas striata* also showed a significant positive effect on grain yield in maize (Prabakaran and Ravi, 1991) and cotton (Radhakrishnan, 1996). The inoculation of phosphorus-solubilizing bacteria with the application of 50 per cent P_2O_5 in the sugarcane gave cane yield and sugar yield, which were found at par with 100 per cent P_2O_5 application (Kathiresan *et al.*, 1995). Radhakrishnan (1996) reported that inoculation of *Azospirillum* and phosphorus solubilizing bacteria resulted in higher root biomass and more bolls in cotton. The combined inoculation of nitrogen fixer *Azospirillum lipoferum*, *Arthrobacter mysorens* and the phosphate solubilizing strain *Agrobacterium radiobacter* showed significant increase on grain yield and ARA in roots of barley (Belimov *et al.*, 1995). Similarly, seed bacterization of bean with *R. leguminosarum* and *P. putida* strain R105 resulted in significant increase in the number of nodules and acetylene reduction activity (ARA) (de Freitas *et al.*, 1993). The interaction of rhizobacteria with AM fungi was found to stimulate mycelial growth and enhanced the mycorrhiza formation on plant roots (Garbaye, 1994; Bianciotto and Bonfante, 2002). The extensive branching of the AM fungal mycelium and host root structures extended beyond the phosphate-depleting zone and access to phosphate ions in soil was elevated (Barea *et al.*, 2005; Smith and Read, 2010).

Future Prospects and Challenges

The extensive characterization of effective beneficial microbes has been carried out from rhizosphere of various crops for their use as bioinoculants for cereals, legumes, vegetables, fruits and flowers (Sindhu *et al.*, 2018). Due to the compatibility and synergistic interaction of selected bacteria, their multi-strain bacterial consortium could be an effective approach for reducing the harmful impact of biotic and abiotic stresses on plant growth leading to improved growth of the inoculated plants (Zhuang *et al.*, 2007; Farwell *et al.*, 2007). Further research on nitrogen fixation, phosphate and potassium solubilization by PGPR strains will increase their use as inoculants inder field conditions. Attempts are required towards optimizing the growth conditions and increased shelf life of PGPR products, which may tolerate adverse environmental condition, leading to improved yield of various crops. Moreover, naturally-occurring PGPR strains could be genetically manipulated with cloned genes leading to increased production of antibiotics, HCN, siderophores or hydrolytic enzymes (Nakkeeran *et al.*, 2005), thereby protecting the plants from the attack of pests and diseases under field conditions.

Conclusion

The presence of microorganisms in the soil is critical to the maintenance of soil functions, in both natural and managed agricultural soils. These soil-inhabiting microbes are involved in key processes such as, decomposition of organic matter and the cycling of elements such as carbon, nitrogen, phosphorus, potassium and sulfur. Moreover, these beneficial microorganisms play key roles in suppressing soil-borne plant diseases and in promoting plant growth (Doran, 2002; Sindhu *et al.*, 2016). For obtaining healthier and more productive plants by inoculation of PGPR strains, a better understanding of plant-microbe interactions is required (Zorner *et al.*, 2018). The integrated nutrient management (INM) should be strengthened as an ecofriendly approach for agricultural sustainability. Similarly, more efforts should be made in characterizing antagonistic microbial strains with improved biological control under field conditions to minimize the use of pesticides. Moreover, biocontrol agents could be used along with reduced application of pesticides under integrated pest management (IPM) strategy to control the infestation of pathogens and insects for sustainable agriculture.

REFERENCES

Abd-Alla, M.H. 1994. Solubilization of rock phosphates by *Rhizobium* and *Bradyrhizobium*. *Folia Microbiol.* 39: 53–56.

Abeysinghe, S. 2009. Effect of combined use of *Bacillus subtilis* CA32 and *Trichoderma harzianum* RU01 on biological control of *Rhizoctonia solani* on *Solanum melongena* and *Capsicum annum*. *Plant Pathol. J.* 8: 9–16.

Adesemoye, A.O. and Kloepper, J.W. 2009. Plant–microbes interactions in enhanced fertilizer-use efficiency. *Appl. Microbiol. Biotechnol.* 85(1): 1–12.

Adesemoye, A.O., Torbert, H.A. and Kloepper, J.W. 2010. Increased plant uptake of nitrogen from ^{15}N-depleted fertilizer using plant growth-promoting rhizobacteria. *Appl. Soil Ecol.* 46(1): 54–58.

Ahemad, M. and Kibret, M. 2014. Mechanisms and applications of plant growth promoting rhizobacteria: Current perspective. *J. King Saud University - Science* 26: 1–20.

Aken, B.V., Correa, P.A., Schnoor, J.L. 2010. Phytoremediation of polychlorinated biphenyls: new trends and promises. *Environ. Sci. Technol.* 44: 2767–2776. https: / /doi.org/10.1021/es902514d

Alvarez-Ayuso, E. 2008. Cadmium in soil-plant systems: an overview. *Int. J. Environ. Pollut.* 33: 275–291. https: / /doi.org/10.1504/IJEP.2008.019399

Anzuay, M.S., Ciancio, M.G.R., Ludueña, L.M., Angelini, J.G., Barros, G., Pastor, N. and Taurian, T. 2017. Growth promotion of peanut (*Arachis hypogaea* L.) and maize (*Zea mays* L.) plants by single and mixed cultures of efficient phosphate solubilizing bacteria that are tolerant to abiotic stress and pesticides. *Microbiol. Res.* 199: 98–109.

Arkhipova, T.N., Veselov, S.U., Melentiev, A.I., Martynenko, E.V. and Kudoyarova, G.R. 2005. Ability of bacterium *Bacillus* subtilis to produce cytokinins and to influence the growth and endogenous hormone content of lettuce plants. *Plant Soil* 272(1-2): 201–209.

Arora, N.K., Khare, E., Verma, A., Sahu, R.K. 2008. *In vivo* control of *Macrophomina phaseolina* by a chitinase and ß-1, 3 glucanse producing pseudomonad NDN1. *Symbiosis* 46: 129–135.

Asghar, H., Zahir, Z., Arshad, M. and Khaliq, A., 2002. Relationship between in vitro production of auxins by rhizobacteria and their growth-promoting activities in *Brassica juncea* L. *Biol. Fertil. Soils* 35(4): 231–237.

Avis, T.J., Gravel, V., Antoun, H. and Tweddell, R.J. 2008. Multifaceted beneficial effects of rhizosphere microorganisms on plant health and productivity. *Soil Biol. Biochem.* 40(7): 1733–1740.

Badri, D.V., Weir, T.L., van der Lelie, D. and Vivanco, J.M. 2009. Rhizosphere chemical dialogues: plant–microbe interactions. *Curr. Opin. Biotechnol.* 20(6): 642–650.

Barazani, O.Z. and Friedman, J. 1999. Is IAA the major root growth factor secreted from plant-growth-mediating bacteria? *J. Chemical Ecol.* 25(10): 2397–2406.

Barbosa, D.D., Brito, S.L., Fernandes, P.D., Fernandes-Júnior, P.I. and Lima, L.M. 2018. Can *Bradyrhizobium* strains inoculation reduce water deficit effects on peanuts?. *World J. Microbiol. Biotechnol.* 34(7): 87–98.

Barea, J.M., Pozo, M.J., Azcon, R. and Azcon-Aguilar, C. 2005. Microbial co-operation in the rhizosphere. *J. Expl. Botany* 56(417): 1761–1778.

Barrios, S., Ouattara, B. and Strobl, E. 2008. The impact of climatic change on agricultural production: Is it different for Africa? *Food Policy* 33(4): 287–298.

Belimov, A.A., Kojemiakov, A.P. and Chuvarliyeva, C.N. 1995. Interaction between barley and mixed cultures of nitrogen fixing and phosphate-solubilizing bacteria. *Plant Soil* 173(1): 29–37.

Berg, G. 2009. Plant–microbe interactions promoting plant growth and health: perspectives for controlled use of microorganisms in agriculture. *Appl. Microbiol. Biotechnol.* 84(1): 11–18.

Bhattacharyya, P.N. and Jha, D.K. 2012. Plant growth-promoting rhizobacteria (PGPR): emergence in agriculture. *World J. Microbiol. Biotechnol.* 28(4): 1327–1350.

Bhattacharyya, P.N., Goswami, M.P. and Bhattacharyya L.H. 2016. Perspective of beneficial microbes in agriculture under changing climatic scenario: A review. *J. Phytol.* 8: 26–41.

Bianciotto, V. and Bonfante, P. 2002. Arbuscular mycorrhizal fungi: a specialised niche for rhizospheric and endocellular bacteria. *Antonie van Leeuwenhoek* 81(1-4): 365–371.

Boiero, L., Perrig, D., Masciarelli, O., Penna, C., Cassán, F. and Luna, V. 2007. Phytohormone production by three strains of *Bradyrhizobium japonicum* and possible physiological and technological implications. *Appl. Microbiol. Biotechnol.* 74(4): 874–880.

Bonaldi, M., Chen, X., Kunova, A., Pizzatti, C., Saracchi, M. and Cortesi, P. 2015. Colonization of lettuce rhizosphere and roots by tagged Streptomyces. *Front. Microbiol.* 6: 25–31.

Buysens, S., Heungens, K., Poppe, J., Hofte, M. 1996. Involvement of pyochelin and pyoverdin in suppression of *Pythium*-induced damping-off of tomato by *Pseudomonas aeruginosa* 7NSK2. *Appl. Environ. Microbiol.* 62: 865–871.

Chabot, R., Antoun, H. and Cescas, M.P. 1996. Growth promotion of maize and lettuce by phosphate-solubilizing *Rhizobium leguminosarum* biovar. *phaseoli*. *Plant Soil* 184(2): 311–321.

Chacko, S., Ramteke, P.W. and John, S.A. 2009. Amidase from plant growth promoting rhizobacterium. *African J. Bacteriol. Res.* 1(4): 46–50.

Chaintreuil, C., Giraud, E., Prin, Y., Lorguin, J., Ba, A., Gillis, M., De laiudie, P. and Dreyfus, B. 2000. Photosynthetic bradyrhizobia are natural endophytes of the African wild rice *Oryza breviligulata*. *Appl. Environ. Microbiol.* 66: 5437–5447.

Chaudhary, D. and Sindhu, S.S. 2015. Inducing salinity tolerance in chickpea (*Cicer arietinum* L.) by inoculation of 1-aminocyclopropane-1-carboxylic acid deaminase-containing *Mesorhizobium* strains. *African J. Microbiol. Res.* 9(2): 117–124.

Chaudhary, D. and Sindhu, S.S. 2017. Amelioration of salt stress in chickpea (*Cicer arietinum* L.) by coinculation of ACC deaminase-containing rhizospheric bacteria with *Mesorhizobium* strains. *Legume Res.* 40(1): 80–86.

Chaudhary, D., Narula, N., Sindhu, S.S. and Behl, R.K. 2013. Plant growth stimulation of wheat (*Triticum aestivum* L.) by inoculation of salinity tolerant *Azotobacter* strains. *Physiol. Mol. Biol. Plants* 19(4): 515–519.

Cheryl, P.L. and Glick, B.R. 1996. Bacterial biosynthesis of indole-3-acetic acid. *Can. J. Microbiol.* 42(3): 207–220.

Choudhary, S.R. and Sindhu, S.S. 2015. Suppression of *Rhizoctonia solani* root rot disease of clusterbean (*Cyamopsis tetragonoloba*) and plant growth promotion by rhizosphere bacteria. *Plant Pathol. J.* 14: 48–57.

Compant, S., Reiter, B., Sessitsch, A., Nowak, J., Clément, C. and Barka, E.A. 2005. Endophytic colonization of *Vitis vinifera* L. by plant growth-promoting bacterium *Burkholderia* sp. strain PsJN. *Appl. Environ. Microbiol.* 71(4): 1685–1693.

Dadarwal, K.R., Yadav, K.S. and Sindhu, S.S. 1997. Biofertilizer production technology: Prospects. In: *Biotechnological approaches in soil microorganisms for sustainable crop production*. (Dadarwal, K.R. ed.) Scientific Publishers, Jodhpur. pp. 323–337.

Datta, B. and Chakrabartty, P.K. 2014. Siderophore biosynthesis genes of *Rhizobium* sp. isolated from *Cicer arietinum* L. *Biotech.* (4): 391–401.

de Souza, J.T., Weller, D.M., Raaijmakers, J.M. 2003. Frequency, diversity and activity of 2, 4-diacetylphloroglucinol producing fluorescent *Pseudomonas* spp. in Dutch take-all decline soils. *Phytopathology* 93: 54–63.

Degenhardt, J., Gershenzon, J., Baldwin, I.T., Kessler, A. 2003. Attracting friends to feast and foes: engineering terpene emission to make crop plants more attractive to herbivore enemies. *Curr. Opin. Biotechnol.* 14: 169–176.

Dell'Amico, E., Cavalca, L. and Andreoni, V. 2008. Improvement of *Brassica napus* growth under cadmium stress by cadmium-resistant rhizobacteria. *Soil Biol. Biochem.* 40(1): 74–84.

Denton, B. 2007. Advances in phytoremediation of heavy metals using plant growth promoting bacteria and fungi. *Basic Biotechnol.* 3: 1–5.

Dimkpa, C.O., Svatoš, A., Dabrowska, P., Schmidt, A., Boland, W. and Kothe, E. 2008. Involvement of siderophores in the reduction of metal-induced inhibition of auxin synthesis in *Streptomyces* spp. *Chemosphere* 74(1): 19–25.

Dominguez-Nuñez, J.A., Benito, B., Berrocal-Lobo, M. and Albanesi, A. 2016. Mycorrhizal fungi: role in the solubilization of potassium. In: *Potassium Solubilizing Microorganisms for Sustainable Agriculture*. Springer, New Delhi. pp. 77-98.

Dong, Y., Iniguez, A.L. and Triplett, E.W. 2003. Quantitative assessments of the host range and strain specificity of endophytic colonization by *Klebsiella pneumoniae* 342. *Plant Soil* 257(1): 49–59.

Doornbos, R.F., van Loon, L.C., Peter, A.H.M. and Bakker, A. 2012. Impact of root exudates and plant defense signaling on bacterial communities in the rhizosphere. *Rev. Sustain. Dev.* 32: 227–243.

Doran, J.W. 2002. Soil health and global sustainability: translating science into practice. *Agric. Ecosyst. Environ.* 88(2): 119–127.

Dubeikovsky, A.N., Mordukhova, E.A., Kochetkov, V.T., Polikarpova, F.Y. and Boronin, A.M. 1993. Growth promotion of blackcurrant softwood cuttings by recombinant strain *Pseudomonas fluorescens* BSP53a synthesizing an increased amount of indole-3-acetic acid. *Soil Biol. Biochem.* 25(9): 1277–1281.

Ehteshami, S.M., Aghaalikhani, M., Khavazi, K. and Chaichi, M.R. 2007. Effect of phosphate solubilizing microorganisms on quantitative and qualitative characteristics of maize (*Zea mays* L.) under water deficit stress. *Pakistan J. Biol. Sci.* 10(20): 3585–3591.

Elkan, G.H. 1992. Biological nitrogen fixation systems in tropical ecosystems: an overview. *Biological nitrogen fixation and sustainability of tropical agriculture. Wiley, Chichester*, pp. 27–40.

Farmer, E.E. 2001. Surface-to-air signals. *Nature* 411: 854–856.

Farwell, A.J., Vesely, S., Nero, V., Rodriguez, H., McCormack, K., Shah, S., Dixon, D.G. and Glick, B.R. 2007. Tolerance of transgenic canola plants (*Brassica napus*) amended with plant growth-promoting bacteria to flooding stress at a metal-contaminated field site. *Environ. Pollut.* 147(3): 540–545.

Felestrino, E.B., Vieira, T.T., Caneschi, W.L., Cordeiro, I.F., Assis, R.A.B., Lemes, C.G.C., Fonseca, N.P., Sanchez, A.B., Cepeda, J.C.C., Ferro, J.A., Garcia, C.C.M., Carmo, F.F., Kamino, L.H.Y. and Moreira, L.M. 2018. Biotechnological potential of plant growth-promoting bacteria from the roots and rhizospheres of endemic plants in ironstone vegetation in southeastern Brazil. *World J. Microbiol. Biotechnol.* 34: 156–170. https: //doi.org/10.1007/s11274-018-2538-0

Fenton, A.M., Stephens, P.M., Crowley, J., O'Callaghan, M., O'Gara, F. 1992. Exploitation of gene(s) involved in 2, 4- diacetyl phloroglucinol biosynthesis to confer a new biocontrol capability to a *Pseudomonas* strain. *Appl. Environ. Microbiol.* 58: 3873–3878.

Freitas, J.R.D. and Germida, J.J. 1990. Plant growth promoting rhizobacteria for winter wheat. *Can. J. Microbiol.* 36(4): 265–272.

Fulchieri, M. and Frioni, L. 1994. *Azospirillum* inoculation on maize (*Zea mays*): effect on yield in a field experiment in central Argentina. *Soil Biol. Biochem.* 26(7): 921–923.

Gaby, J.C. and Buckley, D.H. 2012. A comprehensive evaluation of PCR primers to amplify the *nifH* gene of nitrogenase. *PLoS One* 7: e42149.

Gamalero, E. and Glick, B.R. 2012a. Ethylene and abiotic stress tolerance in plants. In *Environmental adaptations and stress tolerance of plants in the era of climate change.* Springer, New York, NY. pp. 395-412.

Gamalero, E. and Glick, B.R. 2012b. Plant Growth-Promoting Bacteria and Metals Phytoremediation. *Phytotechnologies: Remediation of Environmental Contaminants,* p.361.

Garbaye, J. 1994. Tansley review no. 76 helper bacteria: a new dimension to the mycorrhizal symbiosis. *New phytologist* 128(2): 197–210.

Gerhardt, K.E., Huang, X.D., Glick, B.R. and Greenberg, B.M. 2009. Phytoremediation and rhizoremediation of organic soil contaminants: potential and challenges. *Plant Sci.* 176(1): 20–30.

Ghosh, P.K., Kumar, De. T., Maiti, T.K. 2015. Production and metabolism of Indole acetic acid in root nodules and symbiont (*Rhizobium undicola*) isolated from root nodule of aquatic medicinal legume *Neptunia oleracea* Lour. *J. Botany* 2015, Article ID 575067, 11 pages. http: //dx.doi.org/10.1155/2015/575067

Glick, B.R. 1995. The enhancement of plant growth by free-living bacteria. *Can. J. Microbiol.* 41(2): 109–117.

Glick, B.R. 2010. Using soil bacteria to facilitate phytoremediation. *Biotechnol. Adv.* 28(3): 367–374.

Glick, B.R. 2012. Plant growth-promoting bacteria: mechanisms and applications. *Scientifica* (Cairo) 2012: 963401.

Glick, B.R. and Bashan, Y. 1997. Genetic manipulation of plant growth-promoting bacteria to enhance biocontrol of phytopathogens. *Biotechnol. Adv.* 15(2): 353–378.

Glick, B.R. and Stearns, J.C. 2011. Making phytoremediation work better: maximizing a plant's growth potential in the midst of adversity. *Intern. J. Phytorem.* 13(sup1): 4–16.

Glick, B.R., Patten, C.L., Holguin, G. and Penrose, D.M. 1999. Biochemical and genetic mechanisms used by plant growth promoting bacteria. Imperial College Press, London. pp. 134–179.

Goel, A., Sindhu, S. and Dadarwal, K. 2002. Stimulation of nodulation and plant growth of chickpea (*Cicer arietinum* L.) by *Pseudomonas* spp. antagonistic to fungal pathogens. *Biol. Fertil. Soils* 36(6): 391–396.

Goel, A.K., Sindhu, S.S. and Dadarwal, K.R. 2001. Symbiotic effectiveness of bacteriocin producing and non-producing strains of *Rhizobium* in green gram (*Vigna radiata*). *Indian J. Expt. Biol.* 39: 821–823.

Govindasamy, V., Senthilkumar, M., Mageshwaran, V. and Annapurna, K. 2009. Detection and characterization of ACC deaminase in plant growth promoting rhizobacteria. *J. Plant Biochem. Biotechnol.* 18(1): 71–76.

Gray E.J. and Smith, D.L. 2005. Intracellular and extracellular PGPR: commonalities and distinctions in the plant-bacterium signaling processes. *Soil Biol. Biochem.* 37: 395–412

Griffiths, B. S., Ritz, K., Ebblewhite, N., Dobson, G. 1999. Soil microbial community structure. Effects of substrate loading rates. *Soil Biol. Biochem.* 31: 145–153.

Grover, M., Ali, S.Z., Sandhya, V., Rasul, A. and Venkateswarlu, B. 2011. Role of microorganisms in adaptation of agriculture crops to abiotic stresses. *World J. Microbiol. Biotechnol.* 27(5): 1231–1240.

Guertal, E.A. 2009. Slow-release nitrogen fertilizers in vegetable production: a review. *HortTechnol.* 19(1): 16–19.

Gupta, A., Joia, J., Sood, A., Sood, R., Sidhu, C., Kaur, G. 2016. Microbes as potential tool for remediation of heavy metals: a review. *J. Microb. Biochem. Technol.* 8: 364–372. https: //doi.org/10.4172/1948-5948.1000310

Gupta, G., Parihar, S.S., Ahirwar, N.K., Snehi, S.K. and Singh, V. 2015. Plant growth promoting rhizobacteria (PGPR): current and future prospects for development of sustainable agriculture. *J. Microbial Biochem. Technol.* 7(2): 96–102.

Gupta, R., Bisaria, V.S. and Sharma, S. 2016. Response of rhizospheric bacterial communities of *Cajanus cajan* to application of bioinoculants and chemical fertilizers: A comparative study. *European J. Soil Biol.* 75: 107–114.

Hafeez, F.Y., Safdar, M.E., Chaudhry, A.U. and Malik, K.A. 2004. Rhizobial inoculation improves seedling emergence, nutrient uptake and growth of cotton. *Australian J. Expl. Agri.* 44(6): 617–622.

Hameeda, B., Harini, G., Rupela, O.P., Wani, S.P. and Reddy, G. 2008. Growth promotion of maize by phosphate-solubilizing bacteria isolated from composts and macrofauna. *Microbiol. Res.* 163(2): 234–242.

Han, H.S. and Lee, K.D. 2005. Plant growth promoting rhizobacteria effect on antioxidant status, photosynthesis, mineral uptake and growth of lettuce under soil salinity. *Res. J. Agric. Biol. Sci.* 1(3): 210–215.

Han, J., Sun, L., Dong, X., Cai, Z., Sun, X., Yang, H., Wang, Y. and Song, W., 2005. Characterization of a novel plant growth-promoting bacteria strain *Delftia tsuruhatensis* HR4 both as a diazotroph and a potential biocontrol agent against various plant pathogens. *System. Appl. Microbiol.* 28(1): 66–76.

Hariprasad, P. and Niranjana, S.R. 2009. Isolation and characterization of phosphate solubilizing rhizobacteria to improve plant health of tomato. *Plant Soil* 316(1-2): 13–24.

Harrison, L.A., Letendre, N., Kovacevich, P., Pierson, E., Weller, D. 1993. Purification of an antibiotic effective against *Gaeumannomyces graminis* var. *tritici* produced by biocontrol agent *Pseudomonas aureofaciens*. *Soil Biol. Biochem.* 25: 215–221

Hayat, R., Ali, S., Amara, U., Khalid, R., Ahmed, I. 2010. Soil beneficial bacteria and their role in plant growth promotion: A review. *Ann. Microbiol.* 60: 579–598.

Hinsinger, P. 2001. Bioavailability of soil inorganic P in the rhizosphere as affected by root-induced chemical changes: a review. *Plant Soil* 237(2): 173–195.

Hu, H.Q., Li, X.S. and He, H. 2010. Characterization of an antimicrobial material from a newly isolated *Bacillus amyloliquefaciens* from mangrove for biocontrol of *Capsicum* bacterial wilt. *Biol. Contr.* 54(3): 359–365.

Huang, Y., Xiao, L., Li, F., Xiao, M., Lin, D., Long, X. and Wu, Z. 2018. Microbial degradation of pesticide residues and an emphasis on the degradation of cypermethrin and 3-phenoxy benzoic acid: A review. *Molecules* 23: 13. doi: 10.3390/molecules23092313.

Hyun, R.K. and Song, H.G. 2007. Effects of application of *Rhodopseudomonas* sp. on seed germination and growth of tomato under axenic conditions. *J. Microbiol. Biotechnol.* 17(11): 1805–1810.

Idris, E.E., Iglesias, D.J., Talon, M. and Borriss, R. 2007. Tryptophan-dependent production of indole-3-acetic acid (IAA) affects level of plant growth promotion by *Bacillusamyloliquefaciens* FZB42. *Mol. Plant-Microbe Inter. 20*(6): 619–626.

Ishizuka, J. 1992. Trends in biological nitrogen fixation research and application. In *Biological Nitrogen Fixation for Sustainable Agriculture*. Springer, Dordrecht. pp. 197–209.

James, E.K., Gyaneshwar, P., Mathan, N., Barraquio, W.L., Reddy, P.M., Iannetta, P.P., Olivares, F.L. and Ladha, J.K. 2002. Infection and colonization of rice seedlings by the plant growth-promoting bacterium *Herbaspirillum seropedicae* Z67. *Mol. Plant-Microbe Interact.* 15(9): 894–906.

Jeon, J.S., Lee, S.S., Kim, H.Y., Ahn, T.S. and Song, H.G. 2003. Plant growth promotion in soil by some inoculated microorganisms. *J. Microbiol.* 41(4): 271–276.

Jha, C.K. and Saraf, M. 2015. Plant growth promoting rhizobacteria (PGPR): a review. *J. Agric. Res. Devel.* 5(2): 108–119.

Joseph, B., Ranjan Patra, R. and Lawrence, R. 2012. Characterization of plant growth promoting rhizobacteria associated with chickpea (*Cicer arietinum* L.). *Intern. J. Plant Production* 1(2): 141–152.

Joshi, M., Shrivastava, R., Sharma, A.K., Prakash, A. 2012. Screening of resistant verities and antagonistic *Fusarium oxysporum* for biocontrol of Fusarium Wilt of Chilli. *Plant Pathol. Microbiol.* 3: 134–152.

Joshi, P. and Bhatt, A.B. 2011. Diversity and function of plant growth promoting rhizobacteria associated with wheat rhizosphere in North Himalayan region. *Intern. J. Environ. Sci.* 1(6): 1135–1142.

Juanda, J.I.H. 2005. Screening of soil bacteria for plant growth promoting activities *in vitro*. *J. Agric. Sci.* 4: 27–31.

Kadouri, D., Jurkevitch, E. and Okon, Y. 2003. Involvement of the reserve material poly-β-hydroxybutyrate in *Azospirillum brasilense* stress endurance and root colonization. *Appl. Environ. Microbiol.* 69(6): 3244–3250.

Kandeler, E., Marschner, P., Tscherko, D., Gahoonia, T.S. and Nielsen, N.E. 2002. Microbial community composition and functional diversity in the rhizosphere of maize. *Plant Soil* 238(2): 301–312.

Kang, B.G., Kim, W.T., Yun, H.S., Chang, S.C. 2010. Use of plant growth-promoting rhizobacteria to control stress responses of plant roots. *Plant Biotechnol. Rep.* 4: 179–183.

Kang, B.R., Anderson, A.J. and Young Cheol Kim, Y.C. 2018. Hydrogen cyanide produced by *Pseudomonas chlororaphis* O6 exhibits nematicidal activity against *Meloidogyne hapla*. *Plant Pathol. J.* 34(1): 35–43. **https: //doi.org/10.5423/PPJ. OA.06.2017.0115**

Kang, G., Li, G. and Guo, T. 2014. Molecular mechanism of salicylic acid-induced abiotic stress tolerance in higher plants. *Acta Physiol. Plant.* 36(9): 2287–2297.

Karlidag, H., Esitken, A., Turan, M. and Sahin, F. 2007. Effects of root inoculation of plant growth promoting rhizobacteria (PGPR) on yield, growth and nutrient element contents of leaves of apple. *Scientia Horticult.* 114(1): 16–20.

Kathiresan G., Manickam G. and Parameswaran P. 1995. Efficiency of phosphobacteria addition on cane yield and quality. *Cooperative Sugar* 26: 629–631.

Khan, A.G. 2005. Role of soil microbes in the rhizospheres of plants growing on trace metal contaminated soils in phytoremediation. *J. Trace Elem. Med.Biol.* 18(4): 355–364.

Kim, H.J., Hummel, J.W., Sudduth, K.A. and Birrell, S.J. 2007. Evaluation of phosphate ion-selective membranes and cobalt-based electrodes for soil nutrient sensing. *Transact. ASABE* 50(2): 415–425.

Kim, J.S., Lee, J.E., Nie, H., Lee, Y.J., Kim, S.T. and Kim, S.H. 2018. Physiological and proteomic analysis of plant growth enhancement by the rhizobacteria *Bacillus* sp. JS. *Genes Genom.* 40(2): 129–136.

Kim, K.Y., Jordan, D. and McDonald, G.A. 1998. *Enterobacter agglomerans*, phosphate solubilizing bacteria and microbial activity in soil: effect of carbon sources. *Soil Biol. Biochem.* 30(8-9): 995–1003.

Kinsella, K., Schulthess, C.P., Morris, T.F. and Stuart, J.D. 2009. Rapid quantification of *Bacillus subtilis* antibiotics in the rhizosphere. *Soil Biol. Biochem.* 41: 374–379.

Kloepper, J.W., Lifshitz, R. and Zablotowicz, R.M., 1989. Free-living bacterial inocula for enhancing crop productivity. *Trends Biotechnol.* 7(2): 39–44.

Kloepper, J.W., Ryu, C.M. and Zhang, S. 2004. Induced systemic resistance and promotion of plant growth by *Bacillus* spp. *Phytopathoogy.* 94(11): 1259–1266.

Kloepper, J.W., Schroth, M.N. and Miller, T.D. 1980. Effects of rhizosphere colonization by plant growth-promoting rhizobacteria on potato plant development and yield. *Phytopathology* 70(11): 1078–1082.

Koumoutsi, A., Chen, X.H., Henne, A., Liesegang, H., Hitzeroth, G., Franke, P., Vater, J. and Borriss, R. 2004. Structural and functional characterization of gene clusters directing non ribosomal synthesis of bioactive cyclic lipopeptides in *Bacillus amyloliquefaciens* strain FZB42. *J. Bacteriol.* 186(4): 1084–1096.

Krasensky, J. and Jonak, C. 2012. Drought, salt, and temperature stress-induced metabolic rearrangements and regulatory networks. *J. Expt. Botany* 63(4): 1593–1608.

Kristanti, R.A., Hadibarata, T., Toyama, T., Tanaka, Y. and Kazuhiro, M. 2011. Bioremediation of crude oil by white rot fungi *Polyporus* sp. S133. *J. Microbiol. Biotechnol.* 21: 995–1000. https: //doi.org/10.

Kumar, A., Patel, J.S. and Meena, V.S. 2018. Rhizospheric microbes for sustainable agriculture: An Overview. In *Role of Rhizospheric Microbes in Soil* (pp. 1–31). Springer, Singapore.

Kumar, G. and Ram, M.R. 2014. Phosphate solubilizing rhizobia isolated from *Vigna trilobata*. *Am. J. Microbiol. Res.* 2: 105–109.

Kumar, N.R., Arasu, V.T. and Gunasekaran, P. 2002. Genotyping of antifungal compounds producing plant growth-promoting rhizobacteria, *Pseudomonas fluorescens*. *Curr. Sci.* 82: 1465–1466.

Kumar, P., Dubey, R.C. and Maheshwari, D.K. 2012. *Bacillus* strains isolated from rhizosphere showed plant growth promoting and antagonistic activity against phytopathogens. *Microbiol. Res.* 167(8): 493–499.

Kumar, S., Kaushik, G., Dar, M.A., Nimesh, S., Lopez-Chuken, U.J. and Villarreal-Chiu, J.F. 2018. Microbial degradation of organophosphate pesticides: A review. *Pedosphere* 28(2): 190-208.

Kumari, A., Goyal, R.K., Choudhary, M. and Sindhu, S.S. 2014. Effect of different nitrogen levels and biofertilizers on growth, yield and nitrogen content of *Chrysenthemum*. *Ann. Agric. Res.* 35(2): 156–163.

Lakshminarayana, K., Shukla, B., Sindhu, S.S., Kumari, P., Narula, N. and Sheoran, R.K. 2000. Analogue-resistant mutants of *Azotobacter chroococcum* derepressed for nitrogenase activity and early ammonia excretion having potential as inoculants for cereal crops. *Indian J. Expt. Biol.* 38: 373–378.

Lanteigne, C., Gadkar, V.J., Wallon, T., Novinscak, A., Filion, M. 2012. Production of DAPG and HCN by *Pseudomonas* sp. LBUM300 contributes to the biological control of bacterial canker of tomato. *Phytopathology* 102: 967–973.

Laranjo, M., Alexandre, A. and Oliveira, S. 2014. Legume growth-promoting rhizobia: an overview on the Mesorhizobium genus. *Microbiol. Res.* 169(1): 2–17.

Li, D.M. and Alexander, M. 1988. Co-inoculation with antibiotic-producing bacteria to increase colonization and nodulation by rhizobia. *Plant Soil* 108(2): 211–219.

Liddycoat, S.M., Greenberg, B.M. and Wolyn, D.J. 2009. The effect of plant growth-promoting rhizobacteria on asparagus seedlings and germinating seeds subjected to water stress under greenhouse conditions. *Can. J. Microbiol.* 55(4): 388–394.

Loper, J.E., Gross, H. 2007. Genomic analysis of antifungal metabolite production by *Pseudomonasfluorescens* Pf-5. *Eur. J. Plant Pathol.* 119: 265–278.

Lugtenberg, B. and Kamilova, F. 2009. Plant-growth-promoting rhizobacteria. *Annu. Rev. Microbiol.* 63: 541–556.

Lugtenberg, B.J., Chin-A-Woeng, T.F. and Bloemberg, G.V. 2002. Microbe–plant interactions: principles and mechanisms. *Antonie Van Leeuwenhoek* 81(1-4): 373–383.

Machado, R.G., de Sá, E.L., Bruxel, M., Giongo, A., da Silva, Santos, N. and Nunes, A.S. 2013. Indole acetic acid producing rhizobia promote growth of Tanzania grass (*Panicum maximum*) and Pensacola grass (*Paspalum saurae*). *Intl. J. Agriculture. Biol.* 1: 15–25.

Maheshwari, D.K., Dubey, R.C., Aeron, A., Kumar, B., Kumar, S., Tewari, S. and Arora, N.K. 2012. Integrated approach for disease management and growth enhancement of *Sesamum indicum* L. utilizing *Azotobacter chroococcum* TRA2 and chemical fertilizer. *World J. Microbiol. Biotechnol. 28*(10): 3015–3024.

Malik, D.K. and Sindhu, S.S. 2011. Phytostimulatory effect of IAA producing Pseudomonas strains on nodulation and plant growth of chickpea (*Cicer arietinum* L.). *Physiol. Mol. Biol. Plants* 17: 25–32.

Mantelin, S. and Touraine, B. 2004. Plant growthpromoting bacteria and nitrate availability: impacts on root development and nitrate uptake. *J. Expl. Botany* 55(394): 27–34.

Marschner, P. and Baumann, K. 2003. Changes in bacterial community structure induced by mycorrhizal colonisation in split-root maize. *Plant Soil* 251(2): 279–289.

Martínez-Viveros, O., Jorquera, M.A., Crowley, D.E., Gajardo, G.M.L.M. and Mora, M.L. 2010. Mechanisms and practical considerations involved in plant growth promotion by rhizobacteria. *J. Soil Sci. Plant Nutr.* 10(3): 293–319.

Maurhofer, M., Keel, C., Schnider, U., Voisard, C., Haas, D., Defago, G. 1992. Influence of enhanced antibiotic production by *Pseudomonas fluorescens* strain CHA0 on disease suppressive capacity. *Phytopathology* 82: 190–195.

Mayak, S., Tirosh, T. and Glick, B.R. 2004. Plant growth-promoting bacteria that confer resistance to water stress in tomatoes and peppers. *Plant Sci.* 166(2): 525–530.

McCully, M.E. 2001. Niches for bacterial endophytes in crop plants: a plant biologist's view. *Functional Plant Biol.* 28(9): 983–990.

Meena, V.S., Meena, S.K., Verma, J.P., Kumar, A., Aeron, A., Mishra, P.K., Bisht, J.K., Pattanayak, A., Naveed, M. and Dotaniya, M.L. 2017. Plant beneficial rhizospheric microorganism (PBRM) strategies to improve nutrients use efficiency: A review. *Ecol. Engin.* 107: 8–32.

Mejri, M.B., Baccar, H., Baldrich, E., Del Campo, F.J., Helali, S., Ktari, T., Simonian, A., Aouni, M. and Abdelghani, A. 2010. Impedance biosensing using phages for bacteria detection: Generation of dual signals as the clue for in-chip assay confirmation. *Biosensors Bioelectronics* 26(4): 1261–1267.

Micallef, S.A., Shiaris, M.P. and Colón-Carmona, A. 2009. Influence of *Arabidopsis thaliana* accessions on rhizobacterial communities and natural variation in root exudates. *J. Expt. Botany* 60(6): 1729–1742.

Miransari, M. and Smith, D.L. 2014. Plant hormones and seed germination. *Environ. Expl. Bot.* 99: 110–121.

Mishra, C.N., Tiwarý, V., Satish-Kumar, V.G., Gupta, V., Kumar, A. and Sharma, I. 2015. Genetic diversity and genotype by trait analysis for agromorphological and physiological traits of wheat (*Triticum aestivum* L.). *SABRAO J. Breeding Genet.* 47(1): 40–48.

Mohanram, S. and Kumar, P. 2019. Rhizosphere microbiome : revisiting the synergy of plant-microbe interactions. *Ann. Microbiol.* 69: 307–320.

Müller, H., Westendorf, C., Leitner, E., Chernin, L., Riedel, K., Schmidt, S., Eberl, L. and Berg, G. 2009. Quorum-sensing effects in the antagonistic rhizosphere bacterium *Serratia plymuthica* HRO-C48. *FEMS Microbiol. Ecol.* 67: 468–478.

Nadeem, S.M., Naveed, M., Zahir, Z.A. and Asghar, H.N. 2013. Plant-microbe interactions for sustainable agriculture: fundamentals and recent advances. In: Arora NK (ed.) Plant Microbe Symbiosis: Fundamentals and Advances. Springer, India, pp. 51–103.

Nagaraj, K.M., Bhaskaran, R. and Velazhahan, R. 2004. Involvement of secondary metabolites and extracellular lytic enzymes produced by *Pseudomonas fluorescens* in inhibition of *Rhizoctonia solani*, the rice sheath blight pathogen. *Microbiol. Res.* 159: 73–81.

Nakkeeran, S., Fernando, W.D. and Siddiqui, Z.A. 2005. Plant growth promoting rhizobacteria formulations and its scope in commercialization for the management of pests and diseases. In *PGPR: Biocontrol and Biofertilization.* Springer, Dordrecht. pp. 257–296.

Naznin, H.A., Kimura, M., Miyazawa, M. and Hyakumachi, M. 2012. Analysis of volatile organic compounds emitted by plant growth promoting fungus *Phoma* sp. GS8- 3 for growth promotion effects on tobacco. *Microbe Environ.* 28: 42–49.

Nelson, L.M. 2004. Plant growth promoting rhizobacteria (PGPR): prospects for new inoculants. *Crop Manag.* 3(1): 3–11.

Nisha, K., Kuldeep, Y. and Ashok, A. 2014. Application of AM Fungi with *Bradyrhizobiumjaponicum* in improving growth, nutrient uptake and yield of *Vigna radiata* L. under saline soil. *J. Stress Physiol. Biochem.* 10(3): 42–50.

Nishijima, F., Evans, W.R. and Vesper, S.J. 1988. Enhanced nodulation of soybean by *Bradyrhizobium* in the presence of *Pseudomonas fluorescens. Plant Soil* 111(1): 149–150.

Ongena, M., Jacques, P. 2008. *Bacillus* lipopeptides: versatile weapons for plant disease biocontrol. *Trends Microbiol.* 16: 115–125.

Ortíz-Castro, R., Contreras-Cornejo, H.A., Macías-Rodríguez, L. and López-Bucio, J. 2009. The role of microbial signals in plant growth and development. *Plant Signal. Behav.* 4(8): 701–712.

Park, J.M., Radhakrishnan, R., Kang, S.M. and Lee, I.J. 2015. IAA producing *Enterobacter* sp. I-3 as a potent bio-herbicide candidate for weed control: a special reference with lettuce growth inhibition. *Indian J. Microbiol.* 55(2): 207–212.

Parmar, P. and Sindhu, S.S. 2018. The novel and efficient method for isolating potassium solubilizing bacteria from rhizosphere soil. *Geomicrobiol. J.* 35(10): 1–7. doi: 10.1080/01490451.2018.1514442

Parte, S.G., Mohekar, A.D. and Kharat, A.S. 2017. Microbial degradation of pesticide: A review. *African J. Microbiol. Res.* 11(24): 992–1012.

Patel, V., Sharma, A., Lal, R., Al-Dhabi, N.A. and Madamwar, D. 2016. Response and resilience of soil microbial communities inhabiting in edible oil stress/ contamination from industrial estates. *BMC Microbiol.* 16(1): 50.

Paul, D. and Sarma, Y.R. 2006. Plant growth promoting rhizhobacteria (PGPR)-mediated root proliferation in black pepper (*Piper nigrum* L.) as evidenced through GS root software. *Archives Phytopathol. Plant Prot.* 39(4): 311–314.

Pena-Yam, L.P., Ruíz-Sánchez, E., Barboza-Corona, J.E. and Reyes-Ramírez, A. 2016. Isolation of Mexican *Bacillus* species and their effects in promoting growth of chili pepper (*Capsicum annuum* L. cv Jalapeño). *Indian J. Microbiol.* 56(3): 375–378.

Phour, M. and Sindhu, S.S. 2018. Bioherbicidal potential of rhizosphere bacteria for management of *Phalaris minor* weed. *Res. Crops* 19(3): 380–386.

Pii, Y., Mimmo, T., Tomasi, N., Terzano, R., Cesco, S. and Crecchio, C. 2015. Microbial interactions in the rhizosphere: beneficial influences of plant growth-promoting rhizobacteria on nutrient acquisition process. A review. *Biol. Fertil. Soils* 51(4): 403–415.

Prabakaran, J. and Ravi, K.B. 1991. Interaction effect of *A. brasilense* and *Pseudomonas* sp. A phosphate solubilizer on the growth of *Zea mays*. In: *Microbiology Abstracts, XXXI Annual Conference of the Association of Microbiologists of India,* TNAU, Coimbatore, pp. 23–26.

Prasad, R., Kumar, M. and Varma, A. 2015. Role of PGPR in soil fertility and plant health. In: Plant-growth-promoting rhizobacteria (PGPR) and medicinal plants. Springer International Publishing. pp. 247–260.

Prashar, P., Kapoor, N. and Sachdeva, S. 2014. Rhizosphere: its structure, bacterial diversity and significance. *Rev. Environ. Sci. Bio/Technol.* 13(1): 63–77.

Press, C.M., Loper, J.E. and Kloepper, J.W. 2001. Role of iron in rhizobacteria-mediated induced systemic resistance of cucumber. *Phytopathology* 91: 593–598.

Radhakrishnan K.C. 1996. Role of biofertilizers in cotton productivity. National Seminar Biofertilizer Production Problem and Constraints, TNAU, Coimbatore, pp. 17.

Rajkumar, M. and Freitas, H. 2008. Influence of metal resistant-plant growth-promoting bacteria on the growth of *Ricinus communis* in soil contaminated with heavy metals. *Chemosphere* 71(5): 834–842.

Rajkumar, M., Nagendran, R., Lee, K.J., Lee, W.H. and Kim, S.Z. 2006. Influence of plant growth promoting bacteria and Cr^{6+} on the growth of Indian mustard. *Chemosphere* 62(5): 741–748.

Ramesh, R. and Phadke, G.S. 2012. Rhizosphere and endophytic bacteria for the suppression of eggplant wilt caused by *Ralstonia solanacearum*. *Crop Prot.* 37: 35–41.

Ran, L.X., Li, Z.N., Wu, G.J., Van Loon, L.C. and Bakker, P.H. 2005. Induction of systemic resistance against bacterial wilt in *Eucalyptus urophylla* by fluorescent *Pseudomonas* sp. *Eur. J. Plant Pathol.* 113: 59–70 doi: 10.1007/s10658-005-0623-3

Reed, S.C., Cleveland, C.C. and Townsend, A.R. 2011. Functional ecology of free-living nitrogen fixation: a contemporary perspective. *Annu. Rev. Ecol. Evol. Syst.* 42: 489–512

Reimann, S., Hauschild, R., Hildebrandt, U. and Sikora, R.A. 2008. Interrelationships between *Rhizobium etli* G12 and *Glomus intraradices* and multitrophic effects in the biological control of the rootknot nematode *Meloidogyne incognita* on tomato. *J. Plant Dis. Protect.* 115: 108–113

Richardson, A.E., Barea, J.M., McNeill, A.M., Prigent-Combaret, C. 2009. Acquisition of phosphorus and nitrogen in the rhizosphere and plant growth promotion by microorganism. *Plant Soil* 321: 305–339.

Ryan, P.R., Dessaux, Y., Thomashow, L.S. and Weller, D.M. 2009. Rhizosphere engineering and management for sustainable agriculture. *Plant Soil* 321(1-2): 363–383.

Ryu, C.M., Farag, M.A., Hu, C.H., Reddy, M.S., Pare, P.W. and Kloepper, J.W. 2003. Volatiles produced by PGPR elicit plant growth promotion and induced resistance in *Arabidopsis*. In *Proceedings of the Sixth International Workshop on Plant Growth Promoting Rhizobacteria,* pp. 436–443.

Sacherer, P., Défago, G., Haas, D. 1994. Extracellular protease and phospholipase C are controlled by the global regulatory gene gacA in the biocontrol strain *Pseudomonas fluorescens* CHA0. *FEMS Microbiol. Lett.* 116: 155–160.

Safdar, H., Javed, N., Khan, S.A., Haq, I., Safdar, A. and Khan, N.A. 2012. Control of *Meloidogyne incognita* (Kofoid and White) Chitwood by Cadusafos (Rugby®) on Tomato. *Pakistan J. Zool.* 44(6): 1703–1710.

Saharan, B.S. and Nehra, V. 2011. Plant growth promoting rhizobacteria: a critical review. *Life Sci. Med. Res.* 21(1): 30.

Sahasrabudhe, M.M. 2011. Screening of rhizobia for indole acetic acid production. *Ann. Biol. Res.* 2(4): 460–468.

Saikia, J., Sarma, R.K., Dhandia, R., Yadav, A., Bharali, R., Gupta, V.K. and Saikia, R. 2018. Alleviation of drought stress in pulse crops with ACC deaminase producing rhizobacteria isolated from acidic soil of Northeast India. *Sci. Repor.* 8(1): 35–60.

Sakthivel, K., Manigundan, K., Gautam, R.K., Singh, P.K., Nakkeeran, S. and Sharma, S.K. 2019. *Bacillus* sp. for suppression of eggplant bacterial wilt pathogen in Andaman islands: Isolation and characterization. *Indian J. Expert. Biol.* 57: 131–137.

Sandhya, V.Z.A.S., Grover, M., Reddy, G. and Venkateswarlu, B. 2009. Alleviation of drought stress effects in sunflower seedlings by the exopolysaccharides producing *Pseudomonasputida* strain GAP-P45. *Biol. Fertil. Soils* 46(1): 17-26.

Sang, M.K., Jeong, J.-J., J. Kim, J. and K.D. Kim, K.D. 2018. Growth promotion and root colonisation in pepper plants by phosphate-solubilising *Chryseobacterium* sp. strain ISE14 that suppresses Phytophthora blight. *Annals Appl. Biol.* 172: 208–223. doi: 10.1111/aab.12413

Sangwan, V.P., Sindhu, S.S., Dahiya, O.S. and Kharb, R.P.S. 2012. Improvement of wheat (*Triticumaestivum* L.) yield under field conditions by inoculation of microbial strains. *Microbiol. J.* 2: 86-95.

Sarhan, E.A.D. and Shehata, H.S. 2014. Potential plant growth-promoting activity of *Pseudomonas* spp. and *Bacillus* spp. as biocontrol agents against damping off in alfalfa. *Plant Pathol. J.* 13(1): 8-17.

Sarkar, A., Ghosh, P.K., Pramanik, K., Mitra, S., Soren, T., Pandey, S., Mondal, M.H. and Maiti, T.K. 2018. A halotolerant *Enterobacter* sp. displaying ACC deaminase activity promotes rice seedling growth under salt stress. *Res. Microbiol.* 169(1): 20-32.

Sarwar, M. and Kremer, R.J. 1995. Enhanced suppression of plant growth through production of L-tryptophan-derived compounds by deleterious rhizobacteria. *Plant Soil* 172(2): 261-269.

Schmülling, T., 2002. New insights into the functions of cytokinins in plant development. *J. Plant Growth Regul.* 21(1): 145-152.

Sethy, S.K. and Ghosh, S. 2013. Effect of heavy metals on germination of seeds. *J. Nat. Sci. Biol. Med.* 4: 272–275. https: / /doi.org/10.4103/0976-9668.116964.

Setiawati, A. and Handayanto, E. 2010. Role of phosphate solubilising bacteria on availability phosphorus in Oxisols and tracing of phosphate in corn by using 32P. In *19th world congress of soil science, soil solutions for a changing world, Brisbane, Australia.*

Shaharoona, B., Arshad, M. and Zahir, Z.A. 2006. Effect of plant growth promoting rhizobacteria containing ACCdeaminase on maize (*Zea mays* L.) growth under axenic conditions and on nodulation in mung bean (*Vigna radiata* L.). *Lett. Appl. Microbiol.*42(2): 155–159.

Shaharoona, B., Naveed, M., Arshad, M. and Zahir, Z.A. 2008. Fertilizer-dependent efficiency of Pseudomonads for improving growth, yield, and nutrient use efficiency of wheat (*Triticum aestivum* L.). *Appl. Microbiol. Biotechnol.* 79(1): 147–155.

Shahzad, R., Khan, A.L., Bilal, S., Waqas, M., Kang, S.M. and Lee, I.J. 2017. Inoculation of abscisic acid-producing endophytic bacteria enhances salinity stress tolerance in *Oryza sativa. Environ. Expt. Botany* 136: 68–77.

Sharma, R., Dahiya, A. and Sindhu, S.S. 2018b. Harnessing proficient rhizobacteria to minimize use of agrochemicals. *Intern. J. Curr. Microbiol. Appl. Sci.* 7(10): 3186–3197.

Sharma, R., Sindhu, S. and Sindhu, S.S. 2018a. Suppression of Alternaria blight disease and plant growth promotion of mustard (*Brassica juncea* L.) by antagonistic rhizosphere bacteria. *Appl. Soil Ecol.* 129: 145–150.

Shi, T.Q., Peng, H., Zeng, S-Y., Ji, R-Y., Shi, K., Huang, H. and Ji X-J. 2017. Microbial production of plant hormones: opportunities and challenges. Bioengineered 8(2): 124–128.

Shoji, S. and Kanno, H. 1994. Use of polyolefin-coated fertilizers for increasing fertilizer efficiency and reducing nitrate leaching and nitrous oxide emissions. *Fertilizer Res.* 39(2): 147–152.

Sindhu, S.S. and Dadarwal, K.R. 1985. Protoplast formation and regeneration in *Rhizobium* and *Azospirillum*. *Curr. Sci.* 54: 344–346.

Sindhu, S.S. and Dadarwal, K.R. 2000. Competition for nodulation among rhizobia in legume-*Rhizobium* symbiosis. *Indian J. Microbiol.* 40: 211–246.

Sindhu, S.S. and Sharma, R. 2019. Amelioration of biotic stress by application of rhizobacteria for agriculture sustainability. In: Plant growth promoting rhizobacteria for sustainable stress management. Vol. 2: Rhizobacteria for Biotic Stress Management. Sayyed, R. and Tabassum, B. eds. Springer (in press).

Sindhu, S.S., Brewin N.J. and Kannenberg E.L. 1990. Immunochemical analysis of lipopolysaccharides from free-living and endosymbiotic forms of *Rhizobium leguminosarum*. *J. Bacteriol.* 172: 1804–1813.

Sindhu, S.S., Dadarwal, K.R. and Davis, T.M. 1992. Non-nodulating chickpea breeding line for the study of symbiotic nitrogen fixation potential. *Indian J. Microbiol.* 32(2): 175–180.

Sindhu, S.S., Gupta, S.K. and Dadarwal, K.R. 1999. Antagonistic effect of *Pseudomonas* spp. on pathogenic fungi and enhancement of plant growth in green gram (*Vigna radiata*). *Biol. Fertil. Soils* 29: 62–68.

Sindhu, S.S., Malik, D.K. and Dadarwal, K.R. 2003. Enhancing the potential of biological nitrogen fixation by genetic manipulations of diazotrophic bacteria for sustainable agriculture. In: Plant Genetic Engineering. Volume 1, Applications and Limitations (Singh, R.P. and Jaiwal, P.K., eds.). Sci. Tech. Publ. LCC Houston, USA. PP. 199–228.

Sindhu, S.S., Parmar, P. and Phour, M. 2014a. Nutrient cycling: potassium solubilization by microorganisms and improvement of crop growth. In: Geomicrobiology and Biogeochemistry. Springer, Berlin, Heidelberg, pp. 175–198.

Sindhu, S.S., Parmar, P., Phour, M. and Kumari, K. 2014c. Rhizosphere microorganisms for improvement in soil fertility and plant growth. In: Microbes in the service of mankind: Tiny bugs with huge impact. Nagpal, R., Ashwani Kumar and Singh, R. eds. JBC Press New Delhi. pp. 32–94.

Sindhu, S.S., Phour, M., Choudhary, S.R. and Chaudhary, D. 2014b. Phosphorus cycling: prospects of using rhizosphere microorganisms for improving phosphorus nutrition of plants. In: Geomicrobiology and Biogeochemistry, Springer, Berlin, Heidelberg. pp. 199–237.

Sindhu, S.S., Rakshiya, Y.S. and Malik, D.K. 2010. Rhizosphere bacteria and their role in biological control of plant diseases. In: Biotechnology: Emerging trends (SAyyed, R.J. and Patil, A.S. eds.). Scientific Publishers, Jodhpur. pp. 73–112.

Sindhu, S.S., Sehrawat, A., Sharma, R. and Dahiya, A. 2016. Biopesticides: use of rhizosphere bacteria for biological control of plant pathogens. *Defense Life Sci. J.* 1: 135–148.

Sindhu, S.S., Sehrawat, A., Sharma, R., Dahiya, A. and Khandelwal, A. 2018. Belowground microbial cross-talk and rhizosphere biology. In: Plant-Microbe Interactions in Agro-Ecological Perspectives. Vol. 1: Fundamental mechanisms, methods and functions (Singh, D.P., ed.). Springer Nature, Singapore Pte Ltd. pp. 695–752.

Sindhu, S.S., Sehrawat, A., Sharma, R., Dahiya, A. and Khandelwal, A., 2017. Belowground microbial crosstalk and rhizosphere biology. In: Plant-Microbe Interactions in Agro-Ecological Perspectives. Springer, Singapore. pp. 695–752.

Sindhu, S.S., Sharma, R. Sindhu, S. and Phour, M. 2019a. Plant nutrient management through inoculation of zinc solubilizing bacteria for sustainable agriculture. In: Biofertilizers for sustainable agriculture and environment, Giri, B., Prasad, R., Wu, Q.S. and Verma, A. eds. Soil Biology, Springer (In press).

Sindhu, S.S., Sharma, R., Sindhu, S. and Sehrawat, A. 2019b. Soil fertility improvement by symbiotic rhizobia for sustainable agriculture. In: Soil Fertility Management for sustainable development. Panpatte, D.G. and Jhala, Y.K. eds. Springer Nature, Singapore Pte Ltd. (In press).

Sindhu, S.S., Suneja, S., Goel, A.K., Parmar, N. and Dadarwal, K.R. 2002. Plant growth promoting effects of *Pseudomonas* sp. on coinoculation with *Mesorhizobium* sp. *Cicer* strain under sterile and wilt sick soil conditions. *Appl. Soil Ecol.* 19: 57–64.

Sindhu, S.S., Verma, N., and Goyal, S. 2013. Ecofriendly utilization of crop residues and organic waste material. In : Human and Animal Health: Environmental perspectives. Garg, S.R. ed. Serial Publishing House, New Delhi. pp. 385–404.

Singh, A., Sarma, B.K., Upadhyay, R.S. and Singh, H.B. 2013. Compatible rhizosphere microbes mediated alleviation of biotic stress in chickpea through enhanced antioxidant and phenylpropanoid activities. *Microbiol. Res.* 168: 33–40.

Singh, R., Gautam, N., Mishra, A. and Gupta, R. 2011. Heavy metals and living system: an overview. *Indian J. Pharmacol.* 43: 246–253. https: //doi. org/10.4103/0253-7613.81505.

Smith, S.E. and Read, D.J. 2010. *Mycorrhizal Symbiosis*. Academic press.

Song, X.D., Xue, X.Y., Chen, D.Z., He, P.J. and Dai, X.H. 2014. Application of biochar from sewage sludge to plant cultivation: Influence of pyrolysis temperature and biochar-to-soil ratio on yield and heavy metal accumulation. *Chemosphere* 109: 213–220.

Spaepen, S. and Vanderleyden, J. 2011. Auxin and plant-microbe interactions. Cold Spring *Harb. Perspect. Biol.* 3: a001438.

Spaepen, S., Vanderleyden, J. and Remans, R. 2007. Indole-3-acetic acid in microbial and microorganism-plant signaling. *FEMS Microbiol. Rev.* 31(4): 425–448.

Stepien, P. and Klobus, G. 2005. Antioxidant defense in the leaves of C3 and C4 plants under salinity stress. *Physiol. Plant.* 125(1): 31–40.

Stirk, W.A. and Van Staden, J. 2010. Flow of cytokinins through the environment. *Plant Growth Regul.* 62(2): 101–116.

Stutz, *e.g.*, Defago, G., Kern, H. 1986. Naturally occurring fluorescent pseudomonads involved in suppression of black root rot of tobacco. *Phytopathology* 76: 181–185.

Sundaramoorthy, S. and Balabaskar, P. 2013. Evaluation of combined efcacy of *Pseudomonas uorescens* and *Bacillus subtilis* in managing tomato wilt caused by *Fusarium oxysporum* f. sp. *lycopersici* (Fol). *Plant Pathol. J.* 12(4): 154–161 doi: 10.3923/ppj.2013.154.161

Talebi, R., Ensafi, M.H., Baghebani, N., Karami, E. and Mohammadi, K. 2013. Physiological responses of chickpea (*Cicer arietinum*) genotypes to drought stress. *Environ. Exp. Biol.* 11: 9–15.

Tambekar D.H, Gulhane S.R, Somkuwar D.O, Ingle K.B and Kanchalwar S.P. 2009. Potential *Rhizobium* and phosphate solubilizers as a biofertilizers from saline belt of Akola and Buldhana district. *Indian Res. J. Agric. Biol. Sci.* 5(4): 578–582.

Tan, H., Zhou, S., Deng, Z., He, M. and Cao, L. 2011. Ribosomal-sequence-directed selection for endophytic streptomycete strains antagonistic to *Ralstonia solanacearum* to control tomato bacterial wilt. *Biol. Contr.* 59(2): 245–254.

Timmusk, S. 2003. *Mechanism of action of the plant growth promoting bacterium Paenibacillus polymyxa* (Doctoral dissertation, Acta Universitatis Upsaliensis).

Ting, A.S., Meon, S., Kadir, J., Radu, S. and Singh, G. 2008. Endophytic microorganisms as potential growth promoters of banana. *BioControl* 53(3): 541–553.

Tjamos, E.C., Tsitsigiannis, D.I., Tjamos, S.E., Antoniou, P.P. and Katinakis, P. 2004. Selection and screening of endorhizosphere bacteria from solarized soils as biocontrol agents against *Verticillium dahliae* of solanaceous hosts. *Europeon J. Plant Pathol.* 110: 35–44.

Tjamos, S.E., Flemetakis, E., Paplomatas, E.J. and Katinakis, P. 2005. Induction of resistance to *Verticillium dahliae* in *Arabidopsis thaliana* by the biocontrol agent K-165 and athogenesis-related proteins gene expression. *Mol. Plant Microbe Interact.* 18: 555–561.

Tripathi, A.K., Nagarajan, T., Verma, S.C. and Le Rudulier, D. 2002. Inhibition of biosynthesis and activity of nitrogenase in *Azospirillum brasilense* Sp7 under salinity stress. *Curr. Microbiol.* 44(5): 363–367.

Upadyay, S.K., Maurya, S.K., Singh, D.P. 2012. Salinity tolerance in free living plant growth promoting rhizobacteria. *Indian J. Sci. Res.* 3: 73–78.

van Hulten, M., Pelser, M., Van Loon, L.C., Pieterse, C.M. and Ton, J. 2006. Costs and benefits of priming for defense in *Arabidopsis*. *Proc. Natl. Acad. Sci.* (USA) 103(14): 5602–5607.

Verma, N., Wati, L., Sindhu, S.S. and Goyal, S. 2014. Organic waste utilization for production of compost and biofuel. In: Microbes in the service of mankind:

Tiny bugs with huge impact. Nagpal, R., Ashwani Kumar and Singh, R. eds. JBC Press, New Delhi. pp. 95–124.

Verma, V.C., Singh, S.K. and Prakash, S. 2011. Bio-control and plant growth promotion potential of siderophore producing endophytic *Streptomyces* from *Azadirachta indica* A. Juss. *J. Basic Microbiol.* 51: 550–556.

Vessey, J.K. 2003. Plant growth promoting rhizobacteria as biofertilizers. *Plant Soil* 255: 571–586.

Viveros, M.O., Jorquera, M.A., Crowley, D.E., Gajardo, G.M.L.M. and Mora, M.L. 2010. Mechanisms and practical considerations involved in plant growth promotion by rhizobacteria. *J. Soil Sci. Plant Nutr.* 10(3): 293–319.

Wackett, L.P. 2013. Microbes and antibiotics. *Microbiol. Biotechnol.* 6(6): 740–741.

Walia, A., Sumal, K. and Kumari, S. 2018. Effect of chlorpyrifos and malathion on soil microbial population and enzyme activity. *Acta Sci. Microbiol.* 1(4), 14–22. doi: 10.31080/ASMI.2018.01.0033

Wang, E.T. and Martinez-Romero, E. 2000. *Sesbania herbacea–Rhizobium huautlense* nodulation in flooded soils and comparative characterization of *S. herbacea*-nodulating rhizobia in different environments. *Microbial Ecol.* 40(1): 25–32.

Wang, X. and Liang, G. 2014. Control efficacy of an endophytic *Bacillus amyloliquefaciens* strain BZ6-1 against peanut bacterial Wilt *Ralstonia solanacearum*. *BioMed Res. Intern.* 2014: 465435.

Wani, S.P. 1990. Inoculation with associative nitrogen-fixing bacteria: Role in cereal grain production improvement. *Indian J. Microbiol.* 30(4): 363–393.

Weller, D.M. 2007. *Pseudomonas* biocontrol agents of soilborne pathogens: looking back over 30 years. *Phytopathology* 97: 250–256.

Weller, S.A., Simpkins, S.A., Stead, D.E., Kurdziel, A., Hird, H. and Weekes, R.J., 2002. Identification of *Agrobacterium* spp. present within *Brassica napus* seed by TaqMan PCR–implications for GM screening procedures. *Arch. Microbiol.* 178(5): 338–343.

Wen, X., Wang, M., Ti, J., Wu, Y. and Chen, F. 2017. Bacterial community composition in the rhizosphere of maize cultivars widely grown in different decades. *Biol. Fertil. Soils* 53(2): 221–229.

Whipps, J.M. 2001. Microbial interactions and biocontrol in the rhizosphere. *J. Expt. Botany* 52(suppl_1): 487–511.

Wu, A.B., Li, H.P., Zhao, C.S. and Liao, Y.C. 2005. Comparative pathogenicity of *Fusarium graminearum* isolates from China revealed by wheat coleoptile and floret inoculations. *Mycopathologia* 160(1): 75–83.

Wu, C.H., Wood, T.K., Mulchandani, A. and Chen, W. 2006. Engineering plant-microbe symbiosis for rhizoremediation of heavy metals. *Appl. Environ. Microbiol.* 72(2): 1129–1134.

Wu, P., Zhang, G., Li, J., Lu, H. and Zhao, W. 2012. Effects of Fe^{2+} concentration on biomass accumulation and energy metabolism in photosynthetic bacteria wastewater treatment. *Biores. Technol.* 119: 55–59.

Xiaolin, F. and Zongwen, L. 1998. Increasing fertlizer use efficiency by means of controlled release fertilizer (CRF) production according to theory and techniques of balanced fertilization. *Plant Nutr. Fertilizen Sci.* 3: 115–124.

Yang, J., Kloepper, J.W. and Ryu, C.M., 2009. Rhizosphere bacteria help plants tolerate abiotic stress. *Trends Plant Sci.* 14(1): 1–4.

Yang, Q.Y., Jia, K., Geng, W.Y., Guo, R.J. and Li, S.D. 2014. Management of cucumber wilt disease by *Bacillus subtilis* B006 through suppression of *Fusarium oxysporum* f. sp. cucumerinum in rhizosphere. *Plant Pathol. J.* 13: 160–166.

Yu, J.M., Wang, D., Pierson III, L.S. and Elizabeth A. Pierson, E.A. 2018. Effect of producing different phenazines on bacterial fitness and biological control in *Pseudomonas chlororaphis* 30-84. *Plant Pathol. J.* 34(1): 44–58. https: //doi. org/10.5423/PPJ.FT.12.2017.0277

Yu, X., Liu, X., Zhu, T.H., Liu, G.H. and Mao, C. 2012. Co-inoculation with phosphate-solubilzing and nitrogen-fixing bacteria on solubilization of rock phosphate and their effect on growth promotion and nutrient uptake by walnut. *Eur. J. Soil Biol.* 50: 112–117.

Zaddy, E., Perevolosky, A. and Okon, Y. 1993. Promotion of plant growth by inoculation with aggregated and single cell suspension by *Azospirillum brasilense. Soil Biol. Biochem.* 25: 819–823.

Zahir, Z.A., Arshad, M. and Frankenberger, W.T. 2004. Plant growth promoting rhizobacteria: applications and perspectives in agriculture. *Adv. Agron.* 81: 98–169.

Zahran, H.H. 2001. Rhizobia from wild legumes: diversity, taxonomy, ecology, nitrogen fixation and biotechnology. *J. Biotechnol.* 91: 143–153.

Zeng, Q., Wu, X. and Wen, X. 2017. Identification and characterization of the rhizosphere phosphate-solubilizing bacterium *Pseudomonas frederiksbergensis* JW-SD2 and its plant growth-promoting effects on poplar seedlings. *Ann. Microbiol.* 67(3): 219–230.

Zhang, F., Dashti, N., Hynes, R.K. and Smith, D.L. 1996. Plant-growth promoting rhizobacteria and soybean (*Glycine max* (L.) Merr.) nodulation and nitrogen fixation at suboptimal root zone temperatures. *Ann. Botany* (London) 77: 453–459.

Zhuang, X., Chen, J., Shim, H. and Bai, Z. 2007. New advances in plant growth-promoting rhizobacteria for bioremediation. *Environ. Intern.* 33(3): 406–413.

Zorner, P., Farmer, S. and Alibek, K. 2018. Quantifying crop rhizosphere microbiome ecology: the next frontier in enhancing the commercial utility of agricultural microbes. *Ind. Biotechnol.* (*New Rochelle NY*) 14(3): 116–119.

Chapter 9

Microbial Enzymes with Industrial Potential

*Harnek Singh Saini[1], Ritu Saini[2], Rajesh Kumar[1] and Divya Bhatia[1]**

[1]Department of Biotechnology, University Institute of Engineering and Technology, Kurukshetra University Kurukshetra – 136 119, Haryana
[2]Indian Institute of Wheat and Barley Research, Karnal, Haryana
**Em-ail: divyabhatia228@gmail.com*

ABSTRACT

Enzymes are considered as naturally occurring bio-catalysts that have a broad industrial applications and uses in our daily life. Enzymes, specific catalysts that execute all catabolic and anabolic reactions in human and other living organisms. The enzymes are favored over chemical catalysts in commercial processes owing to their high catalytic efficiency, specific mechanism of activity, stereo-specificity, environment friendly nature and minimum energy need. In recent years, application of enzymes in pharmaceutics and medicine industry has been exponentially increased. In various industries, enzymes are utilized to improve quality, process yield, stability of product and flavor and byproduct transformations.

Introduction

Enzymes owing to their catalytic properties play a crucial role at each levels of biochemical and metabolic reactions. Some enzymes are of utmost importance and are used as catalysts in various industrial processes. The most significant natural enzymes are amylase, cellulase, invertase, lactase and lipase. Microorganisms are the primary source of enzymes, because of their simple isolation and maintenance protocol, short doubling time and relatively easy genetic modification. In addition to this, the microbial enzymes show high activity and stability than enzymes from other natural sources (Anbu *et al.*, 2013; Anbu *et al.*, 2015; Gopinath *et al.*, 2013).

Recently, the microorganisms are intensively explored as a source of industrially relevant enzymes (Jayani *et al.*, 2005). Microbial Enzymes production is preferred because of less process time and space need, easy process optimization and modification and less production cost (Gurung *et al.*, 2013). Enzymes have significant applications in various sectors *e.g.* Food processing and preservation, textiles, healthcare and diagnostics, pharmaceuticals industry, detergents, leather, Forensics, bioremediation, Biofuel production, personal care, pulp and paper processing, medicine and therapy.

α-Amylases

α-Amylase is the first industrially produced enzyme that hydrolyse α-1, 4-glycosidic linkages in present in starch to produce glucose and maltose. Alpha-amylase has been isolated from fungi, yeasts, bacteria and actinomycetes, however, fungal and bacterial enzymes have significantly contributed in various industries (Tanyildizi *et al.*, 2005). In starch processing industry, chemical hydrolysis has been substituted by microbial amylases based hydrolysis of starch (Veille and Zeikus, 2001). In fungi, α-amylase production is restricted to *Aspergillus* sp. and *P. brunneum* (Kathiresan and Maniyannan, 2006). Among all extracellular enzymes produced by *Aspergillus* sp., amylases are the most significant industrial enzymes (Hutcheon *et al.*, 2005). *A. oryzae* has been considered a suitable host owing to secretion of a large number of usefull proteins and enzymes.α-Amylase is mainly derived from the genus *Bacillus* for industrial applications. *B. subtilis*, *B. stearothermophilus*, *B. licheniformis* and *B. amyloliquefaciens* are mainly involved in amylase production (Harshemi *et al*, 2011).

α-Amylases have significant utilization in various industrial processes such as food, processing, fermentation, textile finishing, paper processing, detergent formulation and pharmaceutical industries. In food industry, α-amylases are used for baking and starch hydrolysis purpose (Rodríguez *et al.*, 2006). However, the amylase application has expanded in many fields such as health care, diagnostics, medicine and food products (Pandey *et al.*, 2000). In pulp and paper industry, α-amylases are used for low-viscosity starch production (Gupta *et al.*, 2003). This enzyme is used to digest starch residue of foods to smaller glucose molecules in detergents and for desizing process in textile industry (Mukherjee *et al.*, 2009). The α-amylases remove selectively the size without affecting the fibres (Ahlawat *et al.*, 2009).

Proteases

Proteases are the enzymes which hydrolyze the peptide bonds existing in proteins. Their major applications are in detergent, pharmaceutical and food industries. 60 per cent of industrial enzyme market belongs to proteases (Singh *et al.*, 2016). Proteases are added as main component in detergents formulations to improve the washing (Baweja *et al.*, 2016). Mainly serine proteases isolated from *Bacillus* sp. are used in detergent industry (Gaur and Wadhwa, 2008). Proteases have significant application in degumming and fabric finishing of silk (Srilakshmi *et al.*, 2014).

In food industry, Proteases have a significant role in various process such as meat production, coagulation of milk and as a digestive enzyme (Patel *et al.*, 2013). In addition, proteases are also used to better the taste, nutritional content, solubility and digestive properties of food (Aruna *et al.*, 2014). Proteases are widely used for the manufacturing of bakery products like bread, cookies, cakes and waffles. These enzymes improve the texture and flavor of products by improving the dough consistency and gluten strength (Miguel *et al.*, 2013). In dairy industry, proteases are mainly used in cheese production by producing casein and macro peptides and in tenderization of meat (Rao *et al.*, 1998; Kumar and Takagi, 1999).

In leather industry, use of proteases leads to improved leather quality with less environmental pollution (Singh *et al.*, 2016; Jaouadi *et al.*, 2013). Proteases isolated from *Bacillus subtilis* have been used in digestion of feathers waste (Jisha *et al.*, 2013). Mixtures of proteases from different bacterial sp. are used commercially to clean clogged drain (Ray, 2012). Proteases are also used for degradation of feathered skin and in therapy for skin infections (Brandelli *et al.*, 2010; Mane and Tale, 2015). Microbial proteases have been explored largely due to effortless production, high yield, low maintenance cost and easy modification at genetic level to enhance the catalytic efficiency (Li *et al.*, 2013). So, Microbial proteases are favored over other proteases for commercial application (Souza *et al.*, 2015).

Lipases

Lipases belong to the hydrolase class of enzymes; which hydrolyse long chain triglycerides/Fats. In human body fats/lipids are digested by naturally present lipases. Naturally fats and lipids are digested in the stomach and pancreas by lipases present in body (Fink *et al.*, 1984). Microorganisms synthesize large amount of lipases as compared to the animal and plants (Antczak *et al.*, 2009). Their use as an industrial catalyst is increasing day by day due to the favorable properties like high catalytic efficiency, bio-degradability and high specificity (Arife *et al.*, 2015). Lipases are present in nature, but due to greater stability, wide availability and low production cost, microbial lipases are more significant than lipases extracted from plant and animal sources (Beisson *et al.*, 2000). Most of the commercially used lipases have been isolated from fungi and bacteria. Lipase production have been reported from *Acinetobacter radioresistens* (Zhao et *al.*, 2013), *Pseudomonas* sp. (Latip *et al.*, 2016), *Bacillus stearothermophilus* (Ekinci *et al.*, 2015), *Rhizopus arrhizus* (Wang *et al.*, 2014), *Aspergillus* sp. (Zubiolo *et al.*, 2014), *Rhizopus homothallicus* (Colla *et al.*, 2015), *Penicillium citrinum* (Kumar *et al.*, 2016), *Penicillium restrictum* (Mukhtar *et al.*, 2015). Commetrcially available lipases are mainly extracellular (Robles-Medina *et al.*, 2009). These are mainly used as additives in detergents (Bajpai and Tyagi, 2007), in biodiesel production (El Batal *et al.*, 2016), in production of drug or for coronary vasodilation (Verma *et al.*, 2012), to process the pulp used for paper manufacturing (Verma *et al.*, 2012), in flavor development (Reetz, 2002), in tea processing (Latha and Ramarethinan, 1999) and also in various biotransformations (Sharma *et al.*, 2001).

Cellulase

Among constituent of plant biomass, maximum contribution is of cellulose

(Tomme *et al.*, 1995). Cellulose is digested by bacteria and fungi exocellularly and the hydrolysed products of cellulose may be utilized by other microbes as carbon and energy sources. The most significant industrial application of cellulases are in food industry, in paper industry, fabric finishing and softening, starch hydrolysis, fermentation, in brewery industry (Gao *et al.*, 2008), protection from infections and diseases in agriculture and detergents for maintaining fabrics texture and shine (Cavaco, 1998). Enzymatic hydrolysis of cellulosic biomass produce sugars that are used in synthesis of various significant products of commercial application, like ethanol, organic acids and animal feeds (Cao *et al.*, 2014; Sing *et al.*, 2016). Cellulase are mainly employed in purification of juices from fruits, in feed production for animals and in brewery industries (Sing *et al.*, 2016; Menendez *et al.*, 2015), flavonoids extraction from flowers (Sajith *et al.*, 2016; Cao *et al.*, 2014), in the isolation of phenolic compounds from grape pomace (Menendez *et al.*, 2015; Meyer *et al.*, 1998), to reduce bitterness of citrus fruits (Kuhad *et al.*, 2011; Sajith *et al.*, 2016), for increasing the digestibility and nutritive value of cereal based food (Dhiman *et al.*, 2002; Beauchemin *et al.*, 2003), removal of impurities within raw cotton quality improvement and fabric care is carried out more effectively (Shah, 2013; Kuhad *et al.*, 2011), also used for softening of fabric (Sajith *et al.*, 2016, Galante and Formantici, 2013), for improved washing effect for household purposes (Singh *et al.*, 2016). Cellulases isolated from *Aspergillus* sp., *Trichoderma* sp., and *Bacillus* sp. is prefereably used in juice purification (Singh and Singh, 2015). Cellulase are used in bleaching of pulp (Kuhad *et al.*, 2011), to improve yield and level of vitamin E content in olive oil (Galante *et al.*, 1998). *Aspergillus, Trichoderma* and *Humicola* produce thermostable cellulase which have applications in detergent industry (Sajith *et al.*, 2016). Cellulase in combination with hemicellulase is used for recycling of waste papers (Lee *et al.*, 2007; Ibarra *et al.*, 2012).

In soil, cellulases decompost soil cellulose and improve the soil qualities which ultimately decrease the requirement of chemical fertilizers (Fontaine *et al.*, 2004). Cellulase can also digest cellulose present in biofilms and consequently, reduce dispersion of respective microorganisms (Loiselle and Anderson, 2003).

Pectinase

Pectin, the key constituent of the middle lamella in plants is hydrolysed by the pectic enzymes. Pectic enzymes are categorized into three classes based on the following criteria:

1. On the basis of substrate hydrolysed (pectin, pectic acid or oligo-D-galacturonic acid).
2. On the basis of Mechanism (whether they wprk by translelimination or hydrolysis).
3. Cleavage type (random cleavage/endo-liquifying or exo enzymes).

Depending on their mechanism of action pectin degrading enzymes are named as pectinases, and include polygalacturonases, pectin esterases, pectin lyases and pectate lyases (Alkorta *et al.*, 1998). Depending upon their source of production pectinase may be alkaline or acidic in nature. Alkaline pectinases are mainly synthesized by both prokaryotic while acidic pectinases are produced from eukaryotic microorganisms (Hoondal *et al.*, 2002; Jayani *et al.*, 2005). Pectinases are isolated from different microorganism sources such as bacteria (Kashyap *et al.*, 2000; Aaisha and Barate, 2016, Oumer and Abate, 2018), yeast (Alimardani-Theuil *et al.*, 2011), fungi (Barman *et al.*, 2015; Oumer and Abate, 2018) and actinomycetes (Das *et al.*, 2013). Pectinases are the commercially significant enzymes, with special applications in fruit juice industry (Sandri *et al.*, 2012). The use of pectinases not only reduces the viscosity of the fruit juice but also enhance juice recovery by degrading the structural polysaccharide known as pectin (Joshi *et al.*, 2011). Pectinases have already been used in wine clarification (Cheirsilp and Umsakul, 2008), sugar extraction from date (Bahramian *et al.*, 2011), in production of pastes and purees from vegetables (Tochi *et al.*, 2009), in extraction of vegetable oil (Iconomou *et al.*, 2010), in textile processing and bioscourimg of cotton fibres (Jayani *et al.*, 2005), processing of animal feed (Praveen and Suneetha, 2014), Recycling and deinking of wastepaper (Singh *et al.*, 2012) and biobleaching of mixed hardwood and bamboo kraft pulps (Kaur *et al.*, 2010).

Xylanase

Xylanases hydrolyze the β-1, 4 linkage of polysaccharide xylan present in the complex plant cell wall. Microbial xylanase have application in biofuel and bioenergy (Hatanaka, 2012), paper and pulp industries (Golugiri *et al.*, 2012; Singh *et al.*, 2013). In paper and pulp industries, xylanase are used in bioleaching and biopulping (Walia *et al.*, 2015b; Khonzue *et al.*, 2011). Several microorganisms have been reported to produce 1, 4-b-D endoxylanases and b-xylosidases. Among bacteria *Bacillus, Arthrobacter, Microbacterium, Pseudoxanthomonas* and *Rhodothermus* are major genera that have been used in production of xylanase (Gupta *et al.*, 2001; Chapla *et al.*, 2012). *Streptomyces, Actinomadura, Nonomuraea*, are the some important actinomycetes that have been explored for production of xylanase (Prakash *et al.*, 2013).

Other Significant Enzymes

Enzymes have an indispensable role in nucleic acid manipulation in the field of genetic engineering for research and development. Restriction endonucleases because of their property of site specific cleavage of DNA are suitable for cloning and DNA polymerases from *Thermus aquaticus* are used in polymerase chain reaction for DNA amplification (Hoseini and Sauer, 2015). Nariginase, a debittering enzyme is mainly isolated from fungal sp. like *Aspergillus, Penicillium, Fusarium, Rhizopus* and bacteria like *Bacillus* sp., *Burkholderia, Thermomicrobium roseum, Pseudomonas etc.* (*Puri, 2012*). Superoxide dismutase (SOD) is used to control skin damage caused by pollutions, pathogens and other environmental factors. Some enzymes also have application in beauty enhancement *e.g.* transglutaminase and S glutathione sulfhydryl oxidase are used in curling of hair (Li *et al.*, 2012).

REFERENCES

Aaisha, G.A. and Barate, D.L. 2016. Isolation and identification of pectinolytic Bacteria from soil samples of Akola Region India. *Int. J. Curr. Microbiol. App. Sci.* 5(1): 514-524.

Ahlawat, S., Dhiman, S.S., Battan, B., Mandhan, R.P. and Sharma, J. 2009. Pectinase production by *Bacillus subtilis* and its potential application in biopreparation of cotton and micropoly fabric. *Process Biochem.* 44: 521-526.

Alimardani-Theuil, P., Gainvors-Claise, A. and Duchiron, F. 2011. Yeasts: an attractive source of pectinases from gene expression to potential applications: a review. *Process Biochem.* 46: 1525-1537.

Alkorta, I., Garbisu, C., Llama, M.J. and Serra, J.L. 1998. Industrial applications of pectic enzymes: a review. *Process Biochem.* 33 (1): 21-28.

Anbu, P., Gopinath, S.C.B., Chaulagain, B.P., Tang, T.H. and Citartan, M. 2015. Microbial enzymes and their applications in industries and medicine 2014. *Bio Med. Res. Int.* DOI 10.1155/2015/816419.

Anbu, P., Gopinath, S.C.B., Cihan, A.C. and Chaulagain, B.P. 2013. Microbial enzymes and their applications in industries and medicine. *Bio Med. Res. Int.* DOI 10.1155/2017/2195808.

Antczak, M.S., Kubiak, A., Antczak, T. and Bielecki, S. 2009. Enzymatic biodiesel synthesis-key factors affecting efficiency of the process. *Renewable Energy.* 34: 1185-1194.

Arife, P.E., Barbaros, D., Nimet, B. and Ahmet, A. 2015. Partial purification and characterization of Lipase from *Geobacillus stearothermophilus* AH22. *J. Enzyme Inhib. Med. Chem.* 31(2): 325-31.

Aruna, K., Shah, J. and Birmole, R. 2014. Production and partial characterization of alkaline protease from *Bacillus tequilensis* strains CSGAB 0139 isolated from spoilt cottage cheese. *Int. J. Appl. Biol. Pharm.* 5: 201-221.

Bahramian, S., Mehrdad, A., Mohammad, C. and Gerami, A. 2011. Optimization of enzymatic extraction of sugars from Kabkab Date Fruit. *Middle-East J. of Sci. Res.* 7(2): 211-216.

Bajpai, D. and Tyagi, V.K. 2007. Laundry detergents: an overview. *J. Oleo. Sci.* 56: 327-340.

Barman, S., Sit, N., Badwaik, L.S. and Deka, S.C. 2015. Pectinase production by *Aspirgillus niger* using banana (*Musa balbisiana*) peel as substrate and its effect on clarification of banana juice. *J. Food Sci. Technol.* 52: 3579-3589.

Baweja, M., Tiwari, R., Singh, P.K., Nain, L. and Shukla, P. 2016. An alkaline protease from *Bacillus pumilus* MP 27: functional analysis of its binding model toward its applications as detergent additive. *Frontiers in Microbiol.* 7: 1195.

Beauchemin, K., Colombatto, D., Morgavi, D. and Yang, W.Z. 2003. Use of exogenous fibrolytic enzymes to improve feed utilization by ruminants. *J. of Animal Sci.* 81: 37-47.

Beisson, F., Tiss, A., Riviere, C. and Verger, R. 2000. Methods for Lipase detection and assay: a critical review. *Eur. J. Lipid. Sci. Technol.* 102: 133-153.

Brandelli, A., Daroit, D.J. and Riffel, A. 2010. Biochemical features of microbial keratinases and their production and applications. *Appl. Microbiol. Biotechnol.* 85: 1735-1750.

Cao, Y., Zhang, H. and Xiong, P. 2014. Application of Cellulase assisted method to extract flavonoids from plants. *Agri. Sci.* and *Technol.* 15: 729.

Cavaco paulo, A. 1998. Mechanism of cellulase action in textile processes. *Carbohydr. Polym.* 37: 273-277.

Chapla, D., Patel, H., Madamwar, D. and Shah, A. 2012. Assessment of a thermostable xylanase from *Paenibacillus* sp. ASCD2 for application in prebleaching of eucalyptus kraft pulp. *Waste Biomass Valor.* 3: 269-274.

Cheirsilp, B. and Umsakul, K. 2008. Processing of banana based wine product using pectinase and a-Amylase. *J. of Food Process Eng.* 31(1): 78-90.

Colla, L.M., Ficanha, A.M.M., Rizzardi, J., Bertolin, T.E., Reinehr, C.O. and Costa, J.A.V. 2015. Production and characterization of Lipases by two new isolates of *Aspergillus* through solid-state and submerged fermentation. *BioMed. Research Inter.* DOI10.1155/2015/725959.

Das, A., Bhattacharya, S., Reddy, N.V. and Sundara, R.S. 2013. Improved production and purification of Pectinase from *Streptomyces* sp. GHBA10 isolated from Valapattanam mangrove habitat, Kerala, India. 2: 16-22.

Dhiman, T., Zaman, M., MacQueen, I. and Boman, R. 2002. Influence of corn processing and frequency of feeding on cow performance. *Journal of Dairy Sci.* 85: 217-226.

Ekinci, A.P., Dinçer, B., Baltaþ, N. and Adýgüzel, A. 2015. Partial purification and characterization of lipase from *Geobacillus stearothermophilus* AH22. *Journal of Enzyme Inhibition and Medicinal Chem.* 31: 325-331.

El-Batal, A.I., Farrag, A.A., Elsayed, M.A. and El-Khawaga, A.M. 2016. Biodiesel production by *Aspergillus niger* Lipase immobilized on barium ferrite magnetic nanoparticles. *Bioeng.* 3: 1-15.

Fink, C.S., Hamosh, P. and Hamosh, M. 1984. Fat digestion in the stom-ach: stability of lingual lipase in the gastric environment. *Pediatric Res.* 18(3): 248-254.

Fontaine, S., Bardoux, G. and Benest, D., Verdiera, B., Mariottib, A. and Abbadi, L. 2004. Mechanisms of the priming effect in a savannah soil amended with cellulose. *Soil Sci. Soci.of Ame. J.* 68: 125-131.

Galante, Y., De Conti, A. and Monteverdi, R. 1998. Application of *Trichoderma* enzymes in the food and feed industries, *Trichoderma* and *Gliocladium.* 2: 327-342.

Galante, Y.M. and Formantici, C. 2013. Enzyme applications in detergency and in manufacturing industries. *Current Org. Chem.* 7: 1399-1422.

Gao, J.H., Weng, D., Zhu, M., Yuan, F. and Guan, Y.Xi. 2008. Production and characterization of cellulolytic enzymes from the thermoacidophilic fungal *Aspergillus terreus* M11 under solid-state cultivation of corn Stover. *Bioresour. Technol.* 99: 7623-7629.

Gaur, S. and Wadhwa, N. 2008. Alkaline protease from senesced leaves of invasive weed *Lantana camara*. *Afr. J. Biotechnol.* 24: 4602-4608.

Golugiri, B.R., Thulluri, C., Cherupally, M., Nidadavolu, N., Achuthananda, D., Mangamuri, L.N. and Addepally, U. 2012. Potential of thermo and alkali stable xylanases from *Thielaviopsis basicola* (MTCC1467) in biobleaching of wood Kraft pulp. *Appl. Biochem. Biotechnol.* 167: 2369-2380.

Gopinath, S.C.B., Anbu, P., Lakshmipriya, T. and Hilda, A. 2013. Strategies to characterize fungal lipases for applications in medicine and dairy industry. *BioMed. Res. Int.* DOI 10.1155/2013/154549.

Gupta, R., Gigras, P., Mohapatra, H., Goswami, V.K. and Chauhan, B. 2003. Microbial amylases: a biotechnological perspective. *Process Biochem.* 38: 1599-1616.

Gupta, S., Kuhad, R.C., Bhushan, B. and Hoondal, G.S. 2001. Improved xylanase production from a haloalkalophilic *Staphylococcus* sp. SG-13 using inexpensive agricultural residues. *World J. Microbiol. Biotechnol.*17: 5-8.

Gurung, N., Ray, S., Bose, S. and Rai, V. 2013. A broader view: microbial enzymes and their relevance in industries, medicine and beyond. *BioMed. Res. Int.* DOI 10.1155/2013/32912.

Harshemi, I., Ali, S., Javed, M.M., Hameed, U., Saleem, A., Adnan, F. and Qadeer, M.A. 2011. Introduction of alpha amylase from a randomly induced mutant strain of *Bacillus amyloliquefaciens* and its application as a desizer in textile industry. *Pak. J. Bot.* 42: 473-484.

Hatanaka, K. 2012. Incorporation of fluorous glycosides to cell membrane and saccharide chain elongation by cellular enzymes. *Top. Curr. Chem.* 308: 291-306.

Hoondal, G.S., Tiwari, R.P., Tewari, R., Dahiya, N. and Beg, Q.K. 2002. Microbial alkaline pectinases and their industrial applications: a review. *App. Microbiol. Biotechnol.* 59: 409-418.

Hoseini, S.S. and Sauer, M.G. 2015. Molecular cloning using polymerase chain reaction, an educational guide for cellular engineering. *J. Biol. Eng.* DOI 10.1186/1754-1611-9.

Hutcheon, G.W., Vasisht, N. and Bolhuis, A. 2005. Characterisation of a highly stable alpha-amylase from the halophilic archaeon *Haloarcula hispanica. Extremophil.* 9: 487-495.

Ibarra, D., Monte, M.C., Blanco, A., Martinez, A.T. and Martinez, M.J. 2012. Enzymatic deinking of secondary fibers: cellulases/hemicellulases versus laccase-mediator system. *J. Ind. Microbiol. Biotechnol.* 39: 1-9.

Iconomou, D., Arapoglou, D. and Israilides, C. 2010. Improvement of phenolic antioxidants and quality characteristics of virgin olive oil with the addition of enzymes and nitrogen during olive paste processing. *Grasas Aceites.* 61: 303-311.

Jaouadi, N.Z., Hatem Rekik, H., Badis, A., Trabelsi, S., Belhoul, M., Yahiaoui, A.B., Aicha, H.B., Toumi, A., Bejar, S. and Jaouadi, B. 2013. Biochemical and molecular characterization of a serine keratinase from *Brevibacillus brevis* US575 with promising keratin-biodegradation and hide-dehairing activities. *Plos one* DOI 10.1371/journal.pone.0076722.

Jayani, R.S., Saxena, S. and Gupta, R. 2005. Microbial pectinolytic enzymes: a review. *Process Biochem.* 40: 2931-2944.

Jisha, V.N., Smitha, R.B., Pradeep, S., Sreedevi, S., Unni, K.N., Sajith, S., Priji, P., Josh, M.S., Benjamin, S. 2013. Versatility of microbial proteases. *J. Adv. Enzyme Res.* 1(3): 39- 51.

Joshi, V. K., Parmar, M. and Rana, N. 2011. Purification and characterization of pectinase produced from apple pomace and evaluation of its efficacy for fruit juice extraction and clarification. *Indian J. of Natural Products and Res.* 2(2): 189-197.

Kathiresan, K. and Manivannan, S. 2006. α-Amylase production by *Penicillium fellutanum* isolated from mangrove rhizosphere soil. *Afr. J. Biotechno.* 5: 829-832.

Kaur, A., Mahajan, R., Singh, A., Garg, G. and Sharma, J. 2010. Application of cellulase-free xylano-pectinolytic enzymes from the same bacterial isolate in biobleaching of kraft pulp. *Bioresour. Technol.* 101: 9150-9155.

Khonzue, P., Laothanachareon, T., Rattanaphan, N., Tinnasulanon, P., Apawasin, S., Paemanee, A., Ruanglek, V., Tanapongpipat, S., Champreda, V. and Eurwilaichitr L. 2011. Optimization of xylanase production from *Aspergillus niger* for bio bleaching of eucalyptus pulp. *Biosci. Biotechnol Biochem.* 75(6): 1129-1134.

Kuhad, R.C., Gupta, R. and Singh, A. 2011. Microbial cellulases and their industrial applications. *Enzyme Res.* DOI 10.4061/2011/280696.

Kumar, C.G. and Takag, H. 1999. Microbial alkaline proteases: from a bioindustrial viewpoint. *Biotechnol. Adv.* 17: 561-594.

Kumar, N.V., Rani, M.E., Gunaseeli, R. and Kannan, N.D. 2016. Utilization of medicinal oil effluent for lipase production by *Penicillium citrinum* MKF3. *Environ. Sci.* and *Eng.* 1: 125-132.

Latha, K. and Ramarethinam, S. 1999. Studies on lipid acyl hydrolases during tea processing. *Ann. Plant Physiol.* 13: 73-78.

Latip, W., Rahman, R.N.Z.R.A., Leow, A.T.C., Shariff, F.M. and Ali, M.S.M. 2016. Expression and characterization of thermotolerant lipase with broad pH profiles isolated from an Antarctic *Pseudomonas* sp. strain AMS3. *Peer J. 4: 2420.*

Lee, C., Darah, I. and Ibrahim, C. 2007. Enzymatic deinking of laser printed office waste papers: some governing parameters on deinking efficiency. *Bioresour. Technol.* 98: 1684-1689.

Li, Q., Yi, L., Marek, P. and Iverson, B.L. 2013. Commercial proteases: present and future. *FEBS let.* 587: 1155-1163.

Li, S., Yang, X. and Yang, S. *et al.*, 2012. Technology prospecting on enzymes: application, marketing and engineering. *Comput. Struct. Biotechnol. J.* 2: 1-11.

Loiselle, M. and Anderson, K.W. 2003. The use of cellulase in inhibiting biofilm formation from organisms commonly found on medical implants. *Biofouling.* 19: 77-85.

Mane, P. and Tale, V. 2015. Overview of microbial therapeutic enzymes. *Int. J. Curr. Microbiol App. Sci.* 4: 17-26.

Menendez, E., Garcia-Fraile, P. and Rivas, R. 2015. Biotechnological applications of bacterial cellulases. *AIMS. Bioeng.* 2(3): 163-182.

Meyer, A.S., Jepsen, S.M. and Sorensen, N.S. 1998. Enzymatic release of antioxidants for human low-density lipoprotein from grape pomace. *J. Agric. Food Chem.* 46: 2439-2446.

Miguel ASM, Martins-Meyer T.S., Veríssimo, da C.F.E, Lobo B.W.P., Dellamora-Ortiz G.M. 2013. Enzymes in bakery: current and future trends. In: Muzzalupo, I. (Ed.). Food Industry. InTech, pp. 3.

Mukherjee, A.K., Borah, M. and Rai, S.K. 2009. To study the influence of different components of fermentable substrates on induction of extracellular amylase synthesis by *Bacillus subtilis* DM-03 in solid state fermentation and exploration of feasibility for inclusion of amylase in laundry detergent formulations. *Biochem. Eng. J.* 43: 149-156.

Mukhtar, H., Hanif, M., Rehman, A., Nawaz, A. and Haq, I. 2015. Studies on the lipase production by *Aspergillus niger* through solid state fermentation. *Pak. J. Bot.* 47: 351-354.

Oumer, O.J. and Abate, D. 2018. Screening and molecular Identification of Pectinase producing microbes from coffee Pulp. *BioMed. Res. Int.* DOI10.1155/2018/2961767.

Pandey, A., Nigam, P., Soccol, C.R., Soccol, V.T., Singh, D. and Mohan, R. 2000. Advances in microbial amylases. *Biotechnol. Appl. Biochem.* 31(Part 2): 135-52.

Patel, N.S., Fung, S.M., Zanichelli, A., Cicardi, M. and Cohn, J.R. 2013. Ecallantide for treatment of acute attacks of acquired C1 esterase inhibitor deficiency. *Allergy Asthma Process.* 34(1): 72-77.

Prakash, D., Nawani, N., Prakash, M., Bodas, M., Mandal, A., Khetmalas, M. and Kapadnis, B. 2013. Actinomycetes: a repertory of green catalysts with a potential revenue resource. *BioMed. Res. Int.* DOI 10.1155/2013/264020.

Praveen, K.G. and Suneetha, V. 2014. A cocktail enzyme pectinase from fruit industrial dump sites: a review. *Res. J. Pharm. Biol. Chem. Sci.* 5(2): 1252-1258.

Puri, M. 2012. Updates on naringinase: structural and biotechnological aspects. *Appl. Microbiol. Biotechnol.* 93: 49-60.

Rao, M.B., Aparna., Tanksale, A.M., Ghatge, M.S. and Deshpande, V.V. 1998. Molecular and biotechnological aspects of microbial proteases. *Microbiol. Mol. Biol. Rev.* 62: 597-635.

Ray, A. 2012. Protease enzyme-potential industrial scope. *Int. J. Tech.* 2: 1-4.

Reetz, M.T. 2002. Lipases as practical biocatalysts. *Curr. Opin. Chem. Biol.* 6: 145-150.

Robles-Medina, A., Gonzalez-Moreno, P.A., Esteban-Cerdan, L. and Molina-Grima, E. 2009. Biocatalysis: towards ever greener biodiesel production. *Biotechnol. Adv.* 27: 398-408.

Rodriguez Couto, S. and Angeles Sanroman M. 2006. Application of solid-state fermentation to food industry-a review. *J. Food Eng.* 76(3): 291-302.

Sajith, S., Priji, P., Sreedevi, S. and Benjamin, S. 2016. An overview on fungal cellulases with an industrial perspective. *J. Nutr. Food Sci.* DOI 10.4172/2155-9600.1000461.

Sandri, I.G., Fontana, R.C., Barfknecht, D.M. and Da Silveira, M.M. 2012. Clarification of fruit juices by fungal pectinases. *Food Sci.* and *Technol.* 44(10): 2217-2222.

Shah, S. 2013. Chemistry and applications of cellulase in textile wet processing. *Res. J. Eng. Sci.* 2: 1-5.

Sharma, R., Yusuf, C. and Banerjee, U.C. 2001. Production, purification, chracterization, and applications of lipases. *Biotechno.l Adv.* 19: 627-662.

Singh, A., Yadav, R.D., Kaur, A. and Mahajan, R. 2012. An ecofriendly cost effective enzymatic methodology for drinking of school waste paper. *Bioresour. Technol.* 120: 322-327.

Singh, K. and Singh, R. 2015. Role of enzymes in fruit juices clarification during processing: a review. *Int. J. Biol. Technol.* 6: 114-124.

Singh, R., Kumar, M. and Mittal, A., Mehta, P.K. 2016. Microbial enzymes: industrial progress in 21st century. *3 Biotech.* 6: 174. DOI 10.1007/s13205-016-0485-8

Singh, R., Mittal, A., Kumar, M. and Mehta, P. K. 2016. Microbial protease in commercial applications. *J. Pharm. Chem. Biol. Sci.*4: 365-374.

Singh, V., Pandey, V.C. and Agrawal, S. 2013. Potential of *Laceyella sacchari* strain B42 crude xylanase in biobleaching of kraft pulp. *Afr. J. Biotechnol.* 12(6): 570-579.

Souza, P.M., Bittencourt, M.L., Caprara, C.C., de Freitas, M., de Almeida, R.P., Silveira, D., Fonseca, Y.M., Ferreira Filho, E.X., Pessoa Junior, A. and Magalhaes, P.O. 2015. A biotechnology perspective of fungal proteases. *Brazilian J. Microbiol.* 46: 337-346.

Srilakshmi, J., Madhavi, J., Lavanya, S. and Ammani, K. 2014. Commercial potential of fungal protease: past, present and future prospects. *J. Pharmaceut. Chem. Biol. Sci.* 2: 218-234.

Tanyildizi, M.S., Ozer, D. and Elibol, M. 2005. Optimization of α-amylase production by *Bacillus* sp. using response surface methodology. *Process Biochem.* 40: 2291-2296.

Tochi, B.N., Wang, Z., Xu, S.Y. and Zhang, W. 2009. The Influence of a pectinase and pectinase/hemicellulases enzyme preparations on percentage pineapple juice recovery, particulates and sensory attributes. *Pak. J. Nutrition.* 8: 1184-1189.

Tomme, P., Driver, D.P., Amandoron, E.A., Miller, J.R.C., Warren, R.A.J. and Kilburn, D.G. 1995. Comparison of a fungal (family I) and bacterial (family II) cellulose-binding domain. *J. Bacteriol.* 177: 4356-4363.

Veille, C. and Zeikus, G.J. 2001. Hyperthermophilic enzymes: sources, use, and molecular mechanisms for thermostability. *Microbiol. Mol. Biol. Rev.* 65: 1-43.

Verma, N., Thakur, S. and Bhatt, A.K. 2012. Microbial Lipases: industrial applications and properties (a review). *Int. Res. J. Biol. Sci.*1: 88-92.

Walia, A., Mehta, P., Guleria, S. and Shirkot, C.K. 2015b. Modification in the properties of paper by using cellulase-free xylanase produced from alkalophilic *Cellulosimicrobium cellulans* CKMX1 in biobleaching of wheat straw pulp. *Can. J. Microbiol.* 61: 1-11.

Wang, J.R., Li, Y.Y., Xu, S.D., Li, P., Liu, J.S. and Liu, D.N. 2014. High-Level expression of proform Lipase from *Rhizopus oryzae* in *Pichia pastoris* and its purification and characterization. *Int. J. Mol. Sci.* 15(1): 203-217.

Zhao, X., Xie, W., Lin, Y., Lin, X., Zheng, S. and Han, S. 2013. Combined strategies for improving the heterologous expression of an alkaline lipase. DOI 10.1016/j.procbio.2013.06.021.

Zubiolo, C., Santos, R.C.A. and Carvalho, N.B. 2014. Encapsulation in a sol–gel matrix of lipase from *Aspergillus niger* obtained by bio conversion of a novel agricultural residue. *Bioprocess Biosyst. Eng.* 37: 1781-1788.

Chapter 10

Xylanases: Microbial Sources and Applications

Sri Kant Kaushik[1], *Vikas Beniwal*[1] *and Mukesh Kumar*[2]*

[1]*Department of Biotechnology, Maharishi Markandeshwar University, Mullana – 133 203, Ambala, Haryana*
[2]*Department of Biotechnology Engineering, Ambala College of Engineering and Applied Research, Devsthali – 133 101, Ambala, Haryana*
**E-mail: mukesh.biotech@gmail.com*

ABSTRACT

Xylanases are hydrolytic enzymes catalysing the breakdown of the β-1, 4 linkages of the D-xylopyranose residues of polysaccharide xylan. After cellulose, xylan is the second most abundant polysaccharide found in soft and hard wood. Xylan in combination with lignin and cellulose forms a cross-linked structure imparting resistance to the plants against phytopathogenic microorganisms. Xylanases along with other debranching enzymes are responsible for the hydrolytic cleavage of xylans. The microbial sources of such hydrolysing enzymes include bacteria, actinomycetes, yeast and fungi. Although, there are numerous reports of microbial xylanolytic enzymes, still further inputs are essential to attain a thorough understanding of the process of xylan degradation by xylanases obtained from various microbial sources along with their potential industrial applications such as paper industry, waste paper recycling and pulp bio-bleaching.

Introduction

Most of the lignocelluloses materials are mainly comprised of about 10–25 per cent lignin, 20–30 per cent hemicellulose, and 40–50 per cent cellulose (Iqbal *et al.*, 2011). These materials are the most abundant, recyclable and constitute nearly half of the total biomass (Dhiman *et al.*, 2008). The majority of these lignocelluloses material comes from the agricultural and forest residues. After celluloses, the hemicelluloses are the abundant fraction of polysaccharides in these lignocellulosic wastes. These hemicelluloses are the main constitute of plant cell wall and concentrate mainly in principal and secondary layers where they are in close association with cellulose and lignin providing rigidity to the plant fibre. Chemically, the hemicelluloses are composed of xylose, mannose, galactose, arabinose and other sugars and their uronic acids as monomers. However, these do not occur as homoglycans but rather as heteroglycans containing different sugar residues often as short appendages linked to the main backbone chain. In case of heteroglycans sugar residues the commonly present sugars are pentoses or hexoxes which include D-xylan, D-galactans, D-mannans, L- arabinans *etc.* (Garcia *et al.*, 2001; Dhiman *et al.*, 2008). Around 10–35 per cent of the hemicelluloses present in hardwood are O-acetyl-4-Omethylglucuronoxylans, whereas 10–15 per cent of the hemicelluloses in softwood are made of arabino-4-O-methylglucuronoxylans (Moreira, 2016; Paramjeet *et al.*, 2018).

Among 5-C sugars of hemicellulosic complexes, xylan constitutes the major proportion of biomass and is principally found in the secondary cell walls of angiosperms and gymnosperms. The composition and structure of xylan varies according to the species and it is highly complicated. In some species (*e.g. Chlorophyceae* and *Rhodophyceae*) absence of cellulose is compensated by a highly crystalline fibrillar material of xylan. Xylan backbone consists of β1,4-linked xylopyranosyl residues along with a variation of substituted side groups like arabinose, acetyl, glucuronic acids, p-coumaric acid, and ferulic acid (Scheller and Ulvskov, 2010). In hardwoods and annual plants generally substitution at C-2 and C-3 positions were observed with acetic acid. In some cases, 4-O-methylglucuronic acid substitutes the C-2 position in hardwoods, softwoods and annual plants. Some sugar residues such as arabinose will substitutes the C-3 position in softwoods and annual plants. The arabinosyl residues may be further esterified by feruloyl and coumaroyl residues. The crosslinking and esterification of these substituents makes the xylan structure more complex and stable. Therefore only the synergistic action of different xvlanolytic enzymes having the ability to cleave main chains and side chains could leads to the complete hydrolysis However xylan does not form tightly packed structure therefore can be transformed to the monosaccharides through acid or enzyme catalysed hydrolysis. Although, acid hydrolysis is rapid but leads to formation of toxic compounds.

The hardwood xylans are composed of O-acetyl-4-O-methyl-glucuronoxylan residues connected through β-1, 4-glycosidic linkages, whereas, xylan present in softwood involves arabino-4-O-methylglucuroxvlans. Water solubility of hardwood xylan is highly associated with presence of acetyl group at C-2 and C-3 atoms (Dudkin *et al.*, 1980). On the other hand, softwood xylans are freely soluble in water due to non-acetylation. Beside this, they possess less degree of branching and shorter length at the C-3 position due to α-L-arabinofuranose units joined by α-1, 3 glycosidic linkages. Native xylan has poor degradability and is present as a arduous substrate, enzymatic degradability of xylan increases because of the de-esterification of the native form of substrate by alkaline extraction leading to the elimination of acetyl groups and hydrolysis of cross links (Beg *et al.*, 2000).

Xylanase: Xylan Hydrolyzing Enzyme

In the last few decades, during the process of production of gas using the suspension of wood xylan by river mud microbes, the importance of enzymes in xylan decomposition was perceived by Hopper-Seyler (Bastawde, 1992). Till then a number of reports were published regarding xylanases production (Battan *et al.*, 2006; Dhiman *et al.*, 2008; Paramjeet *et al.*, 2018), classification and substrate type of xylanolytic enzymes (Rabinovich *et al.*, 2002; Shallom and Shoham, 2003) (Table 10.1). Xylanases have been reported to be diverse and ubiquitous in nature as their mode of action and diverse specificity leads to the catabolization of diverse type of xylan molecules (Collins *et al.*, 2005). The complete breakdown of xylan molecule may be achieved due to the synergistic role of each xylan hydrolyzing enzymes. For example, the synergistic approach between acetylxylan esterase and endoxvlanases is responsible for the efficient breakdown of acetylated xylan (Biely *et al.*, 1988). Multicomponent complex xylan-degrading enzyme system of thermophilic actinomycete *T. fusca* is the best example of synergistic approach towards the degradation of xylan. This complex system was made up of β-xylosidase, α-arabinofuranosidase, endoxylanase, and acetyl esterases (Bachmann and McCarthy, 1991). Commercially available xylan was sequentially degraded by other multi-enzyme complex enzymes from *Penicillium* sp. AHT-1 consisting of xylanase, β-xylosidase, and α-L-arabinofuranosidase, This complex multienzyme system works optimally under the ambient conditions (temperature and pH) thus sufficient to completely de- polymerize the substrate (Rahman *et al.*, 2003). At the industrial level, the synergistic co-operation of xylanolytic enzymes from *Streptomyces* sp. TUBB-12-2 in pre-bleaching of kraft pulps was quite effective (Elegir *et al.*, 1994). The synergistic effect of xylanase in combination with enzymes from other groups is also reported (Ahlawat, 2007).

Table 10.1: The Major Xylanases and their Classification (Adapted from Shallom and Shoham, 2003)

Enzymes	*EC Number*	*Family*	*Substrates*
Endo-β-1,4 xylanase	3.3.1.8	GH-5,8,10,11,43	β-1,4 xylan
Exo-β-1,4 xylosidase	3.2.1.37	GH-3,39,43,52,54	β-1,4- xylooligomers xylobiose
Acetyl xylan esterases	3.1.1.72	CE-1,2,3,4,5	2 or 3-O-acetyl xylan
α–glucuronidase	3.2.1.139	GH-67	4-O-methyl-dinna-glucuronic acid (1,2) xylooligomers
α-L-arabinofuranosidases	3.2.1.55	GH- 3,43,51,54,62	α-1,2-arabinofuranosyl (1,2) or (1,3) xylooligomers α-1,5 arabinan

The major enzymes actively involved in the breakdown of xylanases are β-1, 4-endoxylanase (EC 3.2.1.8), β-D -xylosidase (3.2.1.37), α-L-arabinofuranosidase (EC 3.2.1.55), α-L-glucuronidase (3.2.1.139), acetylxylan esterase (EC 3.1.1.72), ferulic acid esterase (3.1.1.73), p-coumaric acidic esterase and phenolic acid esterase. β-xylosidase (EC 3.2.1.37) and endoxylanases (β-1, 4-D- Xylanohydrolase; EC 3.2.1.8) are responsible for the hydrolysis of xylan backbone. Other enzymes such as arabinosidase, acetyl esterase, glucuronosidase, ferulic esterase (Manin *et al.*,

1994; Linden *et al.*; 1994, Wood and Wilson, 1995; Faulds *et al.*, 1995) are involved in the removal of xylan backbone side chains. Cooperative action of these enzymes converts the xylan molecule into constituent sugars. Xylotriose is the smallest oligosaccharide that can be obtained by the action of xylanases (Dobrev *et al.*, 2007).

Families of Xylaneses

The best characterized hemicellulase was β-1, 4-D- xylanase. D-xylanase of this type have been categorized into Endo and Exo (β-1, 4-D-xylan xylanohydrolase) xylanase (Ghose and Bisaria, 1987). Further two major groups of enzymes hydrolyzing the glycosidic bonds: glycosidic hydrolases (GHs) and enzymes hydrolyzing the ester linkages of acetate or ferulic acid side groups: carbohydrate esterases (CEs) have been classified on the basis of the sequence of the amino acid or nucleic acid of their catalytic sites. They have also been grouped into various families depending upon the homology in their primary sequence (Rabinovich *et al.*, 2002). Glycosyl hydrolases (GHs) are further divided into sub-families, Family 10 or (F) and Family 11 or (G) depending on BLAST (Basic Local Alignment Search Tool) performance and pair-wise alignment which confirmed the similarities and relatedness of these families (Collins *et al.*, 2005; Davies and Henrissat, 1995).

Multiple Forms of Xylanases

Enzymes having the potential to degrade the polysacharides such as xylans are generally produced as multiple isoenzymes; *i.e.* enzymes which exist in more than one physical form differing from each other in properties like temperature, optimum pH *etc.* and having same specific catalytic function. These isoenzymes can also differ at structural and molecular level, some have cellulose-binding domains (CBD) and others being the part of multi domain proteins. The multiple forms of xylanases are required to overcome the hindrance in hydrolysis of xylan due to the environmental factors and changes the accessibility of xylosidic linkages (Wong *et al.*, 1988).

This multiple forms of the enzyme are reported to be a result of genetic redundancy, differential mRNA processing, post-translational modifications that lead to the superior and faster hydrolysis (Badhan *et al.*, 2007). Xylanases isoenzyme can also be the product of different alleles of the same gene or may be a result of independent genes. The most potential cause for xylanase multiplicity in *Clostridium thermocellum* has been attributed to substrate diversity for enzyme production. The synergistic action of acetyl xylan esterase and endoxylanases is thought to be responsible for the comprehensive degradation of acetylated xylan (Biely *et al.*, 1985; Wong *et al.*, 1988). Some hydrolyze the backbone xylan chain near substituents much more efficiently than others. The accessibility of some linkages changes during hydrolysis which requires the production of a multiple enzymes with specialized functions. Thus, the production of multiple xylanases is probably one strategy that a microorganism uses to achieve efficient substrate hydrolysis.

Numerous reports are available on the occurrence of multiple forms of xylanase and are commonly observed in *Bacillus, Trichoderma. Stremomyces,* and *Aspergillus* sp (Kuhad *et al.*, 1997; Sa-Pereira *et al.*, 2002). Fifteen different form of xylanase alone are synthesized by the *Aspergillus niger* while 13 and 11 xylanases have been reported from *Trichoderrna viride* (Biely *et al.*, 1985) and from *Butyrivibrio fibrisolvens* (Lin and Thomson, 1991) respectively. In an exceptional case, *Phanerochaete chrysosporium* when grown on avicel has been synthesizing more than 30 different xylanases of 3 different groups which were confirmed through analytical electro-focusing (Dobozi *et al.*, 1992). There is a long list of several microorganisms having the property of synthesizing the multiple form of xylanases for *e.g. Aspergillus awamori, A. kawachii, A. nidulans, A. sojae, Bacillus circulans, B. polymyxa* and many other species of genus *Bacillus, Cellulomonas fimi, Cellulomonas* sp. N.C.I.M 2353, *Clostridium acetobutylicum, C. stercorarium, C. thermolacticum, Fibrobacter succinogenes, Fusarium* oxysporum F3, *Humicola grisea* var. *thermoidea, Melanocarpus albomyces* IIS 68, *Neocallimastix frontalis, Neurospora crassa, Robillarda* sp., *Streptomyces cyaneus, S. exfoliates, S. Thermoviolaceus, Talaromyces byssochlamydoides, T. emersonii, Thermomonospora curvata, Thermotoga maritime, Thermotoga thermarum, Trichoderma harzianum. T koningii, T. lignorum, and T reesei* (Wong *et al.*, 1988; Tenkanen *et al.*, 1992; Eleiger *et al.*, 1995; Kimura *et al.*, 1995; Sunna and Antranikian, 1997: Saraswat and Bisaria, 1997; Chaudhary and Deobagkar, 19978).

Endoxylanases (β-1, 4-D- xylanohydrolase, EC 3.2.1.8)

Among all the xylanases, endoxylanases are the most widely explored and characterized enzymes. As the structure of the xylan is highly complex and further it is substituted by various side groups therefore, a synergistic approach of different forms of xylanase is required to completely hydrolyze xylan to its monomeric units. Amongst all the xylanolytic enzymes, endo-β-1, 4-xylanases (Sunna and Antranikian, 1997) is the most significant and crucial. These are the only hemicellulases that have been well characterized (Kuhad *et al.*, 1997). Hydrolysis of xylan will lead to the formation of short chains of xylo- oligosaccharides of varying length. Biochemical properties, kinetics and affinity towards the substrate of endoxylanases vary according to the source type. Endoxylanases from some fungi apparently has shown molecular weight variations from 16 to 75 kDa along with variations in iso-electric point (pI) and having less affinity towards the substrate molecules (Coughlan *et al.*, 1993). On the basis of molecular structure and their catalytic specificity, endoxylanases are of two categories *i.e.* Non-debranching (*Arabinose non-liberating*) and Debranching (*Arabinose liberating*) enzymes (Dekker and Richards, 1976). Further, Non-debranching endoxylanases can be sub-divided in to two categories, one category which liberates end products in form of xylobiose and xylose and second category releases xylo-oligosaccharides. Category one Non-debranching enzymes can act on L-arabinosyl initiated branch points as β - (1, 4) linkages and can produce only xylobiose and xylose as the major end products. Whereas, second category Non-debranching enzymes are not able to remove branches at α-(1, 2) and α-(1, 3) and can yield mostly short chains of xylo-oligosaccharides. However, xylotriose and xylobiose remain unaffected by these endoxylanases.

Some bacterial endoxylanases also possess cellulose binding domain, the function of which is still unclear. Further the mechanism of action of these endoxylanases is significantly determined by the presence or absence and amount of substituents groups located in the active site region. These active sites are generally responsible for the liberation of xylo-oligosaccharides. However, it is still unclear whether the substitution at active site will lead to the inhibition of xylanase through 'Gross Stearic Hindrance' mechanism or it was just the inability of the active site to accommodate the substrate molecule (Coughlan and Hazlewood, 1993). In general, polymeric xylan gets cleaved only at unsubstituted regions however Frederick *et al.* (1985) reported that some xylanases cleave the main chain only in the vicinity of the substituents.

Biosynthesis of Xylanases

Production of xylanase may be constitutive or inducible in nature and can be of growth associated (Sanghi *et al.*, 2008) and non-growth associated (Archana and Satyabnarayana, 1998; Breccia *et al.*, 1998). However, the regulatory mechanism of xylanase synthesis is still unclear. According to one of the proposed mechanism, only small fragments of xylan fragments can move inside the microbial cells, whereas, the larger xylan molecules are not able to cross the microbial cells. To regulate the xylanase synthesis, small xylan fragments of low molecular weight, are constitutively produced in the medium (Kulkarni *et al.*, 1999). Components of the medium also have significant effect on the inducible nature of xylanase, with maximum production in xylan and xylan rich substrates containing media. A xylan containing substrate primarily turns to be an inducer and the extent of induction is species specific. Xylan was reported as a strong inducer in *Aspergillus awamori* (Siedenberg *et al.*, 1998; Dobrev *et al.*, 2007) and *Streptomyces* sp. QG-11-3 (Beg *et al.*, 2000) weak inducer in case of *Cellulomonas flavigena* (Avalos *et al.*, 1996). Many other compounds like xylo oligosaccharides, lignocellulosic and L-sorbose residues are significantly involved in the induction of production of xylanases. Increased production of monosaccharides and accumulation of oligosaccharides which has been generated due to hydrolysis of metabolizable sugars were responsible for the repression of xylanase production (Coughlan and Hazlewood, 1993). Amino acids rich production media have also demonstrated encouraging effect on *Bacillus* sp. xylanase production (Ikura *et al.*, 1987; Kulkarni *et al.*, 1999), *Staphylococcus* sp. (Gupta *et al.*, 1992) and *Streptomyces* sp. (Beg *et al.*, 2000).

Microbial Sources for Xylanases

Numerous reports are available regarding the different microbial systems having the characteristic feature of producing xylanase (Biely, 1985; Wong *et al.*, 1988; Eriksson *et al.*, 1990; Bastawde, 1992; Kuhad *et al.*, 1997, Annamalai *et al.*, 2009; Walia *et al.*, 2013). Almost every type of microbial sources such as bacterial, fungal, actinomycetes, yeasts, rumen microorganisms (protozoa) of marine environment to terrestrial environment and from algae, snails to crustaceans, insects and seeds of terrestrial plants can synthesize xylanases (Liu *et al.*, 1998; Beg *et al.*, 2000). Both types of xylanase secretion *i.e.* intra-cellularly and extra-cellularly along with the cell wall bound secretion in case of bacterial system have also been reported (Dekker and Richards, 1976; Dekker, 1985). Cell associated endoxylanase from the *Thermoanaerobacterium* sp. Strain JW/SL-YS485 has also been reported (Shao and Wiegel, 1995). Some of the plants and higher animals also have been reported for their xylanase producing features.

Certain microbial species such as *Bacillus* sp., *Streptomyces* sp., *Cellulomonos* sp. *Thermomonospora* sp. and *Chainia* sp. are well known for their natural ability to produce xylanase in higher yields (Kuhad *et al.*, 1997). Anaerobic bacterial genera viz. *Clostridium, Thermoanaerobacter, Acetovibrio,* and *Bacteroides* are known to be the good producer of xylanase enzyme (Eriksson *et al.*, 1990). Other bacterial sources include *Thermobifida halotolerans* YIM 90462, *Actinomadura* sp. strain Cpt20, *Bacillus pumilus* SSP 34 (Zhang *et al.*, 2012; Taibi *et al.*, 2012; Subramaniyan, 2012).

As endoxylanases are the most characterized and studied enzyme therefore, numerous reports were available on its microbial producers. Main producers of endo-xylanases are *Aspergillus* sp. (Frederick *et al.*, 1985; Bailey and Poutanen, 1989; Gawande and Kamat, 1999), *Trichoderma reesi* (Tenkanen *et al.*, 1992), *Cyathus stercoreus* (Saxena *et al.*, 1994), *Fusarium oxysporum* (Singh *et al.*, 1995; Kuhad *et al.*, 1998). *Cellulomonas* sp. N.C.I.M. 2353 (Chaudhary and Deobagkar, 1997), *Aspergillus nidulans* KK-99 (Taneja *et al.*, 2002), *Bacillus* sp. (Gupta *et al.*, 1992), *Streptomyces* sp. (Elegir *et al.*, 1995; Beg *et al.*, 2000), *Staphylococcus* sp. SG-13 (Gupta *et al.*, 2000) and many other microbial systems. Among all the microbial sources, *Bacillus subtilis, Bacillus pumilus, Cellulomonas uda, Clostridium acetobutylicum,* and *Streptomyces xylophagus* are the producers of industrially important β-1, 4- endoxylanases (Kuhad *et al.*, 1997).

Fungal systems are generally known for their capability of producing xylanases working optimally under acidic pH and have been studied in detail (Coughlan *et al.*, 1993). *Ascomycetes, Deuteromycetes,* and *Basidiomycetes* are some of the common groups known for xylanase synthesis. Production and characterization of xylanases from several *Trichoderma, Aspergillus* and *Fusarium* strains have also been reported by many other workers (Dekker, 1985; Singh *et al.*, 1995). Among the fungal strains, Brown rot fungi and white rot fungi both have the ability to degrade the xylan rich substrates. Brown rot fungi characteristically left the lignin contents undigested and acted specifically over hemicellulose and cellulosic material. Whereas white rot fungi as the name suggest convert the woody material to fibrous, pale coloured material which eventually converted to the whitish mass. Some of the best xylanase producers among brown rot fungi are *Tyromyces balsemeus, Poria placenta, Gloeophyllum trabeum, Lenzites trabeum, Lentinus lepidius, Coniophora puteana* (Eriksson *et al.*, 1990). The fungal strains such as *Trametes versicolor, Phanerochaete chrysosporium, Phlebia radiata, Dichomitus squalens, Heterobasidium annosum, Pleurotus ostreatus, Phellinus pini, Polyporus anceps etc.* comes under the category of white rot fungi. Few soft rot fungal species are also known for their ability to synthesize endoxylanases. Some of the example includes *Chaetomium cellulolyticum, Aspergillus niger, Fusarium oxysporum, Trichoderma viride, Thielavia terrestris, Penicillium jenthillenum, Dactylomyces crustaceous*. Among all the fungal system, xylanases from *A. niger* have been best characterized (Eriksson *et al.*, 1990). Some fungal sources also known for their ability to synthesize multienzyme complex of xylanolytic system *i.e.* they have the property to secrete multiple different form of xylanase enzyme. Multienzyme complex of endoxylanases has been reported from *Curvularia inaequalis, Trichoderma reesei* and *Aspergillus nidulans* under this category (Gupta *et al.*, 2000, Beg *et al.*, 2000; Taneja *et al.*, 2002).

Anaerobic fungal systems basically inhabitants of the alimentary canal of herbivorous animals also have the ability to hydrolyze the xylan rich substrates due to their characteristic feature of producing an array of all the xylan hydrolysing enzymes. Some of the well-studied examples of anaerobic fungi are *Neocallimastix frontalis, N. patriciarum, Piromyces (Piromonas) communis* and *Caecomyces* sp. CR4 (Matsui and Ban-Tokuda, 2008).

Among other sources, yeasts such as *Cryptococcus, Aureobasidium,* and *Trichosporon* have been reported to produce xylanases. Mainly xylanolytic activity has been observed in *T. cutaneum and C. albidus* (Eriksson, 1990). Xylanolytic activity has also been reported from halophilic bacteria viz *Halorhabdus utahensis* (Waino and Ingvorsen, 2003), and *Halobacterium* strain CL8 (Wejse *et al.,* 2003). A list of various types of hemicellulose producing microorganisms with certain properties is given in Table 10.2.

Table 10.2: List of different Hemicellulase Producing Microorganisms

Microorganism	*Temp. Optima (°C)*	*pH Optima*	*pI*	*Molecular Weight (kDa)*
Aeromonas caviae ME 1	50	7	7.1	20
Acedobacterium capsulatum	65	5	7.3	41
Aspergillus awamori	55	5.5-6	5.7-6.7	39
	50	5.0	3.7	23
	45-50	4.0	3.3-3.5	26
Aspergillus fischeri	60	6	-	31
Aspergillus fumigates AR1	65	6.5	-	>212
Aspergillus niger	65	5.5	-	24
Aspergillus nidulans	56	0.6	3.4	34
Aspergillus kawachii	50-60	2.0-5.5	3.5-6.7	26-35
Aspergillus sojae	60	5.5	3.5	32.7
	50	5.5	-	35
Aspergillus syndowii	60	5.5	-	30
Aureobasidium pullulans ATCC 42023	35	3-4.5	-	21
Aureobasidium pullulans Y-2311-1	54	4.8	9.4	25
Bacillus amyloliquefaciens	80	6.8-7.0	10.1	18.5-19.6
Bacillus sp. strain AR009	60	9.0	-	23
	75	9.0-10.0	-	48
Bacillus circulans WL-12	-	5.5-7.0	4.5	85
	-	5.5-7.0	9.1	15
Bacillus sp. CCMI 966	60	6.0	-	340
Bacillus sp. NG-27	70	7,8.4	-	-

Microorganism	Temp. Optima (°C)	pH Optima	pI	Molecular Weight (kDa)
Bacillus sp. NCIM 59	50-60	6.0	4,8	15.8, 35
Bacillus polymyxa	45	6.0-7.0	9.3	34
	50-62	4.0-6.0	> 9.3	34
	55	6.0-7.0	9.0	22
Bacillus pumilus	55	8.0	4.8	40
Bacillus polymyxa CECT 153	50	6.5	4.7	61
Bacillus sp 11-1S	80	4.0	-	-
Bacillus sp. strain 41-1(36)	50	9.0	5.3	3.6
Bacillus sp. strain BP-23	50	5.5	9.3	32
Bacillus sp. strain BP-7	55	6.0	7.0-9.0	22-120
Bacillus sp. strain K-1	60	5.5	-	23
Bacillus sp. strain TAR-1	75	6.0	4.1	40
Bacillus sp. W1	65	6.0	8.5	21.5
	70	7.0-9.0	3.6	49.5
Bacillus sp. strain SPS-0	75	6.0	-	99
Bacillus sp. W1 (JCM 2888)	65	6.0	8.5	21.5
	70	7.9	3.7	49.5
Bacillus sp. W2	65	6.0	8.3	22.5
	70	7.0-9.5	3.7	50
Bacillus stearothermophilus T-6	75	6.0	9.0	43
Bacillus strain XE	75	6.0	7.7	22
Cellulomonas flavigena PN-120	55	9.0	6.14	56
Cellulomonas fimi	45	5.0	-	14
	40	6.0	-	22
	40	5.5-6.5	-	150
Cellulomonas sp. N.C.I.M.2353	55	6.5	8	22, 33, 53
Cephalosporium sp. strain RYM-202	7.5-8.0	50	6.3	35
	7.5-8.0	50	4.4	24
Chaetomium cellulolyticum	-	6.0-7.0	8.9	25
	-	6.0-7.0	8.4	47
	-	6.0-7.0	5.0	57
Clostridium thermolacticum	80	6.5	4.45	39
	80	6.5	4.55	55
	80	6.5	4.65	65
Cryptococcus flavous	55	4.5	10.0	31
Cryptococcus albidus	25	5.0	-	48
Erwinia chrysanthemi	55	5.5	8.8	42

Microorganism	Temp. Optima (°C)	pH Optima	pI	Molecular Weight (kDa)
Fibrobactor succinogenes	39	7.0	-	537
	55	6.3	-	66
Fusarium oxysporum F3	60	6.0	9.5	20.8
	55	6.0	8.45-8.7	23.5
Humicola insolens	55-60	6.0-6.65	9.0	6.0
	55-60	6.0-6.5	7.7	21
Humicola lanuginsoa	65	6.0	4.1	22.5
Halophilic bacterium strain CL8	60	6.0	5.0	43
	65	6.0	3.4	62
Geotrichum candidum	50	4.0	3.4	60-67
Irpex lacteus	-	4.6-5.2	7.6-8	38
Lentinula edodes	60	4.5-5.0	3.6	41
Micrococcus sp. AR-135	55	7.5-9.0	-	56
Microtetraspora flexuosa S IIX	80	6.0	8.4	26.3
	80	6.0	9.45	16.8
Myrothecium verrucaria	45	5.5	4.35	15.9
Neocallimastix frontalis	55	5.5	-	45
	55	6.0	-	70
Neurospora crassa	-	4.8	4.8	33
	-	4.8	4.5	30
Paecilomyces varioti	65	5.5-7.0	3.9	25
Penicillum purpurogenum	-	7.0	4.5-5.5	33
	-	3.5	4.5-5.5	23
Pyromyces sp. E2	50	6.0	9.06	12.5
	50	6.0	9.06	12.5
Recombinant E. coli PATBX4.5	60	6.0	4	35
	60	6.0	8	14.5
Robillarda sp. Y-20	50	4.5-6.0	9.7	17.6
	50	4.5-6.0	3.5	59
Schizophyllum commune	-	5.0	4.5	21
Schizophyllum commune	45-50	5.0	-	33
Staphylococcus sp. SG-13	50	7.5, 9.2	-	60
Streptomyces lividans 66	57	6.0	> 10.2	22
Streptomyces sp. No 3137	60-65	5.5-6.5	7.1	50
	60-65	5.0-6.0	10.06	25
	60-65	5.0-6.0	10.26	25
Streptomyces sp. strain C5-A13	55	8.5-9.5	5	19.2

Microorganism	*Temp. Optima (°C)*	*pH Optima*	*pI*	*Molecular Weight (kDa)*
Streptomyces sp. B-12-2	55	6.0	7.5	26.4
	60	6.0	8.3	23.8
	60	7.0	5.4	36.2
	60	7.0	5.0	26.2
	60	6.0	4.8	40.5
Streptomyces sp. strain S38	60	6.0-6.5	9.8	24.5
	60	6.0-7.5	5.2	37.5
	60	6.0-8.0	4.7	38
Streptomyces sp. A451	50	5.6	8.6	22.8
	50	5.4	8.9	33.1
Streptomyces viridosporus T7A	65-70	7.0-8.8	10.2-10.5	59
Streptomyces thermoviolaceus OPC-520	60-70	7.0	4.2, 8	33, 54
Streptomyces chattanogensis CECT 3336	50	6.0	9	48
Streptomyces sp. QG-11-3	60	8.6	-	-
Streptomyces T-7	60	4.5-5.5	7.8	20.643
Thermoanaerobacterium sp. strain JW/SL- YS485	80	6.5	-	350
Thermoascus aurantiacus	-	5.1	7.1	32
Thermomyces lanuginosus DSM 5826	60-70	7.0	4.1	25.5
Thermomyces lanuginosus SSBP	70-75	6.5	3.8	23.6
Thermomonospora curvata	75	6.8-7.8	4.2-8.4	15-36
Thermotoga thermarum	80	6.0	-	266
	90-100	7.0	-	35
Thermotoga maritime MSB8	92-105	5.4, 6.2	5.6	40, 120
Trichoderma harzianun	50	5.0	-	20
Trichoderma koningii	-	4.9-5.5	7.3	18
	-	4.9-5.8	7.2	29
Trichoderma longibrachiatum	45	5.0-6.0	-	37.7
Trichoderma reesei	45, 40	5.0-5.5 4.0-4.5	9, 5.5	20, 19
Trichoderma viride	53	5.0	9.3	2

Industrial Applications of Xylanases

During last few years numerous investigations were reported over the bulk use and different biotechnology applications of the microbial xylanolytic enzymes at industrial levels such as in paper and pulp industries, textile industry, fruit juice industry, waste paper recycling, food and feed industry (Wong and Saddler, 1992; Kuhad and Singh, 1993).

The major application of xylanolytic enzymes is in paper and pulp industry for bio-bleaching; to reduce the consumption of polluting oxidizing chemicals for imparting whiteness to the paper (Taneja *et al.*, 2002; Ninawe and Kuhad, 2006). Xylanase is very much popular in paper industry for pre-treatment of pulp and during bio-bleaching because it was observed that enzyme treatment enhanced separation of pulp fiber *i.e.* fibrillation as well as facilitated enzymatic deinking of recycled fibers. Conventionally, oxidizing chemicals such as chlorine water (Cl_2) and chlorine dioxide (ClO_2) were used to remove the lignin from paper pulp and process was termed as *'Sulphate or Kraft'* process. Use of xylanase during kraft processing reduces the use of organo-chlorine pollutants (such as dioxin) from paper making process. Further bio-bleaching improves various physical properties of the paper (Battan *et al.*, 2007). Apart from paper and pulp, xylanases are used concurrently with cellulase and pectinase for clarifying must and juices (Puchart *et al.*, 1999), animal feed industry, for increasing the specific bread volume, liquefying vegetables and fruits (Biely, 1985), for facilitating the process of composting and forage crops pretreatment for the improvement of digestibility of ruminant feeds. Xylanase in combination with other hydrolytic enzymes such as pectinases and cellulases was widely applied for degumming and debarking process. Whereas mixture of hydrolytics enzymes was efficiently used in debarking as the initial step in processing of wood by Wang and Saddler (1992). Some useful fermentation substrates such as xylose can be produced from agricultural wastes and food industries as they contained large amount of xylan and further these fermentation substrates could be efficiently converted to single cell proteins, xylitol, ethanol and other organic solvents and acids (Kuhad and Singh, 1993; Olsson and Hahn Hagerdal, 1996; Dominguez, 1998). Some of the products such as xylitol can be used as sweetening agent and also used in toothpaste formulations. Further it has antimicrobial properties and having odontological applications such as teeth hardening, remineralisation (Parajo *et al.*, 1998; Roberto *et al.*, 2003). Xylitol produced founds its applications in several foodstuffs like ice cream, chewing gum, soft drinks and candy.

Xylanase enzyme also has other applications such as production of plant protoplasts through improving cell wall maceration (Wong *et al.*, 1998). This technique can be efficiently applied for the separation of gluten from starch in wheat and other cereals.

Xylanase was also helpful for extracting the fermentable sugars from barley for improvement in the extraction of flavours, spices oils and pigments. Other applications include processing of aromatizing wines, nuts, rhizosecretion in transgenic tobacco (Borisjuk *et al.*, 1999), and induction in fatty acylation and glycosylation of phytosterols (Moreau *et al.*, 1994). Xylanase also improves the cleaning ability of detergents that are specially applied in cleaning fruits, vegetables, grass stains and soils (Sun and Cheng, 2002). Microbial enzymes successfullyemployed at industrial levels for commercial purposes are compiled in Table 10.3.

Table 10.3: List of Commercially available Xylanases (Beg *et al.*, 2001)

Supplier	*Product Trade Name*	*Application*
Meito Sankyo, Nogaya, Japan	Xylanase	Pulp bleaching
Voest Alpine, Austria	VAI Xylanase	Pulp bleaching
Sandoz, Charlotte, N.C. and Basle, Switzerland	Cartazyme	Pulp bleaching
Alko Rajamaki, Finland	Ecopulp	Pulp bleaching
Thomas Swan, UK	Ecozyme	Pulp bleaching
Biocon India, Bangalore	Bleachzyme F	Pulp bleaching
Rohm, Germany	Rholase 7118	Pulp bleaching
Novo Nordisk, Denmark	Pulpzyme HA, Pulpzyme Hb, Pulpzyme HC, BiofeedBeta, Biofeed Plus, Ceremix	
Genercor, Finland; Ciba Giegy, Switzerland	Irgazyme 40-4X, Albazyme 40-4X, Irgazyme 10A, Albazyme 10A, Multifect xylanase	Pulp bleaching
Rohn Enzyme 0Y; Primalco, Finland	Ecopulp X-100, Ecopulp X-200, Ecopulp X-200/4, Ecopulp TX-100, Ecopulp TX-200, Ecopulp X-100	Feed Brewing
Solvay Interox, USA	Optipulp L-8000	Food
Enzyme Development USA	Enzeko xylanase	Baking, Food, Feed
Iogen, Canada	GS – 35, HS70	Pulp bleaching
Sankyo, Japan	Sanzyme PX, Alpelase F, Sanzyme X	Feed Food
Clarient, UK	Cartazyme HS 10, Cartazyme HT, Cartazyme SR 10, Cartazyme PS 10, Cartazyme 9407 E, Cartazyme NS 10, Cartazyme MP	Pulp bleaching

Paper Industry

Xylanase and other enzymes have successfully proven their significant use in paper industry for applications like bleaching and pitch removal, increasing of freeness of pulp fibres respectively as proposed by researchers in their hypothesis decades ago (Li and Hardin, 1998; Duarte *et al.*, 2003; Sandrim *et al.*, 2005; Ahlawat *et al.*, 2007).

These enzymes are also being used in enzymatic treatment cellulosic pulp during pulp production for residual xylan removal. (Duarte *et al.*, 2003; Roncero *et al.*, 2003; Sandrim *et al.*, 2005). The scott mill in Miranda (Spain) used xylanse pre-bleaching process development by Sandoz corp in the pulp production using xylanases. Thereafter, the xylanase mill trials started with notable frequency. In 1990s, 10 mills were involved in the xylanase mediated pulp pre-bleaching on a marketable basis (Bajpai, 1997). Xylanases, mannoses and other related enzymes have the ability to decompose the hemicelluloses part of the pulp keeping the cellulose unaffected. Process of enzyme-assisted bio-bleaching was generally carried out at higher temperature (~ 60 °C) and at alkaline pH (~ 9.0), therefore alkalothermophilic xylanases are required at industrial level.

However fungal xylanases are operative inside the acidic pH range of 4.0 - 6.0 (Da-Silva *et al.*, 1994; Beg *et al.*, 2000a) and bacteria are effective over a wide-ranging pH of 5.0-9.0 (Kulkarni and Rao, 1996; Khanongnuch *et al.*, 1999). Therefore, bacterial xylanse having low molecular weight and cellulose freeby nature was selectively used for the removal of hemicellulosic components with minimum damage to the cellulose pulp.

Bio-bleaching and bio-processing of pulps using a combination of xylanases with other enzymes such as laccases and pectinases further increases the significance of enzymes at industrial level (Ahlawat *et al.*, 2007). Once hemicellulose(s) were modified with enzymatic pretreatment; lignin layer gets easily penetrated by degradative action of laccase (Viikari *et al.*, 1994; Bajpai, 2004). Kapoor *et al.* (2007) described different and interactive properties of a xylanase and laccase in waste pulps and soda bleaching for eco-friendly paper production. Most of the bio-bleaching studies are confirmed to kraft and sulfite pulps (Oakley *et al.*, 2003), however some reports revels the use of non-woody pulp with significant increase in brightness. Increased brightness of 19.6, 11.6 and 7.9 per cent in banana, silk cotton, and cotton respectively was reported using *Bacillus subtills* C01 xylanase pre-treated pulp (Ayyachamy and Vatsala, 2007; Ahlawat *et al.*, 2007). Described the synergistic effect of using pectinase with xylanase for pretreatment of kraft pulp and reduction in Cl_2 consumption during bio-bleaching.

Presently a substantial number of Japanese, North American, South American, and European, industries are using xylanase for bleaching process. Pulpzyme HA, (Novo Nordisk A/S) from *Trichoderma reesei* became as the first commercially obtainable xylanase for bio-bleaching of wood pulps and was used to reduce the dosage of active chlorine in the first bleaching stage. Numerous biotech companies throughout the world are making various xylanse preparations, such as Ecopulp (Alko), Irgazyme (Genencor international), VAI xylanse (Voest alpine) and Cartazyme (Sandoz). Canada, the biggest manufacturer of pulp, is delivering greater than 10 per cent xylanase based bleached pulps. However, in India, xylanase commercialization in pulp bleaching is yet to be started, but it is anticipated to be a part of paper mills very soon.

Waste Paper Recycling

Xylanase also finds its applications in waste paper management. Waste paper treatment is achieved using two stage process of beating and pulping. In the first stage, moistening of loose fibers is attained through the fiber separation or fiber dispersing and the process is known as hydrating process. Enzymatic treatment process in its first step involves preliminary saturation of paper and xylanase incubation, the next step include mechanical pulp removal with subsequent pulp heating for enzyme deactivation and removal of fibers. Treatment of xylanase results in removal of a number of reducing sugars from paper pulp waste. This process is temperature dependent as majority of the pulp fibre xylan is hydrolysed at higher temperature. Xylanase treatment causes the waste paper pulp fibres to swell and separate. Hydrolysis and detachment of xylan residues results in more dispersed and smooth fibers in pre-treated pulp as compared to untreated pulp.

REFERENCES

Ahlawat, S., Battan, B., Dhiman, S. S., Sharma, J., and Mandhan, R. P. 2007. Production of thermostable pectinase and xylanase for their potential application in bleaching of kraft pulp. *J. Ind. Microbiol. Biotechno. 34*(12), 763-770.

Archana, A., and Satyanarayana, T. 1998. Cellulase-free xylanase production by thermophilic *Bacillus licheniformis* A99. *Indian J. Microbiol. 38*(3), 135-139.

Ayyachamy, M., and Vatsala, T. M. 2007. Production and partial characterization of cellulase free xylanase by *Bacillus subtilis* C 01 using agriresidues and its application in biobleaching of nonwoody plant pulps. *Lett. Appl. microbiol. 45*(5), 467-472.

Bachmann, S. L., and McCarthy, A. J. 1991. Purification and cooperative activity of enzymes constituting the xylan-degrading system of *Thermomonospora fusca. Appl. Environ. Microbiol. 57*(8), 2121-2130.

Badhan, A. K., Chadha, B. S., Kaur, J., Saini, H. S., and Bhat, M. K. 2007. Production of multiple xylanolytic and cellulolytic enzymes by thermophilic fungus *Myceliophthora* sp. IMI 387099. *Biores.Technol. 98*(3), 504-510.

Bailey, M. J., and Poutanen, K. 1989. Production of xylanolytic enzymes by strains of *Aspergillus. Appl. Microbiol. Biotechnol. 30*(1), 5-10.

Bajpai, P. 1997. Microbial xylanolytic enzyme system: properties and applications. In *Advances in applied microbiology* (Vol. 43, pp. 141-194). Academic Press.

Bajpai, P. 2004. Biological bleaching of chemical pulps. *Crit.Rev. Biotechnol. 24*(1), 1-58.

Bastawde, K. B. 1992. Xylan structure, microbial xylanases, and their mode of action. *World J. Microbiol. Biotechnol. 8*(4), 353-368.

Battan, B., Sharma, J., and Kuhad, R. C. 2006. High-level xylanase production by alkaliphilic *Bacillus pumilus* ASH under solid-state fermentation. *World J. Microbiol. Biotechnol. 22*(12), 1281-1287.

Battan, B., Sharma, J., Dhiman, S. S., and Kuhad, R. C. 2007. Enhanced production of cellulase-free thermostable xylanase by *Bacillus pumilus* ASH and its potential application in paper industry. Enzyme Microb. Technol. 41(6-7), 733-739.

Beg, Q. K., Bhushan, B., Kapoor, M., and Hoondal, G. S. 2000. Production and characterization of thermostable xylanase and pectinase from *Streptomyces* sp. QG-11-3. *J. Ind. Microbiol. Biotechnol.24*(6), 396-402.

Beg, Q. K., Bhushan, B., Kapoor, M., and Hoondal, G. S. 2000a. Enhanced production of a thermostable xylanase from *Streptomyces* sp. QG-11-3 and its application in biobleaching of eucalyptus kraft pulp. *Enzyme Microb. Technol. 27*(7), 459-466.

Beg, Q. K., Kapoor, M., Mahajan, L., and Hoondal, G. S. 2001. *Appl. Microbiol. Biotechnol.* 56, 326–338. *CrossRef Google Scholar*.

Biely, P. 1985. Microbial xylanolytic systems. *Trends Biotechnol. 3*(11), 286-290.

Biely, P., MacKenzie, C. R., and Schneider, H. 1988. Production of acetyl xylan esterase by *Trichoderma reesei* and *Schizophyllum commune. Canadian J. Microbial. 34*(6), 767-772.

Borisjuk, N. V., Borisjuk, L. G., Logendra, S., Petersen, F., Gleba, Y., and Raskin, I. 1999. Production of recombinant proteins in plant root exudates. *Nature Biotechnol. 17*(5), 466.

Breccia, J. D., Siñeriz, F., Baigori, M. D., Castro, G. R., and Hatti-Kaul, R. 1998. Purification and characterization of a thermostable xylanase from *Bacillus amyloliquefaciens*. *Enzyme Microb. Technol. 22*(1), 42-49.

Chaudhary, P., and Deobagkar, D. N. 1997. Purification and characterization of xylanases from *Cellulomonas* sp. NCIM 2353. *Biotechnol. Appl. Biochem. 25*(2), 127-133.

Collins, T., Gerday, C., and Feller, G. 2005. Xylanases, xylanase families and extremophilic xylanases. *FEMS Microbiol. Rev. 29*(1), 3-23.

Coughlan, M. P., and Hazlewood, G. P. 1993. beta1, 4-Dxylandegrading enzyme systems: biochemistry, molecular biology and applications. *Biotechnol. Appl. Biochem*. 17(3): 259-289.

Coughlan, M. P., Tuohy Filho, M. G., EXF, P. J., Claeyssens, M., Vrsanska, M., and Hughes, M. M. 1993. Enzymological aspects of microbial hemicellulases with emphasis on fungal systems. Hemicellulose and hemicellulases. *Portland Press, London*. 53-84.

Da Silva, R., Yim, D. K., and Park, Y. K. 1994. Application of thermostable xylanases from *Humicola* sp. for pulp improvement. *J. Ferment. Bioeng. 77*(1), 109-111.

Damiano, V. B., Bocchini, D. A., Gomes, E., and Da Silva, R. 2003. Application of crude xylanase from *Bacillus licheniformis* 77-2 to the bleaching of eucalyptus Kraft pulp. *World J. Microbiol. Biotechnol. 19*(2), 139-144.

Davies, G., and Henrissat, B. 1995.Structures and mechanisms of glycosyl hydrolases. *Structure, 3*(9), 853-859.

Dekker, R. F. 1985. Biodegradation of the hemicelluloses. *Biosynthesis and biodegradation of wood components,* Academic Press. 505-533.

Dekker, R. F., and Richards, G. N. 1976. Hemicellulases: their occurrence, purification, properties, and mode of action. *Adv. Carbo. Chem. Biochem.* (Vol. 32, pp. 277-352).

Dhiman, S. S., Sharma, J., and Battan, B. 2008. Industrial applications and future prospects of microbial xylanases: A Review. *BioRes. 3*(4), 1377-1402.

Dhiman, S. S., Sharma, J., and Battan, B. 2008. Pretreatment processing of fabrics by alkalothermophilic xylanase from *Bacillus stearothermophilus* SDX. *Enzyme Microb.Technol. 43*(3), 262-269.

Dobozi, M. S., Szakács, G., and Bruschi, C. V. 1992. Xylanase activity of Phanerochaete chrysosporium. *Appl. Environ. Microbiol. 58*(11), 3466-3471.

Dobrev, G. T., Pishtiyski, I. G., Stanchev, V. S., and Mircheva, R. 2007. Optimization of nutrient medium containing agricultural wastes for xylanase production by *Aspergillus niger* B03 using optimal composite experimental design. *Biores. Technol. 98*(14), 2671-2678.

Dominguez, J. M. 1998. Xylitol production by free and immobilized *Debaryomyces hansenii*. *Biotechnol. Lett.* *20*(1), 53-56.

Duarte, M. C. T., da Silva, E. C., Ponezi, A. N., Portugal, E. P., Vicente, J. R., and Davanzo, E. 2003. Xylan-hydrolyzing enzyme system from *Bacillus pumilus* CBMAI 0008 and its effects on *Eucalyptus grandis* kraft pulp for pulp bleaching improvement. *Biores. Technol.* *88*(1), 9-15.

Dudkin, M. S., Parfent'eva, M. A., and Shkantova, N. G. 1980. Structure of the xylan of the stems of *Brassica napus*. *Chem. Nat. Comp.* *16*(5), 432-434.

Elegir, G., Sykes, M., and Jeffries, T. W. 1995. Differential and synergistic action of Streptomyces endoxylanases in prebleaching of kraft pulps. *Enzyme Microb. Technol.* *17*(10), 954-959.

Elegir, G., Szakács, G., and Jeffries, T. W. 1994. Purification, characterization, and substrate specificities of multiple xylanases from *Streptomyces* sp. strain B-12-2. *Appl. Environ. Microbiol.* *60*(7), 2609-2615.

Eriksson, K. E. L., Blanchette, R., and Ander, P.1990. Microbial and enzymatic degradation of wood and wood components. *Berlin, Germany CrossRef Google Scholar*.

Faulds, C. B., Ralet, M. C., Williamson, G., Hazlewood, G. P., and Gilbert, H. J. 1995. Specificity of an esterase (XYLD) from *Pseudomonas fluorescens* subsp. cellulose. *Biochem. Biophys. Acta.* *1243*(2), 265-269.

Frederick, M. M., Kiang, C. H., Frederick, J. R., and Reilly, P. J. 1985. Purification and characterization of endoxylanases from *Aspergillus niger*. I. Two isozymes active on xylan backbones near branch points. *Biotechnol. Bioeng.* *27*(4), 525-532.

Garcia, R. B., Nagashima Jr, T., Praxedes, A. K., Raffin, F. N., Moura, T. F., and do Egito, E. S. T. 2001. Preparation of micro and nanoparticles from corn cobs xylan. *Pol. Bull.* *46*(5), 371-379.

Gawande, P. V., and Kamat, M. Y. 1999. Production of *Aspergillus* xylanase by lignocellulosic waste fermentation and its application. *J. Appl. Microbiol.* *87*(4), 511-519.

Ghose, T. K., and Bisaria, V. S. 1987. Measurement of hemicellulase activities: Part I Xylanases. *Pure Appl. Chem.* *59*(12), 1739-1751.

Gupta, N., Vohra, R. M., and Hoondal, G. S. 1992. A thermostable extracellular xylanase from alkalophilic *Bacillus* sp. NG-27. *Biotechnol. lett.* *14*(11), 1045-1046.

Gupta, S., Bhushan, B., and Hoondal, G. S. 2000. Isolation, purification and characterization of xylanasefrom *Staphylococcus* sp. SG13 and its application in biobleaching of kraft pulp. *J. Appl. Microbiol.* *88*(2), 325-334.

Ikura, Y., and Horikoshi, K. 1987. Stimulatory effect of certain amino acids on xylanase production by alkalophilic *Bacillus* sp. *Agric. Biol. Chem.* *51*(11), 3143-3145.

Iqbal, H. M. N., Asgher, M., and Bhatti, H. N. 2011. Optimization of physical and nutritional factors for synthesis of lignin degrading enzymes by a novel strain of *Trametes versicolor*. *BioRes.* *6*(2), 1273-1287.

Kapoor, M., Kapoor, R. K., and Kuhad, R. C. 2007. Differential and synergistic effects of xylanase and laccase mediator system (LMS) in bleaching of soda and waste pulps. *J. Appl. microbial.* *103*(2), 305-317.

Khanongnuch, C., Lumyong, S., Ooi, T., and Kinoshita, S. 1999. A non-cellulase producing strain of *Bacillus subtilis* and its potential use in pulp biobleaching. *Biotechnol. Lett.* *21*(1), 61-63.

Kimura, I., Sasahara, H., and Tajima, S. 1995. Purification and characterization of two xylanases and an arabinofuranosidase from *Aspergillus sojae*. *J. Ferment. Bioeng.* *80*(4), 334-339.

Kuhad, R. C., and Singh, A. 1993. Lignocellulose biotechnology: current and future prospects. *Crit. Rev. Biotechnol.* *13*(2), 151-172.

Kuhad, R. C., Manchanda, M., and Singh, A. 1998. Optimization of xylanase production by a hyperxylanolytic mutant strain of *Fusarium oxysporum*. *Proc. Biochem.* *33*(6), 641-647.

Kuhad, R. C., Singh, A., and Eriksson, K. E. L. 1997. Microorganisms and enzymes involved in the degradation of plant fiber cell walls. In *Biotechnology in the pulp and paper industry* (pp. 45-125). Springer, Berlin, Heidelberg.

Kulkarni, N., and Rao, M. 1996. Application of xylanase from alkaliphilic thermophilic *Bacillus* sp. NCIM 59 in biobleaching of bagasse pulp. *J. Biotechnol.* *51*(2), 167-173.

Kulkarni, N., Shendye, A., and Rao, M.1999. Molecular and biotechnological aspects of xylanases. *FEMS Microbial. Rev.* *23*(4), 411-456.

Li, Y., and Hardin, I. R. 1998. Enzymatic Scouring of Cotton-Surfactants, Agitation, and Selection of Enzymes. *Textile Chem. Color.* *30*(9).

Lin, L. L., and Thomson, J. A. 1991. An analysis of the extracellular xylanases and cellulases of *Butyrivibrio fibrisolvens* H17c. *FEMS Microbiol. Lett.* *84*(2), 197-204.

Linden, J. C., Decker, S. R., and Samara, M. 1994. Role of acetyl esterase in biomass conversion. *Am. Chem. Soc. Symp. Series. (USA).*

Liu, W., Zhu, W., Lu, Y., Kong, J., and Ma, G.1998. Production, partial purification and characterization of xylanase from *Trichosporon cutaneum* SL409. *Proc. Biochem.* *33*(3), 331-336.

Manin, C., Shareek, F., Morosoli, R., and Kluepfel, D. 1994. Purification and characterization of an α-L-arabinofuranosidase from *Streptomyces lividans* 66 and DNA sequence of the gene (abfA). *Biochem. J.* *302*(2), 443-449.

Matsui, H., and Ban-Tokuda, T. 2008. Studies on carboxymethyl cellulase and xylanase activities of anaerobic fungal isolate CR4 from the bovine rumen. *Curr. Microbiol.* *57*(6), 615-619.

Moreau, R. A., Powell, M. J., Whitaker, B. D., Bailey, B. A., and Anderson, J. D. 1994. Xylanase treatment of plant cells induces glycosylation and fatty acylation of phytosterols. *Physiol. Plant. 91*(4), 575-580.

Moreira, L. 2016. Insights into the mechanism of enzymatic hydrolysis of xylan. *Appl. Microbiol. Biotechnol. 100*(12), 5205–5214.

Ninawe, S., and Kuhad, R. C. 2006. Bleaching of wheat straw-rich soda pulp with xylanase from a thermoalkalophilic *Streptomyces cyaneus* SN32. *Biores. Technol. 97*(18), 2291-2295.

Oakley, A. J., Heinrich, T., Thompson, C. A., and Wilce, M. C. 2003. Characterization of a family 11 xylanase from *Bacillus subtillis* B230 used for paper bleaching. *Acta. Crystallogr. D. Biol. Crystal. 59*(4), 627-636.

Olsson, L., and Hahn-Hägerdal, B. 1996. Fermentation of lignocellulosic hydrolysates for ethanol production. *Enzyme Microbiol. Technol. 18*(5), 312-331.

Parajo, J. C., Dominguez, H., and Domínguez, J. 1998. Biotechnological production of xylitol. Part 1: Interest of xylitol and fundamentals of its biosynthesis. *Biores. Technol. 65*(3), 191-201.

Paramjeet, S., Manasa, P., and Korrapati, N. 2018. Biofuels: Production of fungal-mediated ligninolytic enzymes and the modes of bioprocesses utilizing agro-based residues. *Biocatal. Agric. Biotechnol.*

Pérez-Avalos, O., Ponce-Noyola, T., Magaña-Plaza, I., and de la Torre, M. 1996. Induction of xylanase and β-xylosidase in *Cellulomonas flavigena* growing on different carbon sources. *Appl. Microbiol. Biotechnol. 46*(4), 405-409.

Puchart, V., Katapodis, P., Biely, P., Kremnický, L., Christakopoulos, P., Vršanská, M., and Bhat, M. K. 1999. Production of xylanases, mannanases, and pectinases by the thermophilic fungus *Thermomyces lanuginosus*. *Enzyme Microb. Technol. 24*(5-6), 355-361.

Rabinovich, M. L., Melnik, M. S., and Bolobova, A. V. 2002. Dedicated to the memory of IV Berezin and RV Feniksova Microbial Cellulases. *Appl. Biochem. Microbiol. 38*(4), 305-322.

Rahman, A. S., Sugitani, N., Hatsu, M., and Takamizawa, K. 2003. A role of xylanase, α-L-arabinofuranosidase, and xylosidase in xylan degradation. *Canadian J.Microbiol. 49*(1), 58-64.

Roberto, I. C., Mussatto, S. I., and Rodrigues, R. C. 2003. Dilute-acid hydrolysis for optimization of xylose recovery from rice straw in a semi-pilot reactor. *Indust. Crops. Prod. 17*(3), 171-176.

Roncero, M. B., Torres, A. L., Colom, J. F., Vidal, T. T. C. F. 2003. TCF bleaching of wheat straw pulp using ozone and xylanase. Part A: paper quality assessment. *Biores. Technol. 87*(3), 305-314.

Sandrim, V. C., Rizzatti, A. C. S., Terenzi, H. F., Jorge, J. A., Milagres, A. M. F., and Polizeli, M. L. T. M. 2005. Purification and biochemical characterization of two xylanases produced by *Aspergillus caespitosus* and their potential for kraft pulp bleaching. *Proc. Biochem. 40*(5), 1823-1828.

Sanghi, A., Garg, N., Sharma, J., Kuhar, K., Kuhad, R. C., and Gupta, V. K. 2008. Optimization of xylanase production using inexpensive agro-residues by alkalophilic *Bacillus subtilis* ASH in solid-state fermentation. *World J.Microbiol. Biotechnol. 24*(5), 633-640.

Sá-Pereira, P., Costa-Ferreira, M., and Aires-Barros, M. R. 2002. Enzymatic properties of a neutral endo-1, 3 (4)-β-xylanase Xyl II from *Bacillus subtilis. J. Biotechnol. 94*(3), 265-275.

Saraswat, V., and Bisaria, V. S. 1997. Biosynthesis of xylanolytic and xylan-debranching enzymes in *Melanocarpus albomyces* IIS 68. *J. Ferment. Bioengg. 83*(4), 352-357.

Saxena, A., Kuhad, R. C., Saxena, R. K., and Gupta, R. 1994. Production and characterization of a xylanase from *Cyathus stercoreus. World J. Microbiol. Biotechnol. 10*(3), 293-295.

Scheller, H. V., and Ulvskov, P. 2010. Hemicelluloses. *Annual review of plant biology, 61*.

Shallom, D., and Shoham, Y. 2003. Microbial hemicellulases. *Curr. Opinion Microbiol. 6*(3), 219-228.

Shao, W., and Wiegel, J. 1995. Purification and characterization of two thermostable acetyl xylan esterases from *Thermoanaerobacterium* sp. strain JW/SL-YS485. *Appl. Environ. Microbiol. 61*(2), 729-733.

Siedenberg, D., Gerlach, S. R., Schügerl, K., Giuseppin, M. L. F., and Hunik, J. 1998. Production of xylanase by *Aspergillus awamori* on synthetic medium in shake flask cultures. *Proc. Biochem. 33*(4), 429-433.

Singh, A., Kuhad, R. C., and Kumar, M. 1995. Xylanase production by a hyperxylanolytic mutant of *Fusarium oxysporum. Enzyme Microb. Technol.* 17: 551-553.

Subramaniyan, S. 2012. Isolation, purification and characterisation of low molecular weight xylanase from *Bacillus pumilus* SSP-34. *Appl. Biochem. Biotechnol. 166*(7), 1831-1842.

Sun, Y., and Cheng, J. 2002. Hydrolysis of lignocellulosic materials for ethanol production: a review. *Biores. Technol. 83*(1), 1-11.

Sunna, A., and Antranikian, G. 1997. Xylanolytic enzymes from fungi and bacteria. *Crit. Rev. Biotechnol. 17*(1), 39-67.

Taibi, Z., Saoudi, B., Boudelaa, M., Trigui, H., Belghith, H., Gargouri, A., and Ladjama, A. 2012. Purification and biochemical characterization of a highly thermostable xylanase from *Actinomadura* sp. strain Cpt20 isolated from poultry compost. *Appl. Biochem. Biotechnol. 166*(3), 663-679.

Taneja, K., Gupta, S., and Kuhad, R. C. 2002. Properties and application of a partially purified alkaline xylanase from an alkalophilic fungus *Aspergillus nidulans* KK-99. *Biores.Technol. 85*(1), 39-42.

Tenkanen, M., Puls, J., and Poutanen, K. 1992. Two major xylanases of *Trichoderma reesei*. *Enzyme Microb. Technol.* *14*(7), 566-574.

Viikari, L., Kantelinen, A., Sundquist, J., and Linko, M. 1994. Xylanases in bleaching: from an idea to the industry. *FEMS Microbiol. Rev.* *13*(23), 335-350.

Waino, M., and Ingvorsen, K. 2003. Production of β-xylanase and β-xylosidase by the extremely halophilic archaeon *Halorhabdus utahensis*. *Extremoph.* *7*(2), 87-93.

Walia, A., Mehta, P., Chauhan, A., and Shirkot, C. K. 2013. Optimization of cellulase-free xylanase production by alkalophilic *Cellulosimicrobium* sp. CKMX1 in solid-state fermentation of apple pomace using central composite design and response surface methodology. *Annals of Microbiol.* *63*(1), 187-198.

Wang, K. K. Y., and Saddler, N. J. 1992. Trichoderma xylanases, their properties and purification. *Crit. Rev. Biotechnol.* *12*, 413-435.

Wejse, P. L., Ingvorsen, K., and Mortensen, K. K. 2003. Purification and characterisation of two extremely halotolerant xylanases from a novel halophilic bacterium. *Extremoph.* *7*(5), 423-431.

Wong, K. K., Tan, L. U. L., and Saddler, J. N. 1988. Multiplicity of beta-1, 4-xylanase in microorganisms: functions and applications. *Microbiol. Rev.* *52*(3), 305.

Wood, T. M., and Wilson, C. A. 1995. α-(4-O-Methyl)-d-glucuronidase activity produced by the rumen anaerobic fungus *Piromonas communis*: A study of selected properties. *Appl. Microbiol. Biotechnol.* *43*(5), 893-900.

Zhang, J., Tang, M., and Viikari, L. 2012. Xylans inhibit enzymatic hydrolysis of lignocellulosic materials by cellulases. *Biores. Technol.* *121*, 8-12.

Chapter 11

Application of Microbes in Pharmaceutical Industries

Diksha Sharma[1], Mona Piplani[2], Archana Sharma[1], Mamta Dhamija[3] and Prabodh Chander Sharma[1]

[1]*Institute of Pharmaceutical Sciences, Kurukshetra University, Kurukshetra – 136 119, Haryana*
[2]*School of Pharmacy, Maharaja Agrasen University, Baddi – 174 103, Haryana*
[3]*ProAdWise Communications Pvt. Limited, Mumbai, Maharashtra*
**E-mail: sharma_prabodh@rediffmail.com*

ABSTRACT

Variety of species of microorganisms, animals and plants found in the nature are significant sources of novel therapeutic agents due to their incredible chemical diversity. Now days, microorganisms are playing a vital role in pharmaceutical and nutraceutical industries at global level. They have been employed for production of variety of biochemicals such as proteins, enzymes, biosurfactants, antioxidants, immunosuppressants, vitamins, pigments, organic acids, nanoparticles and antibiotics, *etc.* having remarkable therapeutic activity. They are also being utilized for the treatment as well as diagnosis of various diseases. The present chapter describes different applications of microorganisms in pharmacy and pharmaceutical industry.

Keywords: Microbes, Pharmaceutical, Nanoparticles, Antimicrobial, Biosurfactant.

Introduction

Pharmaceutical agents are required for the treatment of diseases and are essential for maintaining the good health of public. Various health challenges are accepted and resolved by advancements in drug therapies which are important for healthy civilization. Immune system of our body helps us to fight against the infectious microorganisms entering inside the body by generating macrophages and some other mechanisms. On the other hand, microorganisms play a vital role in the treatment of various diseases and hence, widely utilized in pharmaceutical and medical industries. The scientific researchers across the world are taking considerable interest in carrying out the investigations on microorganisms for finding their significant role in the treatment of specific diseases. Microbes are ubiquitous in nature. Although they are very small in size (ranging from 0.15-700 μm) but they have incredible impact in pharmaceuticals. Louis Pastuer was the scientist who first time utilized microorganisms in medicine in the mid 19th century. Great inventions from vaccines to devices have been carried out by advancement of microbiology in healthcare industries, particularly in pharmaceutical and medical industries. The natural products having industrial applications can be produced from primary and secondary metabolites of microorganisms (Gupta, 2014). The number of natural compounds has exceeded one million after the technical improvement in screening program and separation as well as isolation techniques. Among these compounds around 5 per cent are of microbial origin (Kalkan, 2013). For last several decades, microorganisms are being utilized in huge production of various biochemicals ranging from alcohol to antibiotics. As a natural source of drugs, microorganisms also have great potential for the cure of different ailments *e.g.* anemia, cancer, diarrohea, obesity, atopic dermatitis, Crohn's diseases, diabetes, irritable bowel disease, *etc.* Besides the antibacterial, antifungal and antiviral activity, secondary metabolites from microorganisms also serve as catalysts or producers for the production of antibiotics, proteins, enzymes, biosurfactants, vitamins, pigments, organic acids, nanoparticles as well as nutraceuticals having significant therapeutic potential. The various pharmaceutical applications have been depicted in Figure 11.1.

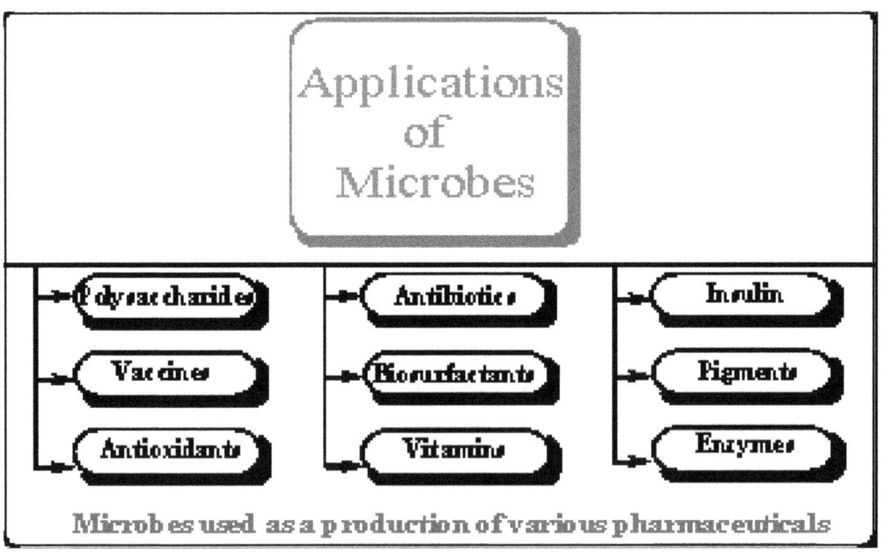

Figure 11.1: Various Applications of Microbes in Pharmaceuticals.

Moreover, having great diversity in structural and functional characteristics, the polysaccharides obtained from microbes are employed for food, pharmaceutical along with medical applications. Recombinant and hybridoma technology are utilized for modification in microorganisms so as to improve the rate of production of various metabolites as well as synthesis of new metabolites. The study of microorganisms and human cytology motivated the researchers to invent various pharmaceutical agents which could be utilized for the treatment of various communicable diseases (Ali, 2013). In the present chapter, we have attempted to review the various uses of microorganisms in the field of pharmaceuticals.

Applications of Microbes

The most significant and inherent features of bioactive microbial metabolites are their origin, interaction with environment in addition to their exclusive chemical structures. The various applications of microbes are given as follows:

Biosurfactants

Biosurfactants, the surface active agents obtained from microbes has gained considerable interest in pharmaceuticals and cosmetics due to their special characteristics *i.e.* biodegradability, less toxicity, biocompatibility, specification as well as effectiveness. These features make them commercially significant alternatives to chemically synthesized surfactants (Kosaric, 2001). Chemical surfactants are acquired from petroleum feedstock whereas biosurfactants are obtained from microbial fermentation process utilizing cheaper raw materials such as agrobased substrates. Biosurfactants can be utilized as wetting agents, foaming agents, emulsifiers as well as de-emulsifiers. In addition to this, they also possess antibacterial, antifungal, antiviral, anticancer, anti-HIV, immunomodulatory and antiadhesive properties which makes them pertinent molecule for application in combating many diseases. The various groups of biosurfactants such as glycolipids, including mannosylerythritol lipid-B, succinoyl trehelose lipid-3, sophorose lipid, succinoyl trehelose lipid-1 and polyol lipid have been found to provoke cell differentiation rather than cell proliferation in the human pro-myelocytic leukemia cell line (Rath, 2011). A dose dependent cytotoxic effect of sophorolipids has also been observed on cancer cells of H7402, A549, HL60 and K562. Biosurfactants derived from bacteria such as *Lactococcus lactis*-53 as well as *Streptococcus thermophilus*-A exhibited antibacterial activity against various bacterial and yeast strains. Lipopeptide iturin is produced by *Bacillus subtilis* and found to exhibit antifungal activity. The sophorolipids and its structural analogues produced by *Candida bombicola* displayed anti-HIV and spermicidal activity. It also acts as a good immunomodulator. Liposomes based on biosurfactants are also found to exhibit modified efficiency of gene transfection when compared with commercially available cationic liposomes. The current application of biosurfactant is its association with microbubbles which is employed as an ultrasound contrasting agent (Fathabad, 2010).

Biotransformation of Compounds

Biotransformation is the process of converting one form of organic compound to another to diminish the diligence and toxicity of the chemical compounds. Microorganisms are also utilized for synthesis of value added compounds having several applications in pharmaceutical industries (Smitha, 2017). Microbial synthesis has several advantages over synthetic methodologies as it avoids use of heavy metal, organic solvents, strong acids and bases *etc.* and is highly substrate specific. Owing to well characterized physiology and genetics, rapid cell growth rate and accessibility of profuse genetic tools, *Escherichia coli* and *Saccharomyces cerevisiae* are widely utilized platform organisms. Transformation of steroids, sterols and non-steroid compounds are carried out by various microorganisms. Testosterone has been found to produce from dehydroepiandrosterone employing *Corynebacterium* species and prostaglandins are found to produce from unsaturated fatty acids by microbial transformation utilizing pathogenic fungi such as *Cryptococcus neoformans* (Du, 2011). Some species of *Streptomyces* and *Bacillus* are also utilized for transformation of antibiotics. In addition to this, recently metal nanoparticles have also been synthesized with help of microorganisms and extensively studied. This synthesis of nanoparticles has been perceived as an efficient way for utilizing microorganisms as convenient nanofactories in future. Biological gold nanoparticles synthesised from extract of red seaweed *Corallina officinalis* exhibited cytotoxic activity against MCF7 human breast cancer cell lines. Similarly gold nanoparticles have been synthesized using leaf extract of olax scandens which demonstrated remarkable inhibition of cancer cells of lung (A549), breast (MCF-7) and colon (COLO 205) (Singh, 2016).

Antibiotics Production

The natural or synthetic chemical substances having tendency to inhibit or entire cell metabolic activity are known as antimicrobial agents/antibiotics. These are molecules having low molecular weight and are produced as secondary metabolites mainly by soil microorganisms *i.e. Actinomycetes, Pseudomonas aeruginosa, Staphylococcus* species *etc.* Antibiotic term was first used by Selman Waksman and his collaborators in 1942 (Shetty, 2014). They found that antibiotic is a substance produced by a type of microorganisms and in much diluted form antagonistic to the growth of other microorganisms. With the advancements of medicinal chemistry, presently most of the antibiotics are semisynthetic modification of natural compounds. The beta lactam antibacterial agents such as penicillins (produced by fungi), cephalosporins and carbapenems come into this category (Singh, 2012). *Bacillus* species is found abundantly in the soil and due to its production of bacitracin as well as resistant endospore formation it acts as growth inhibitor for other microorganisms. Likewise, the genus streptomyces is utilized to synthesize greater part of antimicrobial agents having clinical importance such as streptomycin, erythromycin, tetracycline, amphotericin and rifamycin *etc.* From the industrial point of view, the rate of production of antibiotics has been improved by employing random mutagenesis, protoplast fusion and development of DNA recombinant technology (Mahajan, 2015).

Insulin Production

Human insulin is generated' by pancreatic beta cells and is found to consist

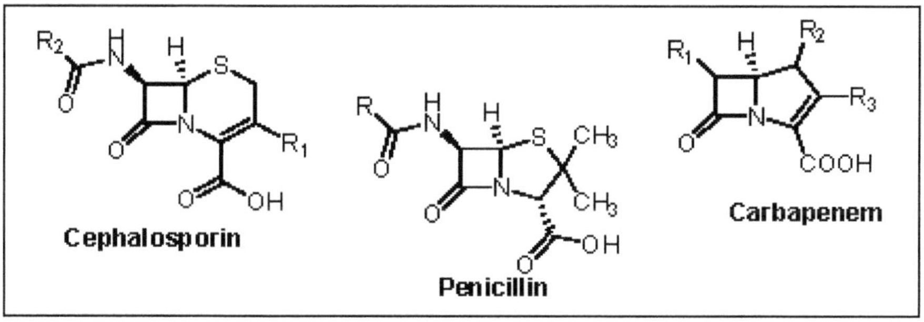

Figure 11.2: Structures of Antibiotics Obtained from Microorganisms.

of 51 amino acids. It plays a prime role in metabolism of fats and carbohydrate in the human body. *E. coli* as well as *Saccharomyces cerevisiae* has been utilized for production of recombinant human insulin for therapeutic use in human. *S. cerevisiae* is the ideal and significant yeast for profitable generation of insulin, although a few different species of yeast are also accessible for insulin production *e.g. Hansenulla polymorpha, Pichia pastoris etc.* These species grow quickly, extremely simple to deal with and acquiescent to a variety of genetic manipulations like bacteria *E. coli* (Baeshen, 2014).

Source of Polysaccharides

A novel class of microbial products is microbial polysaccharides which are of great interest in pharmaceuticals. These products are mainly used in pharmaceuticals, biomedical and food industries and are produced by sugar fermentation of some microorganisms such as *Spingomonas paucimobilis, Xanthomonas campestris* and *Leuconostoc mesentroides*. Among all, some of the polysaccharides such as dextran is

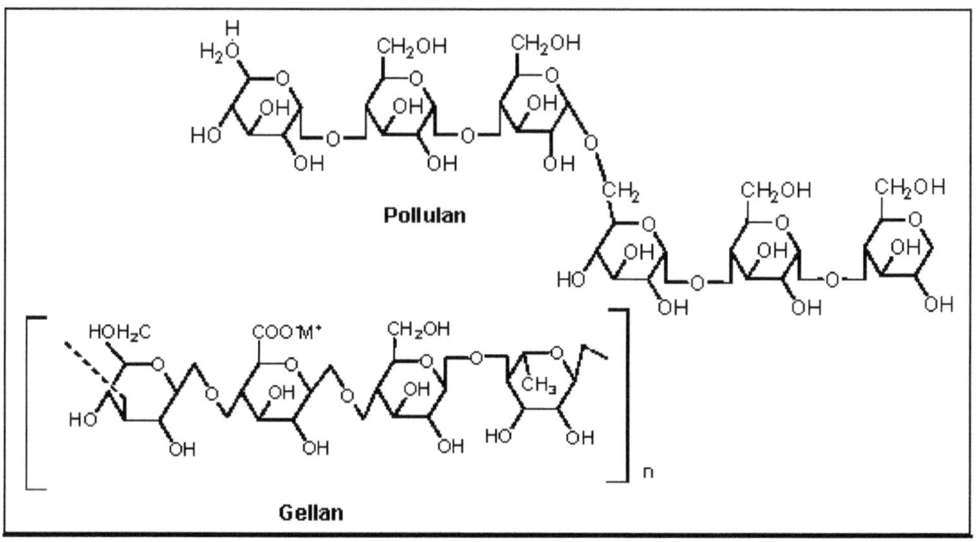

Figure 11.3: Structures of some Polysaccharides Obtained from Microorganisms.

obtained by lactic acid bacteria. Dextran-derived oligosaccharides have been shown to possess prebiotic effects (Meybodi, 2015). Xanthan is obtained from *Xanthomonas campestris* which is mainly used as a stabilizer, thickener and emulsifier. Pollulan is obtained by *Aureobasidium pullulans* and it can be used in denture adhesive and in sugar coated pharmaceutical composition such as granules, pills, tooth powders, shampoos, cosmetics and lotions, tablets, non-toxic conjugates for vaccines and sustained release formulations due to its great transparent film-foaming ability, water solubility, moisture absorptivity and tackiness properties. Gellan is obtained by *Pseudomonas elodea* used as an emulsion based gels and suspending agent for vitamins, minerals and protein. Curdlan is mainly obtained by *Alcaligenes faecais* which is found to exhibit a linear effect on reducing oil and moisture transfer. Levan, obtained by *Zymomonas mobilis* can be hydrolysed by gastric acid and used as a prebiotic. These substances are mainly applicable to act as a gelling and/or flocculating agent which modify the rheology properties and enhance the emulsion stability in food, pharmaceutical and biomedical industries (Singh, 2008).

Source of Pigments

There are two major sources of natural pigments such as plants and microorganisms. But the major drawbacks of the natural pigments obtained from plants are: low water solubility, heat or adverse pH, instability against light and its availability is not regular throughout the year. On the other hand, the pigments obtained from microorganisms are stable and easily available due to cultivation technology. Hence, the microbial pigments are of great interest as compared to natural pigments. Microbial pigments have numerous beneficial properties like antiproliferative, anticancer, antibiotic, immunosuppressive, biodegradability *etc.* By using fermentation technology some microbes *i.e.* yeast, fungi, bacteria, mould *etc.* have been employed for the production of variety of pigments in industries. There are some important pigments which are produced by microorganisms such as: Beta-carotene is produced from *Rhodococcus maris* which is utilized for the cure of a variety of ailments like erythropoietic protoporphyria. It can be utilized to decrease the risk of breast cancer and is also used as an antioxidant. Canthaxanthin is obtained by *Bradyrhizobium* sp. and *Haloferax alexandrine* which is mainly used in pharmaceutical preparations. Indigoidine, obtained from *Corynebacterium insidiosum,* is used in protection from oxidative stress. Prodigiosin, produced from *Alteromonas rubra, Serratia rubidaea, Streptoverticillium rubrireticuli, Rugamonas rubra, Vibrio gaogenes* and *Serratia marcescens* is used as an anticancer, immunosuppressant, antifungal, algicidal, antiproliferative and was observed to be an active constituent for treatment and prevention of diabetes mellitus (Venil, 2013). It has also been evaluated in the standard 60 cell line panels of human tumor cells which are derived from brain, renal, colon, ovarian, lung cancer, leukemia and melanoma. Pyocyanin is obtained by *Pseudomonas aeruginosa* and is utilized in reducing local inammation, oxidative metabolism. Violacein is obtained by *Chromobacterium violaceum, Janthinobacterium lividum* displayed anticancer, antitumoral, antiparasitary, antiprotozoan, antiviral, antibacterial and antioxidant activities. Zeaxanthin, produced from *Paracoccus zeaxanthinifaciens Flavobacterium sp.,* and *Staphylococcus aureus* is employed in the treatment of various diseases which chiefly affects the eyes. Xanthomonadin

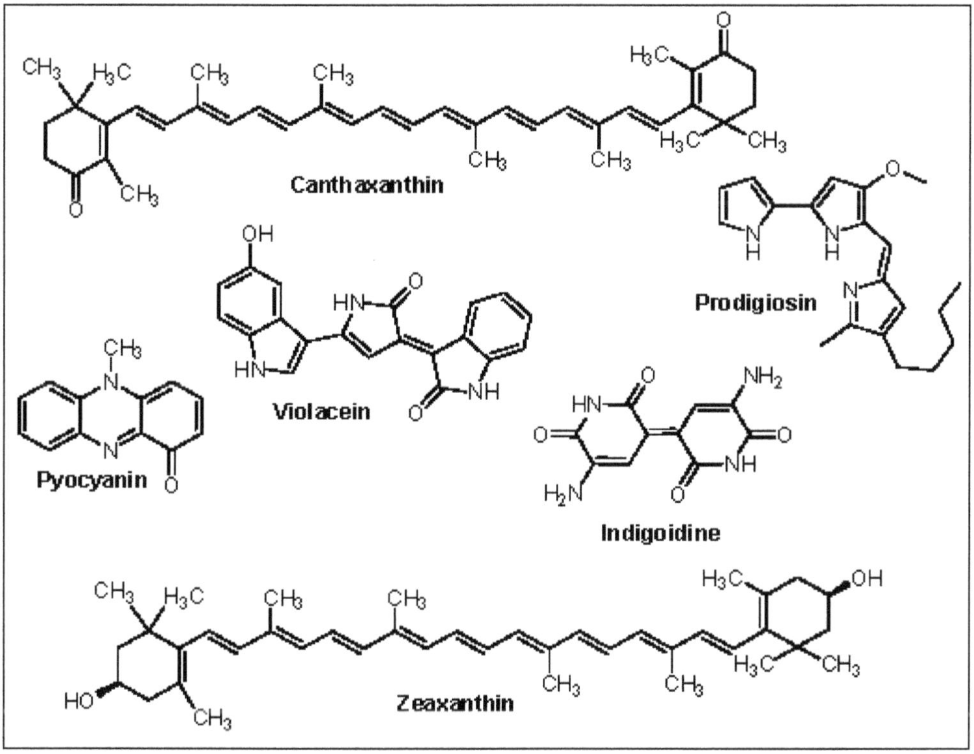

Figure 11.4: Structures of Various Pigments obtained from Microorganisms.

is obtained by *Xanthomonas oryzae* which has been utilized as chemotaxonomic and diagnostic markers. Several polysaccharides such as anthocyanins are water soluble, flavonoid pigments which showed an extensive variety of activities that affect positively the health properties and modulate immune response, reduce inflammatory disorder, decrease the risk of cancer and antioxidant activity (Kumar, 2015).

Source of Vaccine

Edward Jenner and Louis pastuer established the basic principle for the preparation of microbial vaccines *i.e.* isolation, inactivation and injection of causative infectious agent. This principle guided the development of number of vaccines in the last centuary. Now days, 20 vaccines have been licensed throughout the world. These vaccines have been proved to be very beneficial for the treatment of various diseases which were considered to be lethal just a century ago. Vaccines are the biochemical or biological formulations that improve immunity to a particular disease and are used for therapy of various infectious diseases. They have been proved as promising weapon against cancer and autoimmune diseases. For the treatment of various infections caused by *Staphylococcus aureus, Neisseria meningitidis*, and *Streptococcus pneumonia* some vaccines containing multivalent polysaccharides have been developed. After the introduction of vaccines small pox have been eradicated in 1977 (Verdier, 2002). Each and every existing vaccines belong to one of the following types (a) subunit natural (*e.g.* chemically inactivated bacterial toxin-based vaccine against *Clostridium tetani*) (b) whole microorganism killed (*e.g.* Rabies vaccine), (c) whole microorganism attenuated (Bacillus Calmette Guerin (BCG) vaccine against *Mycobacterium tuberculosis*), and subunit polysaccharide glycoconjugate (*e.g.* pneumococcal glycoconjugate vaccine). Moreover, due to viability of quick production and comparatively less cost of the various vaccines containing evolved promoters, cytokines, immunostimulatory sequences, *etc.*, for comparison testing, DNA vaccines are of particular importance (Noyes, 2015). By DNA shuffling, viral vaccine vectors can be improved to obtain required properties of stability, expression level and tropism. There is an incredible opportunity for DNA shuffling as a device for enhancing the effectiveness in addition to triumph rate of the production of new entire organism, bacterial, viral and recombinant protein vaccines (Patten, 2008). DNA shuffling process can be applied to pathogenic viruses followed by their selection for desirable attenuation characteristics while retaining the property of immunogenicity required for a vaccine (Kaushik, 2008).

Source of Enzyme

Enzymatic procedures in the chemical industry depend mainly on cost competitiveness. Due to this, the utilization of microbial enzymes in different fields *e.g.* industrial and pharmaceutical have expanded to a great extent during the previous few years. Various medically important enzymes inhibitors having high catalytic efficiency, greater affinity and specificity have been isolated from microbes which are consistent and economically viable. The most important enzymes are: Lipase, obtained by *Aspergillus niger, Aspergillus oryzae, Aspergillus avus,* and *Candida antarctica* used to treat disorders of the pancreas, digestive aids, dyspepsia, gastrointestinal disturbances, malignant tumor, cutaneous manifestations of digestive allergies and various other such type of infections (Zaidi, 2014). Yeast lipases are used in herbicides and can also be employed to catalyze butanolysis in anhydrous solvents to obtain enantiopure 2-chloro-and 2-bromo-propionic acids in various pharmaceuticals. Protease, produced from *Aspergillus niger, Aspergillus avus, Bacillus subtilis, Pseudomonas* sp., *Serraita* sp., and *Streptomyces* sp., is displayed as an antibacterial agents and extensively utilized in the preparation of chiral alcohols, carboxylic acids, amines or epoxides along with synthesis of peptides in organic solvents (Adrio, 2014). Amylase is obtained from *Aspergillus sp., Bacillus subtilis, B. licheniformi, B. amyloilquifaciens* and *Streptomyces* sp. Amylases exhibited remarkable application in a wide number of industrial processes such as fermentation, food and pharmaceutical industries which is helpful in easy digestion. By inhibiting the amylases present in gastrointestinal tract the degradation of poly, oligo, and disaccharides into glucose can be inhibited, which is the essential therapeutic strategy for the therapy of type 2 diabetes. Amylase inhibitors play an important role in controlling obesity and can also be useful for weight loss because some versions of amylase inhibitors exhibit great potential to reduce carbohydrate absorption in humans and treatment of rumen acidosis (Jayaraj, 2013). Chitinase, obtained by *Streptococcus pneumonia, Bacillus anthracis,* and *Clostridium perfringens* has showed

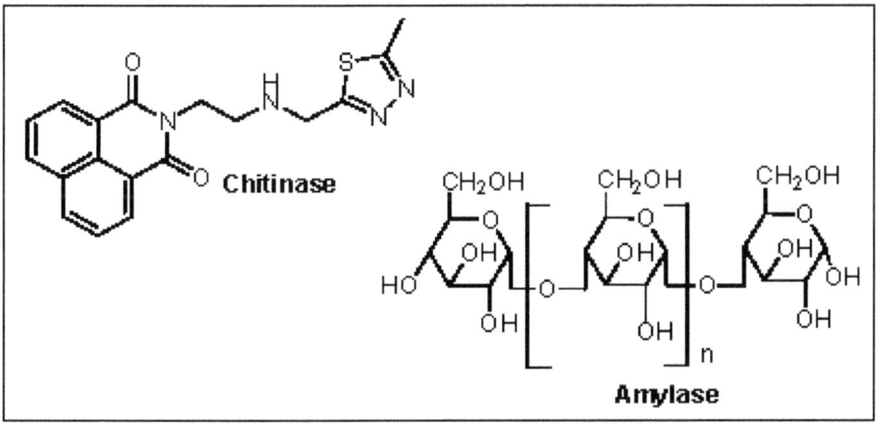

Figure 11.5: Structures of some Enzymes Obtained from Microorganisms.

great antimicrobial activity and utilized for the treatment of a number of infections. It also exhibits activity against new drug-resistant bacterial strains. Tyrosinase is obtained by *Streptomyces glausescens* and *Erwinia herbicola* which is utilized for the treatment of cancer and parkinson's disease and this enzyme is very important for melanin biosynthesis (Gupta, 2014). Melanin absorbs UV rays of sunlight and thus plays a significant role in protecting the skin from dreadful effect of UV light. It also removes reactive oxygen species. As per literature survey, numerous enzymes have been found to possess great potential activities against various diseases namely cardiovascular disease, antiinflammatory, antioxidant, skin ulcer, anticoagulant, antiviral and gout. Metabolic and digestive enzymes either alone or in association with different remedies are utilized for the treatment of numerous diseases *i.e.* cardiovascular diseases, digestive disorders, skin ulcers, leukemia, celiac disease, pompe's disease, parkinson's disease, fabry's disease, inflammation and pancreatic disorders *etc.* They are additionally utilized in biochemical investigation, monitoring and diagnosis of various alarming diseases (Mane, 2015).

Source of Antioxidants

It has been evident in last few decades that the free radicals serves as causative agents for occurrence of different diseases such as cardiovascular diseases, diabetes, autoimmune disorders, cancer, aging and neurodegenerative diseases *etc.* (Gupta, 2013). Since 1980s, the isolation of microbial antioxidants became center of attention for many researchers. A connection between microorganisms and antioxidants has been established through various studies. The various toxic oxidative reactions are catalysed by free radicals which causes the production of toxic lipid peroxides. The microbes belonging to probiotic group protect the body from the dangerous free radicals generally connected with alzheimer's disease, cancer, diabetes, cardiovascular disease and age-related functional decline. Due to toxins, antibiotics, sugar, stress, ageing and other chemicals, the level of hazardous free radicals augment in the body. The microorganism *Streptococcus thermophilus* can be used for the protection of body from these free radicals because it possesses the powerful antioxidant activity. Malnutrition because of short term fasting can be recovered by the probiotic bacteria namely *S. thermophilus* and it also diminishes the related intestinal atrophy in animal studies (Debbab, 2010). Moreover, the anti-oxidant activity of ethyl acetate extracts of numerous *Penicillium* and *Aspergillus species,* including *Rhizopus oryzae,* were examined and it was inferred that the extracts of two *Penicillium* and four *Aspergillus* species were found to protect linoleic acid in better way compared to control. The ansamycin antibiotics include mycotrienin II, trienomycin A, and trienomycin B, which possess antioxidant activity. It was found that among these three compounds mycotrienin II was proved to be most active compound, but was taken as a moderate antioxidant when compared with BHT (butylated hydroxytoluene) (Gupta, 2014).

Source of Vitamin

Vitamins, defined as a fundamental micronutrients needed in minor amount that can be easily synthesized by plants and animals but cannot be produced by mammals, play a vital role in metabolism of all living organisms. Vitamins produced by microbes are widely utilized in pharmaceuticals, foods and cosmetics such as vitamin-E (α-tocopherol), obtained from fresh water microalgae *Euglena gracilis Z.* and marine microalgae *Dunaliella tertiolecta* was found to possess highest antioxidant activity. These are useful in prevention of light induced pathologies of the skin, eyes and degenerative disorders *viz.* cardiovascular diseases, atherosclerosis as well as cancer (Survase, 2006). Vitamin- K is the vital cofactor that is required for the conversion of glutamic acid residues of some precise proteins present in blood as well as bone to g-carboxyglutamic acid which is produced from mutated strain of *Bacillus subtilis*. Ascorbic acid (vitamin C) is obtained by *Cynobacterium sp.* exhibited antioxidant activity. Riboflavin (vitamin B2) is obtained from *Eremothecium ashbyii, Ashbya gossypii, Bacillus sp.* and is helpful in the treatment of migraine and malaria. In the world of science and medicine, Vitamin B12 (Cobalamin) is one of the most fascinating molecules which were originally found to possess anti-pernicious anemia

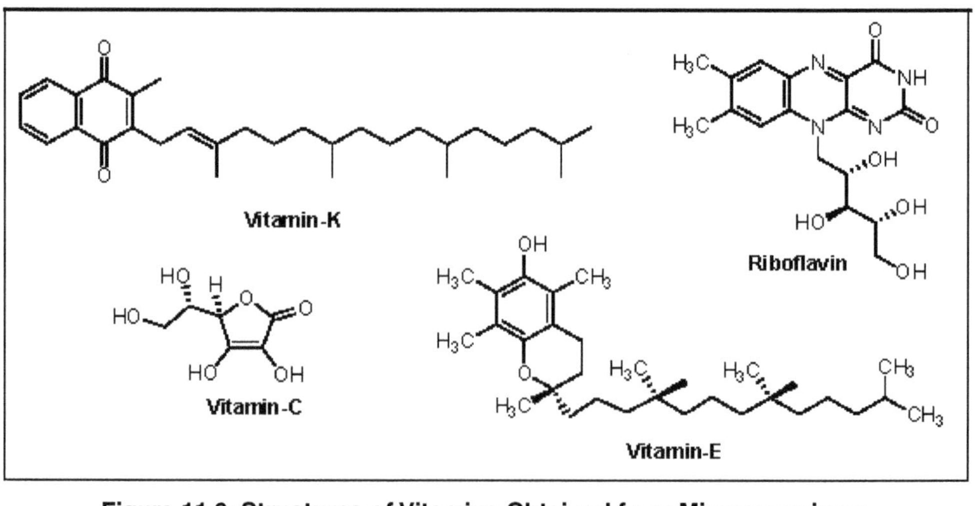

Figure 11.6: Structures of Vitamins Obtained from Microorganisms.

factor in the early 1920s. Vitamin B12 is obtained by *Propionibacterium shermanii, Pseudomonas denitrificans*. Nicotinic acid, obtained by *Rhodococcus rhodochrous* is mainly used in pharmaceutical industry. The same *R. rhodochrous* enzyme can be employed for the synthesis of p-aminobenzonitrile to p-aminobenzoic acid (Acevedo-Rocha, 2018).

Conclusion

The microbes are providing cost effective and valuable pharmaceuticals which are non toxic and are of wide range of uses such as biosurfactants, vitamins, enzymes, antioxidants, *etc*. Keeping in view, its present awareness and future prospective, the continuous growth of new microbial metabolites with modified qualities is being carried out. Cloning and genetic engineering recommend alternative approaches to modify quality and production of secondary metabolites by incorporating suitable biosynthetic pathways. Hence, based on present study, it may be concluded that microbes are the blessings for existing world having an ample range of applications.

REFERENCES

Acevedo-Rocha, C.G., Gronenberg, L.S., Mack, M., Commichau, F.M., Genee, H.J. 2018. Microbial cell factories for the sustainable manufacturing of B vitamins. *Curr. Opin. Biotechnol.* 56: 18-29.

Adrio, J.L., Demain, A. L. 2014. Microbial Enzymes: Tools for Biotechnological Processes. *Biomol.* 4: 117-139.

Ali, M.Z. 2013. An assignment on role of microorganisms in pharmaceutical industry. DOI: 10.13140/RG.2.1.4671.0883.

Baeshen, N.A., Baeshen, M.N., Sheikh, A., Bora, R.S., Ahmed, M.M.M., Ramadan, H.A., Saini, K.S. and Redwan, E.M. 2014. Cell factories for insulin production. *Microbial Cell Factories*. 13: 141.

Debbab, A., Aly, A.H., Lin, W.H., Proksch, P. 2010. Bioactive compounds from marine bacteria and fungi. *Microb. Biotechnol.* 3(5): 544-563.

Du, J., Shao, Z. 2011. Engineering microbial factories for synthesis of value added products. *J. Ind. Microbiol. Biotechnol.* 38(8): 873-890.

Fathabad, *e.g.* 2010. Biosurfactants in pharmaceutical industry. *Am. J. Drug Discov. Dev.* 1-11.

Gupta, C., Prakash, D. and Gupta, S. 2014.Natural useful therapeutic products from microbes. *J. Microbiol. Exp.* 1(1): 30-37.

Gupta, C., Prakash, D., Gupta, S. 2013. Functional foods enhanced with microbial antioxidants. *Academic Journal of Nutrition*. 2(2): 10-18.

Jayaraj, S., Suresh, S., Kadeppagari, R.K. 2013. Amylase inhibitors and their biomedical applications. *Starch/Starke.* 65: 535–542.

Kalkan, S.O. 2013. Aspect of Microbial Sources for Pharmaceuticals; Past, Present and Future. *The Open Conference Proccedings Journal*. In proceedings of ICNP 2013. 4: 258.

Kaushik, D.K., Sehgal, D. 2008. Developing antibacterial vaccines in genomics and proteomics era. *Scand. J. Immunol.*67: 544–552.

Kosaric, N. 2001. Biosurfactants and Their Application for Soil Bioremediation. *Food Technol. Biotechnol.* 39 (4): 295–304.

Kumar, A., Vishwakarma, H.S., Singh, J., Dwivedi, S., Kumar, M. 2015. Microbial pigments: production and their applications in various industries. *Int. j. pharm. chem. biol. sci.*5 (1): 203-212.

Mahajan, G. and Balachandran, L. 2015, Biodiversity in Production of Antibiotics and Other Bioactive Compounds. *Adv. Biochem. Eng. Biotechnol.* 147: 37–58.

Mane, P., Tale, V. 2015. Overview of microbial therapeutic enzymes. *Int. J. Curr. Microbiol. App. Sci.* 4(4): 17-26.

Meybodi, N.M., Mohammadifar, M.A. 2015. Microbial Exopolysaccharides: A Review of Their Function and Application in Food Sciences. *J. Food Qual. Hazards Control*.2: 112-117.

Nabih A Baeshen, Mohammed N Baeshen, Abdullah Sheikh, Roop S Bora, Mohamed Morsi M Ahmed, Hassan A I Ramadan, Kulvinder Singh Saini and Elrashdy M Redwan. Cell factories for insulin production. Microbial Cell Factories 2014, 13: 141. Insulin

Noyes, A., Huffman, B., Berrill, A., Merchant, N., Godavarti, R., Titchener-Hooker, N., Coffman, J., Sunasara, K., Mukhopadhyay, T. 2015. High throughput screening of particle conditioning operations: II. Evaluation of scale-up heuristics with prokaryotically expressed polysaccharide vaccines. ý*Biotechnol. Bioeng.* 112: 1568–1582.

Patten, P.A., Howardt, R.I., Stemmer, W.P.C. 2008. Applications of DNA shuffling to pharmaceuticals and vaccines. *Curr. Opin. Biotechnol.* 8: 724-733.

Rath, K., Mishra, B. and Vuppu, S. 2011. Recent potential usage of surfactant from microbial origin in pharmaceutical and biomedical arena: a perspective. *International Research Journal of Pharmacy*. 2(8): 11-15.

Shetty, P.R., Buddana, S.K., Tatipamula, V.B., Naga, Y.V.V., Ahmad, J. 2014. Production of polypeptide antibiotic from *Streptomyces parvulus* and its antibacterial activity. Brazilian Journal of Microbiology 45(1): 303-312.

Singh, A.P., Singh, R.B. and Mishra, S. 2012. Microbial and biochemical aspects of antibiotic producing microorganisms from soil samples of certain industrial area of India- An overview. *The Open Nutraceuticals Journal*. 5: 107-112.

Singh, P., Kim, Y-J., Zhang, D., Yang, D-C. 2016. Biological synthesis of nanoparticles from plants and microorganisms. Trends in Biotechnology. 34(7): 588-599.

Singh, R.S., Saini, G.K., Kennedy, J.F. 2008. Pollulan: Microbial source, Production and application. ý*Carbohydr. Polym.* 73: 515–531.

Smitha, M.S., Singh, S., Singh, R. 2017. Microbial transformation: a process for chemical alterations. *J. Bacteriol. Mycol. Open Access*. 4(2): 47-51.

Survase, S.A., Bajaj, I.B., Singhal, R.S. 2006. Biotechnological production of vitamins. *Food Technol. Biotechnol.* 44 (3): 381-396.

Venil, C.K., Zakaria, Z.A., Ahmad, W.A. 2013. Bacterial pigments and their applications. *Process Biochem.* 48: 1065-1079.

Verdier, F. 2002. Non-clinical vaccine safety assessment. *Toxicol.* 174: 37–43.

Zaidi, K.U., Ali, A.S., Ali, S.A., Naaz, I. 2014. Microbial Tyrosinases: Promising Enzymes for Pharmaceutical, Food Bioprocessing, and Environmental Industry. *Biochem. Res. Int.* http: //dx.doi.org/10.1155/2014/854687.

Chapter 12

Gut Microbiome as a Determinant of Health and Disease: An Appraisal

Priyanka Sharma

Department of Biotechnology,
Goswami Ganesh Dutta Sanatan Dharma College, Chandigarh – 160 030
E-mail: drpriyankamailbox@gmail.com

ABSTRACT

Gut microbiome represents the largest microbial ecosystem inside humans. It has enormous diversity and functional capacity. It plays a crucial role in metabolism, maintains the homeostasis of the immune system and affects the physiology of the host in numerous ways. There are several illustrations to state that the gut microbiome determines health and diseased phenotype of the host. Researchers have even dissected specific microbiome signature of some diseases. This highlights the importance of studies that analyze the host response provoked by modulations in the gut microbiome size and composition. The present review analyzes the implications of gut microbiome perturbations for health and disease and also present therapeutic potential of restoration of gut microbiome for dysbiosis related health disorders.

Keywords: *Gut microbiome, Microbial ecosystem, Homeostasis, Dysbiosis, Therapeutic potential.*

Introduction

Human gut is microbiologically highly specialized and metabolically most active region of the human body. GI tract is inhabited by several distinct microbiological communities localized at various distinct niches in the gut ecosystem. They are together referred to as "gut microbiome". It represents a dynamic microbiological system co-evolving in the GI tract as an integral part of human body. Around 100 trillion microorganisms inhabit the human gut symbiotically. They aid in the normal gut functioning, digests food and promote host immunity in exchange for constant supply of food. (Chow *et al.*, 2010; Shreiner *et al.*, 2015).

Evidences have been found that human microbiome is involved in metabolic functions like fermentation and absorption of carbohydrates which cannot be digested by humans and ultimately, energy harvesting from digested food. It also aids in the storage of energy harvested. For instance, the capacity of digesting xyloglucans present in common vegetables like lettuce and onions has been found to be a trait of certain rare species of Bacteroides residing as a component of gut microbiome in human body (Clemente *et al.*, 2012; Shreiner *et al.*, 2015).

Host health or diseased state has been found to be in sync with the complex interplay between the human gastrointestinal tract and its indigenous microbiota. The structure and metabolome of the gut microbiota has been found to exert profound effects on most physiological functions of the body. Gut microbiome is metabolically active and the metabolites of the gut microbiome have been shown to exert tremendous effect on the signalling of neural and immune system of the body. Gut-associated microorganisms have been found to aid in the liberation of short chain fatty acids (SCFA) from the indigestible dietary fibres. SCFA serves as important energy source for intestinal epithelium and are known to play a critical role in modulating immune responses in the gut (Shreiner *et al.*, 2015; Sommer *et al.*, 2017).

Interactions between gut microbiome and host immune system are complex and bidirectional. Gut microbiota provides signals to the immune system for maturation of immune cells. In vivo studies have demonstrated that gut microbiome secretes a wide range of biologically important metabolites like Histamines which have immunoregulatory effects. These metabolites strengthen the intestinal mucosal immune system for maintaining intestinal integrity (Clemente *et al.*, 2012; Sommer *et al.*, 2017).

Gut microbiome has been proven to influence immune homeostasis. The human gut ecosystem comprises of a discrete community of commensal, symbiotic and pathogenic microorganisms. Immune system of a healthy host is capable of distinguishing the commensal microorganisms of the gut microbiome from the invading pathogenic microorganisms. This ability of the immune system is, in turn, attributed to the structure and metabolome of gut microbiome. The balance between the immune tolerance and inflammation is the regulated by the interactions between the gut microbiota and the host immune system involving various signalling pathways and metabolites. The ability of the host body to sustain beneficial microorganisms inhabiting the gut and resist the opportunistic and pathogenic microflora contributes to the host health and the pathogenesis of any disorder.

Gut microbiota in a eubiotic status in the numerous microhabitats along the length of GI tract plays a crucial role in the well being of its host and significantly reduce the risk of the diseases (Schmidt *et al.*, 2018). It has been demonstrated that the modulations in the population size and composition of the microflora residing the gut has direct impact on the health of its host. A wide range of gastroenterologic diseases (like diabetes, obesity, Crohn's disease, Irritable bowel disease, *etc.*), autoimmune diseases, asthma and other allergic disorders have been attributed to disturbed host-microbiome interactions (O'Hara and Shanahan, 2006; Clemente *et al.*, 2012).

Several studies analyzing the modifications in the structure and function of gut microflora in conjunction of the host health led to the development of various strategies for diagnosis, cure and prevention of various health disorders. The present paper reviews the effects of composition of gut microbiome and interaction between the host and its gut microbiome as a determinant of health and diseased state of the body.

Healthy Gut Microbiota

Human microbiome constitutes both prokaryotic and eukaryotic microbes. Gut microbiome is composed of complex, interactive, multikingdom, multispecies microbial communities. Bacteria, archaea, viruses and eukaryotic microbes have been isolated, identified and characterized from the gut microbiome (Shreiner *et al.*, 2015) but the gut microbiome has been found to be dominated by bacterial members; specially, by the members of division Bacteroidetes and Firmicutes (Turnbaugh *et al.*, 2006; Clemente *et al.*, 2012).

Although there exist a huge variability among the constituent members of the gut microbiome, it can be mostly classified into three variants or 'enterotypes' based on dominant genera. These three 'enterotypes' are Bacteroides, Prevotella and Ruminococcus (Arumugam *et al.*, 2011). These microorganisms have tremendous effect on physiological state of the body.

So far, the bacterial partners of the gut microbiome community have been studied majorly and eukaryotic and viral members have not been much explored. The facts that most of the viral sequences identified so far were novel and our nucleic acid extraction methods are biased towards encapsulated viruses only, clearly suggests that the characterization of the human virome is at its initial stages (Clemente *et al.*, 2012; Schmidt *et al.*, 2018).

Development and Establishment of Gut Microbiome

The gut also known as, gastrointestinal tract (GI tract or GIT)/Alimentary canal/Digestive tract harbour a complex community of microbial life. The human GI tract is absolutely sterile at the time of infant birth. The gut microbiome is established immediately after birth. It co-evolves and adapts throughout the life of the host and even takes over the host body when it dies. Immediately after birth, infants are exposed to a myriad of microorganisms from surrounding environments and are colonized by the microbes that they first encounter. It was demonstrated that the early microbial colonizers are generally aerotolerant microbes because the gut initially contains oxygen, but later on, they are replaced by anaerobic microorganisms that are typical of the adult gut microbiota (Palmer *et al.*, 2007). These initial colonizers are notably replaced by phylotypes specific to the infant and distinct from maternal microbiome by 11 months of age (Vaishampayan *et al.*, 2010). This marks the beginning of the convergence of infant's gut microbiome toward an adult like microbiota. It takes about 2.5 years for the infant microbiome to fully resemble the adult microbiome (Palmer *et al.*, 2007; Koenig *et al.*, 2011). There are major shifts in taxonomic composition with life events like antibiotic use and the introduction of solid foods. After that the phylogenetic diversity gradually increases over time owing to ongoing exposure to parents, dietary habits and other environmental sources depending upon the lifestyle (Koenig *et al.*, 2011). Adult microbiome has been found to be mostly dominated by Bacteroidetes and Firmiculates. Once the microbiome has reached its maturity, it is known to remain mostly stable throughout life in healthy individuals (Clemente *et al.*, 2012).

The nature of the gut microbiome and hence, also the immunity of an infant is greatly influenced by its mode of delivery. The Infants who have born vaginally harbour microbiome resembling that of their mother's vagina. On another hand, the young ones born by caesarean section have microbiota characteristic of skin, *i.e.*, it is found to be dominated by taxa such as *Staphylococcus* and *Propionibacterium* species (Dominguez-Bello *et al.*, 2010). Lower bacterial cell counts were observed in the faecal samples of the children born by caesarean section and a higher number of antibody secreting cells were found in their body (Huurre *et al.*, 2008; Adlerberth and Wold, 2009; Dominguez-Bello *et al.*, 2010).

Every individual has its own unique microbiota. Gut microbiome is highly diverse even among the members of the same species. Various studies have been conducted with the object of determining the variability among the human microbiome. Large-scale projects like the Human Genome Project and MetaHIT have deeply dissected the bacterial component of the human microbiome (Turnbaugh *et al.*, 2007; Peterson *et al.*, 2009; Qin *et al.*, 2010). Although, a large variability was displayed in the composition of healthy individual's microbiome; with twins sharing less than 50 per cent of their species-level bacterial taxa and even fewer viral sequences (Reyes *et al.*, 2010), a huge resemblance has been found in the bacterial communities at different body sites which lacked any considerable inter-individual variability (Costello *et al.*, 2009; Grice *et al.*, 2009; Nasidze *et al.*, 2009; Robinson *et al.*, 2010; Ravel *et al.*, 2011).

Gut Microbiome and Genetics

Among the various factors that shape the gut microbiome the primary factor is the host genetics. Specific host genomic sites have been shown to influence the interpersonal variability in the microbiome community (Schmidt *et al.*, 2018).

In spite of all these variabilities, there exists a core share of functionalities among the various members of the gut microbiome community and this forms the basis of metagenomic studies aiming to establish the link between diseases and the microbiota (Clemente *et al.*, 2012).

Temporal Consistency in Gut Microbiome

Numerous studies on the temporal dynamics of the human microbiome have established that diet has profound effects on the nature, structure and composition of gut microbiome. These studies illustrated that composition of microbiome in a healthy host remains stable provided that the variables like diet, lifestyle, diseases and environment are kept constant. Once establishment in a healthy human host, human virome showed minimal variation in more than 95 per cent of the viral sequences over a period of one year (Scanlan and Marchesi, 2008; Costello *et al.*, 2009; Reyes *et al.*, 2010; Caporaso *et al.*, 2011).

Impact of Diet

Dietary interventions brought about significant changes in the microbiota of mice within 24 hours on shifting them from high fat and high sugar western diet to a low fat and plant polysaccharide rich diet. Similar changes were observed in another study on humans when the diet was changed from high fat/low fibre to low fat/high fiber diet. Wu and his coworkers, in 2011, established a correlation between the diet and the enterotypes. It was found that those individuals which ate a diet rich in animal fat were found to have a Bacteroides dominated enterotype, whereas those on a carbohydrate-rich diet were found to have Prevotella-dominated enterotype (Turnbaugh *et al.*, 2009b; Wu *et al.*, 2011).

Effect of Antibiotics

There are compelling evidences suggesting external perturbations like antibiotic overuse can result in dysbiosis. Antibiotic use has devastating effects on the gut microbiome diversity. The extent of the effects of antibiotic on nature of gut microbiome and the particular taxa affected vary among individuals. Some taxa do not restore even months after antibiotic treatment while others might develop antibiotic resistance. The disrupted gut microbiome reduces the resistance of the body to pathogenic microorganisms which may out-compete the commensal microbiota and colonize the gut. This results in the permanent changes to the gut microbiome and varying states of diseases (Sullivan *et al.*, 2001; Jernberg *et al.*, 2007; Dethlefsen *et al.*, 2008; Sommer *et al.*, 2009).

Functions Attributed to Gut Microbiome and Relationship between the Host and its Microbiome

Gut microbiome performs many significant roles in the human body ranging from digestion and absorption of nutrients to promoting health and boosting immunity. As described earlier, human gut harbours a consortium of diverse group of microorganisms. There exists a dynamic interaction between the various constituent members of the gut microbiome. This complex interaction of the bacterial, viral and eukaryotic members of the gut microbiome with one another and also with the host immune system plays a fundamental role in the well-being of the host (Chow *et al.*, 2010; Clemente *et al.*, 2012).

The potential functions of gut microbiota in the host body can be attributed to one of the three mechanisms:

1. **Catabolic/bioconversion pathways catalyzed by gut microbiome**: example, fermentation of starch to short chain fatty acids and lignans to phytoestrogens by the process of hydrolysis; deconjugation of proteins to amines and phenols; reduction of xenobiotic compounds like toxins and allopathic drugs by esterification, *etc.*

2. **Synthetic metabolic activities of the gut microbiome**: example, synthesis of vitamins and cofactors, polymers and bioactive signalling molecules like amines.
3. **Epithelial and immune response developed due to host-microbe interactions**: example, altered mucous expression, responses due to sialation, altered expression of nutrient receptors, development of (+/-) proliferative response; immune response, (+/-) inflammatory response (Bassis *et al.*, 2013; Walker 2016).

Dysbiosis and Disease

Health abnormalities are rarely caused by a single factor. Interplay of different elements result in complex and influential interactions which induce a disease phenotype. Human gut is constantly in contact with the external environment receiving many foreign materials like food, air, microbes and other substances from the environment. The GI tract, being exposed to the several foreign particles is associated with the 'Common Mucosal Immune System' which is bestowed with the responsibility of maintaining homeostasis within the GI tract. The 'Common Mucosal Immune System' comprises of GALT (Gut Associated lymphoid tissue), along with the immune tissues of respiratory and genitourinary tract (Swiatczak and Cohen, 2015; Ishiguro *et al.*, 2018).

It is hypothesized that the common mucosal immune system has developed an intricate association with the gut microbiome over the course of evolution. The host microbiome plays a vital role in the development of the innate and the adaptive immunity (O'Hara and Shanahan, 2006; Ley *et al.*, 2008; Chow *et al.*, 2010). There are compelling evidences to suggest that commensal microorganisms of the gut microbiome are capable of modulating host immune system to promote its own fitness. Gut microbiome can suppress inflammatory responses, promote immunological tolerance and program the adaptive immune system by impacting T-cell differentiation. It is hardly any surprise that the human immune system recognizes its gut microbiome and their metabolites as self, and therefore, does not induce any immune response against them while differentiating them from the intruding pathogenic microorganisms and their compounds and eliminating them through appropriate immune response (Hooper *et al.*, 2003; Lee and Mazmanian, 2010; Lathrop *et al.*, 2011).

Helicobacter pylori colonizes the human gut within first 10 years of life and can exist asymptomatically throughout the host life. Evolutionary studies indicate that it has been associated with human microbiome for at least 58,000 years. There are illustrations to state that *H. pylori* can induce a continuous innate response resulting in incidental protection against other pathogens. Nearly all adults in developing countries, like India, harbor *H. pylori*, but the bacterial population is scarce in the microbiome of Western countries owing to increased antibiotic use. The elevated rates of asthma and other allergic disorders in the west have been linked to the decreased prevalence of *H. pylori* in the gut microbiome. Similarly, many other diseases like adenocarcinoma of the esophagus, gastro-esophageal reflux disease

(GERD), and Barrett's esophagus have also been linked to reduced prevalence of cag A+ strains of *H. pylori* (Banatvala *et al.*, 1993; Brown, 2000; Anderson, 2005; Perry *et al.*, 2006; Linz *et al.*, 2007; Atherton and Blaser, 2009; Blaser, 2011). In another recent study, monkeys infected with *H. pylori* were found resistant to tuberculosis (Perry *et al.*, 2010).

There are evidences that the susceptibility of the host to viral infections is promoted by certain specific configurations of the microbiota. Like *H. pylori*, Herpes virus also has evolutionary beneficial relationship with humans. Herpes virus infection can reduce susceptibility to certain bacterial infections. Similarly, Helminthes have been shown to confer resistance against autoimmune diseases (Barton *et al.*, 2007).

Various studies have demonstrated a disease-related dysbioses in a wide array of health disorders ranging from metabolic to neurological diseases while others emphasized on the post-perturbation resilience as the hallmark of host health (Sommer *et al.*, 2017). A delicate balance is maintained by the common mucosal immune system wherein the immune system selectively eliminates the infectious microorganisms but at the same time, ensures immune tolerance to the healthy gut microbiota.

Many diseases have been linked to specific microbiome signature. Recently proposed microbiome "Global Positioning System" distinguish the healthy and diseased state of the host based on the multi-omic readouts (Gilbert *et al.*, 2016). Many researchers have described the profound effects caused by gut microbiome imbalances on the health of its host. Some researches even highlighted the underlying molecular mechanisms linking bacterial, viral and eukaryotic communities of the gut microbiome with health abnormalities (Table 12.1).

Tools and Techniques to Study Gut Microbiome

Several techniques are used to study the signalling between the gut microbiome and the host's body that can trigger alterations in host function. These techniques can be broadly classified into two; analogous to anatomy (those that evaluate the community structure and composition of the microorganisms of the gut microbiome) and analogous to physiology (those that provide the insights into the functional potential). The Human microbiome project provided the advances like the high throughput sequencing technique for nucleic acids; which allows the direct DNA sequencing of the entire microbiome and is likely to promote the understanding of the role of indigenous microbiota in health and its disorders. The *In situ* microbial functional analysis have been carried out by metatranscriptomic analysis (sequence analysis of expressed genes by reverse transcriptase) which quantitatively assays the RNA transcripts present in a microbiome and provides the knowledge of the microbial metagenome that is being expressed at a particular time under given conditions. Mass spectrometry techniques have been successfully applied to quantitate the relative abundance of proteins and other microbial metabolites in a given microbiome. (Lagier *et al.*, 2012; Di *et al.*, 2013; Aleen-Vercoe, 2013; Sommer, 2015).

Table 12.1: Health Disorders and the Associated Microbiome

Health Disorder	*Cause*	*Clinical Manifestations*	*Microbiome Condition Contributing to Disease*	*Microbiome Condition Cntributing to Health*	*References*
Crohn's disease	Unknown etiology involving interaction of virome, genome and microbiota	Inflammatory response of intestinal mucosa	1. Decline in the population size of Firmicutes and B. Fragilis. 2. Reduced diversity, greater abundances of Enterobacteriaceae, Pasteurellaceae, Fusobacteriaceae, Neisseriaceae, Veillonellaceae, and Gemellaceae. 3. Reduced abundances of Bifidobacteriaceae, Erysipelotrichaceae, Clostridiales, and Bacteroidales.	Restoration of a diverse microbiota with FMT.	Guslandi *et al.*, 2000; Loftus, 2004; Halfvarson *et al.*, 2006; Sartor, 2006; Anderson *et al.*, 2011; Shreiner *et al.*, 2015
Irritable bowel syndrome (IBS)	Acute gastrointestinal infection	Chronic abdominal pain, flatulence, bloating, and/ or alterations in bowel habits	1. Decreased proportion of *Eubacterium*, *Roseburia*, and especially, *Faecalibacterium.Prausnitzii* 2. Raised Basidiomycota/Ascomycota ratio. 3. Depletion of *Saccharomyces cerevisiae* 4. Increased proportion of *Candida albicans.*	Low FODMAP (fermentable oligosaccharides, disaccharides, monosaccharides and polyols) diet induced microbiome shift that improved symptoms.	Halmos *et al.*, 2014; Shreiner *et al.*, 2015; Sokol *et al.*, 2016; Ishiguro *et al.*, 2018
Autoimmune disorders	Inappropriate function of adaptive immune system	Immune system attacks body's own tissues.	1. Decreased proportion of species belonging to Clostridia clusters XIVa and IV occurs in Multiple sclerosis (MS) 2. Increased rare gut bacterial taxa (Actinobacteria), decreased gut microbial diversity has been observed in patients with rheumatoid arthritis (RA)	Introduction of bacterial polysaccharides from *Bacteroides fragilis* protects against the disease development	Lee and Mazmanian, 2010; Ishiguro *et al.*, 2018

Health Disorder	*Cause*	*Clinical Manifestations*	*Microbiome Condition Contributing to Disease*	*Microbiome Condition Cntributing to Health*	*References*
Lactose maldigestion and intolerance	Decreased production or absence of β-galactosidase	Bloating, gas, cramps and diarrhoea due	—	Ingestion of milk or yoghurt containing live bacteria	Marteau *et al.*, 1990; de Vrese *et al.*, 2001
Antibiotic-associated Diarrhoea	Antimicrobial therapy disrupts the composition of the GI tract microflora hampering its function and also making it susceptible to infection by opportunistic pathogens.	Diarrhoea	Shift in microbiome community.	*Lactobacillus rhamnosus* GG, *S. boulardii*, and *Enterococcus faecium* SF68 proved effective at reducing the incidence of diarrhoea.	Wunderlich *et al.*, 1989; McFarland *et al.*, 1995; Vanderhoof *et al.*, 1999
Clostridium difficile -Intestinal infection and relapse	Toxin A and B produced by *Clostridium*	Diarrhoea and colitis. *C. difficile* spores induce relapses.	1. Probiotic intake was found to reduce the incidence of *C. difficile* toxin-positive patients. 2. Intervention with *Saccharomyces boulardii* an adjunct to the antibiotic therapy prevented relapse in a placebo-controlled trial.	-	McFarland *et al.*, 1994; Plummer *et al.*, 2004
Acute diarrhoea and Gastroenteritis	Rotavirus infection	Severe watery diarrhoea, fever, vomiting. Some may even experience abdominal pain, cramping, nausea, gas and dehydration.		Ingestion of probiotic strains *Bifi dobacterium bifi dum, Lb. rhamnosus* GG, *Streptococcus thermophilus* proved beneficial in curing the condition.	Saavedra *et al.*, 1994; Kaila *et al.*, 1995; Guandalini *et al.*, 2000

Health Disorder	*Cause*	*Clinical Manifestations*	*Microbiome Condition Contributing to Disease*	*Microbiome Condition Cntributing to Health*	*References*
Asthma and Allergy	Th2 asthma, characterized by the presence of allergen-specific Th2 cells that induce production of allergen-specific IgE; secrete cytokines that attract inflammatory eosinophils to the site. The second form of asthma is driven by Th17, which produces pro-inflammatory cytokine IL17	Shortness of breath, chest tightness or pain, coughing or wheezing.	1. Increased count of Proteobacteria in microbiome is observed. 2. Decreased abundance of four bacterial genera, *Lachnospira, Faecalibacterium, Rothia,* and *Veillonella.* 3. Decreased prevalence of *H. Pylori*	hookworm eggs reduces and/or eliminates allergic responses	Banatvala *et al.*, 1993 Anderson, 2005; Feary *et al.*, 2011; Clemente *et al.*, 2012
Obesity	A physiological state provoked by western diet and genetic makeup of an individual.	1. Increased capacity for harvesting energy from food. 2. Low-level inflammation marked by raised level of mast cells, t cells, macrophages and induction of cytokines such as TNF-a, IL-1b, and CCL2	1. Increased population size of hydrogen-utilizing methanogenic Archaea and hydrogen-producing Prevotellaceae. 2. Increased population of Firmicutes and reduction in size of Bacteroidetes. 3. *A. muciniphila* counts decreased in obese and type 2 diabetic mice	Treatment with *A. muciniphila* reversed high-fat diet-induced metabolic disorders, including adipose tissue inflammation, fat mass gain, metabolic endotoxemia, and insulin resistance	Ley *et al.*, 2005; Turnbaugh *et al.*, 2006, 2008, 2009a; Zhang *et al.*, 2009; Ley, 2010; Gregor and Hotamisligil, 2011; Clemente *et al.*, 2012; Dao *et al.*, 2016
Diabetes	Development of insulin resistance and/or Low to little insulin production	Excessive thirst or urination, hunger, fatigue, weight loss, blurred vision.	1. Greater abundance of *Bacteroides ovatus* and firmicute strain CO19 and reduced diversity of microbiome over time. 2. Depletion of some butyrate-producing bacteria and increased proportion of opportunistic pathogens.	-	Giongo *et al.*, 2011; Ishiguro *et al.*, 2018

Restoring Gut Microbiota

Microbiota Transplantation

Restoring gut microbiome by transfer of foreign microbiota into the host's gut has proven to be a valuable tool in promoting health in certain dysbiosis caused diseases. Several studies showed that certain health abnormalities like obesity can be reverted back to the healthy state by transplantation of microbiota from healthy to the unhealthy person (Turnbaugh *et al.*, 2006; Wen *et al.*, 2008; Bakken, 2009; Vijay-Kumar *et al.*, 2010).

Faecal microbiota transplantation (FMT) also known as "Stool transplantation" is defined as the transfer of stool from a healthy donor to the GI tract of the patient, in order to promote health by establishing desired microbiome state. FMT proved more efficient than traditional antibiotics in the treatment of *Clostridium difficile* infection (RCDI). FMT resulted in an increased proportion of alpha diversity in recipient's microbiome and a gradual shift in the microbiome structure towards donor composition (Seekatz *et al.*, 2014).

A study involving transplantation of gut microbiome from a healthy donor to a *Clostridium difficile*-associated disease (CDAD) patient demonstrated a shift in the gut microbiome composition of the recipient from a depleted Firmicutes and Bacteroidetes-deficient configuration to a healthy microbiome composition resembling the donor, *i.e.*, dominated by Bacteroides sp. The imbalance in the gut microbial community, when corrected, also resulted in the gradual disappearance of the all the symptoms associated with CDAD in the patient. (Khoruts *et al.*, 2010).

Nutritional Supplements (Probiotic, Prebiotic and Synbiotic Approach)

Probiotics are the live non pathogenic microbial strains that supplement the gut microbiota while the prebiotics are the fermented nutrients that promote the growth of already existing gut microbiome. Synbiotic, on the other hand, is a complex of non pathogenic microbe/s along with prebiotic used by these microbes to replace a missing function in a microbiome. Application of dietary intervention has been studied in numerous health disorders. In a recent study, it was found that fiber rich diet supplemented with prebiotics resulted in a shift in the gut microbiome towards increased Prevotella/Bacteroides ratios and contributed to weight loss in obese Children (Zhang *et al.*, 2015). Another study showed that the treatment with inulin-type fructans triggered an increase in *Bifidobacterium* and *Anaerostipes* spp. (Vandeputte *et al.*, 2017). Oral administration of *Lactobacillus reuteri* GMNL-263 has been found to improve insulin resistance and ameliorate hepatic steatosis in high-fructose-fed rats (Jain and Sharma, 2012; Hussey and Bergman, 2014). In order to promote beneficial microflora in the human microbiome, a complete redesigning of diet (referred to as "Nutritional therapy") could be very promising.

Surgical Intervention

Bariatric surgery, specifically the Roux-en-Y gastric bypass, has been found to be a successful method of weight loss in obese patients. Surgical interventions involved in the method resulted in changes in microbiome composition and also induced physiological changes. Post-operative human gut showed enrichment for bacteria in the phylum Gammaproteobacteria and decreased proportion of Firmicutes and Bacteroidetes (Hussey and Bergman, 2014).

Food Habits

Food habits of an individual have profound effect on the diversity of its gut microbiome. Studies have revealed that the mammals dependent on plant based diet tend to harbor greater microbial diversity than those on carnivorous diet. In a study it was demonstrated that switching a lean mouse from normal to high fat diet resulted in an increase in the population of a specific class of bacteria belonging to Firmiculates phyla and induced obesity. The condition reverted when the mice was switched back to the normal diet (Hussey and Bergman, 2014).

Future Aspects and Challenges

The complexity of disease model has raised multi-folds over the past few decades from 'Koch's postulate of pathogenicity' proposed in 1890 to the present day demonstrations of multifaceted interactions among the members of the microbiome community, its effects on microbial metabolome and its influence on the host physiology. Future treatments that target gut microbiome focuses on the assessment of anatomical and physiological aspects of gut microbiome with precision and to establish a direct link between gut microbiome and various health conditions.

Elucidating the gut microbiome variability among the individuals of the same species awaits extensive research on the geographically diverse populations to know the effect of various host extrinsic factors (such as medication, diet, physical activity and cultural habits) and intrinsic factors (such as genetics, sex, innate and adaptive immunity) before implicating any individual component for any health disorder (Schmidt *et al.*, 2018). A deeper insight into the role that the gut microbiome plays in the health, the knowledge of the structure and composition of the various taxa of microorganisms at distinct microhabitats in the human microbiome and the development of the tools to use specific microorganisms to replace or replenish the gut microbiome and restore health are the thrust areas for the future research in the field.

Machine learning tools should be applied to improve characterization of gut microbiome variability and its different functional profiles. It is also crucial to apply artificial intelligence tools to build predictive models for diagnosis of various Gut microbiome-linked health disorders and its treatment by correction of dysbiosis (Clemente *et al.*, 2012). Integrative studies promoting the understanding of the various contributing elements that provoke the development of a health abnormality and how these influential elements interact to induce the disease phenotype are mandatory to facilitate the prediction and treatment of such diseases.

Conclusion

The immunity, susceptibility and resistance of the body to different health conditions, even the acute and chronic ones, has been attributed to the ability of the host to establish and maintain the structure and function of microbiota at several distinct microhabitats of the GI tract. A closer study on the interaction of the common mucosal immune system with the host microbiome and the understanding of structure and metabolome of the gut microbiome suggests that this knowledge could be developed as a potent tool in determining the presence of a pathogenic disorder in host and also in preventing it's occurrence by developing body's immunity. It has also been suggested that the focus of the various attempts made to correct dysbiosis should be kept on the entire microbial community (bacterial, eukaryotic and viral constituents of gut microbiome) rather the individual component of the gut microbiome. Under the light of growing understanding of healthy gut microbiome structure and the its impact on human physiology, it is imperative to apply machine intelligence to establish the link between the human microbiome, especially gut microbiome (which forms the majority of the human body's microbiome) to pathogenesis of different disorders.

REFERENCES

Adlerberth, I., and Wold, A.E. 2009. Establishment of the gut microbiota in Western infants. *Acta Paediatr*. 98: 229–238.

Allen-Vercoe E. 2013. Bringing the gut microbiota into focus through microbial culture: recent progress and future perspective. *Curr Opin Microbiol*.16: 625-9. doi: 10.1016/j.mib.2013.09.008.

Anderson, C.A., Boucher, G., Lees, C.W., Franke, A., D'Amato, M., Taylor, K.D., Lee, J.C., Goyette, P., Imielinski, M., Latiano, A., *et al.*, 2011. Meta-analysis identifies 29 additional ulcerative colitis risk loci, increasing the number of confirmed associations to 47. *Nat. Genet*. 43: 246–252.

Anderson, H.R. 2005. Prevalence of asthma. *BMJ* 330: 1037–1038.

Arumugam, M., Raes, J., Pelletier, E., Le Paslier, D., Yamada, T., Mende, D.R., Fernandes, G.R., Tap, J., Bruls, T., Batto, J.M., *et al.*, MetaHIT Consortium. 2011. Enterotypes of the human gut microbiome. *Nature*. 473: 174–180.

Atherton, J.C., and Blaser, M.J. 2009. Coadaptation of *Helicobacter pylori* and humans: ancient history, modern implications. *J. Clin. Invest*. 119: 2475–2487.

Bakken, J.S. 2009. Fecal bacteriotherapy for recurrent *Clostridium difficile* infection. *Anaerobe*. 15: 285–289.

Banatvala, N., Mayo, K., Megraud, F., Jennings, R., Deeks, J.J., and Feldman, R.A. 1993. The cohort effect and Helicobacter pylori. *J. Infect. Dis*. 168: 219–221.

Barton, E.S., White, D.W., Cathelyn, J.S., Brett-McClellan, K.A., Engle, M., Diamond, M.S., Miller, V.L., and Virgin, H.W., 4th. 2007. Herpesvirus latency confers symbiotic protection from bacterial infection. *Nature*. 447: 326–329.

Bassis, C.M., Young, V.B. and Schmidt, T.M. Methods for characterizing microbial communities associated with the human body. In: Fredricks DN, ed. The human microbiota: how microbial communities affect health and disease. Wiley, 2013: 51-74doi: 10.1002/9781118409855.ch2.

Blaser, M. 2011. Antibiotic overuse: Stop the killing of beneficial bacteria. *Nature.* 476: 393–394.

Brown, L.M. 2000. *Helicobacter pylori*: epidemiology and routes of transmission. *Epidemiol. Rev.* 22: 283–297.

Caporaso, J.G., Lauber, C.L., Costello, E.K., Berg-Lyons, D., Gonzalez, A., Stombaugh, J., Knights, D., Gajer, P., Ravel, J., Fierer, N., *et al.*, 2011. Moving pictures of the human microbiome. *Genome Biol.* 12: R50.

Chow, J., Lee, S.M., Shen, Y., Khosravi, A., and Mazmanian, S.K. 2010. Host bacterial symbiosis in health and disease. Adv. Immunol. 107: 243–274.

Clemente, J.C., Ursell, L.K., Parfrey, L.W., and Knight, R. 2012. The impact of the gut microbiota on human health: an integrative view. *Cell. 148*(6): 1258-70.

Costello, E.K., Lauber, C.L., Hamady, M., Fierer, N., Gordon, J.I., and Knight, R. 2009. Bacterial community variation in human body habitats across space and time. Science 326: 1694–1697.

Dao, M.C., *et al.*, 2016. Akkermansia muciniphila and improved metabolic health during a dietary intervention in obesity: relationship with gut microbiome richness and ecology. *Gut.* 65 (3): 426–436.

De Vrese, M., Stegelmann, A., Richter, B., Fenselau, S., Laue, C. and Schrezenmeir, J. 2001. Probiotics-compensation for lactase insufficiency. *American Journal of Clinical Nutrition*, 73: 421S–429S.

Dethlefsen, L., Huse, S., Sogin, M.L., and Relman, D.A. 2008. The pervasive effects of an antibiotic on the human gut microbiota, as revealed by deep 16S rRNA sequencing. *PLoS Biol.* 6: e280.

Di Bella, J.M., Bao, Y., Gloor, G.B., *et al.*, 2013. High throughput sequencing methods and analysis for microbiome research. *J Microbiol Methods.* 95: 401-14. doi: 10.1016/j.mimet.2013.08.011.

Dominguez-Bello, M.G., Costello, E.K., Contreras, M., Magris, M., Hidalgo, G., Fierer, N., and Knight, R. 2010. Delivery mode shapes the acquisition and structure of the initial microbiota across multiple body habitats in newborns. Proc. Natl. Acad. Sci. USA 107, 11971–11975.

Feary, J., Britton, J., and Leonardi-Bee, J. 2011. Atopy and current intestinal parasite infection: a systematic reviewand meta-analysis. *Allergy.* 66: 569–578.

Gilbert, J.A., Quinn, R.A., Debelius, J., Xu, Z.Z., Morton, J., Garg, N., Jansson, J.K., Dorrestein, P.C., and Knight, R. 2016. Microbiome-wide association studies link dynamic microbial consortia to disease. *Nature.* 535: 94–103.

Giongo, A., Gano, K.A., Crabb, D.B., Mukherjee, N., Novelo, L.L., Casella, G., Drew, J.C., Ilonen, J., Knip, M., Hyo¨ ty, H., *et al.*, 2011. Toward defining the autoimmune microbiome for type 1 diabetes. *ISME J.* 5: 82–91.

Gregor, M.F., and Hotamisligil, G.S. 2011. Inflammatory mechanisms in obesity. *Annu. Rev. Immunol.* 29: 415–445.

Grice, E.A., Kong, H.H., Conlan, S., Deming, C.B., Davis, J., Young, A.C., Bouffard, G.G., Blakesley, R.W., Murray, P.R., Green, E.D., *et al.*, NISC Comparative Sequencing Program. 2009. Topographical and temporal diversity of the human skin microbiome. *Science*. 324: 1190–1192.

Guandalini, S., Pensabene, L., Zikri, M.A., Dias, J.A., Casali, L.G., Hoekstra, H., Kolacek, S., Massar, K., Micetic- Turk, D., Papadopoulou, A., de Sousa, J.S., Sandhu, B., Szajewska, H. and Weizman, Z. 2000 *Lactobacillus* GG administered in oral rehydration solution to children with acute diarrhea: a multicentre European trial. *Journal of Pediatrics, Gastroenterology and Nutrition*. 30: 54–60.

Guslandi, M., Mezzi, G., Sorghi, M. and Testoni, P.A. 2000 *Saccharomyces boulardii* in maintenance treatment of Crohn's disease. *Digestive Disease Science*. 45: 1462–1464.

Halfvarson, J., Jess, T., Magnuson, A., Montgomery, S.M., Orholm, M., Tysk, C., Binder, V., and Ja¨ rnerot, G. 2006. Environmental factors in inflammatory bowel disease: a co-twin control study of a Swedish-Danish twin population. *Inflamm. Bowel Dis.* 12: 925–933.

Halmos E.P., Power V.A., Shepherd S.J., *et al.*, 2014. A diet low in FODMAPs reduces symptoms of irritable bowel syndrome. *Gastroenterology*. 146(1): 67–75. e5.

Hooper, L.V., Stappenbeck, T.S., Hong, C.V., and Gordon, J.I. 2003. Angiogenins: a new class of microbicidal proteins involved in innate immunity. *Nat. Immunol.* 4: 269–273.

Hussey, S. and Bergman, M. 2014. The Gut Microbiota and Effects on Metabolism, Eds: Linda M. McManus, Richard N. Mitchell, Pathobiology of Human Disease, Academic Press, pp. 508-526

Huurre, A., Kallioma¨ ki, M., Rautava, S., Rinne, M., Salminen, S., and Isolauri, E. 2008. Mode of delivery - effects on gut microbiota and humoral immunity. *Neonatology*. 93: 236–240.

Ishiguro, E., Haskey, N., Campbell, K. 2018. Gut Microbiota in Health and Disease. In: Gut Microbiota, Academic Press, pp. 57-90.

Jain, P., Sharma, P. 2012. Probiotics and Their Efficacy in Improving Oral Health: A Review. *Journal of Applied Pharmaceutical Science*. 2(11): 151-163

Jernberg, C., Lo¨ fmark, S., Edlund, C., and Jansson, J.K. 2007. Long-term ecological impacts of antibiotic administration on the human intestinal microbiota. *ISME J.* 1: 56–66.

Kaila, M,. Isolauri, E., Saxelin, M., Arvilommi, H. and Vesikari, T. 1995. Viable versus inactivated *Lactobacillus* strain GG in acute rotavirus diarrhoea. *Archives of Diseases in Children*. 72: 51–53.

Khoruts, A., Dicksved, J., Jansson, J.K., and Sadowsky, M.J. 2010. Changes in the composition of the human fecal microbiome after bacteriotherapy for recurrent Clostridium difficile-associated diarrhea. *J. Clin. Gastroenterol*. 44: 354–360.

Koenig, J.E., Spor, A., Scalfone, N., Fricker, A.D., Stombaugh, J., Knight, R., Angenent, L.T., and Ley, R.E. 2011. Succession of microbial consortia in the developing infant gut microbiome. Proc. Natl. Acad. Sci. USA. 108(Suppl 1): 4578–4585. Published online July 28, 2010. 10.1073/pnas.1000081107.

Lagier, J.C., Armougom, F., Million, M., *et al.*, 2012. Microbial culturomics: paradigm shift in the human gut microbiome study. *Clin Microbiol Infect.*18: 1185-93. doi: 10.1111/1469-0691.12023.

Lathrop, S.K., Bloom, S.M., Rao, S.M., Nutsch, K., Lio, C.W., Santacruz, N., Peterson, D.A., Stappenbeck, T.S., and Hsieh, C.S. 2011. Peripheral education of the immune system by colonic commensal microbiota. *Nature*. 478: 250–254.

Lee, Y.K., and Mazmanian, S.K. 2010. Has the microbiota played a critical role in the evolution of the adaptive immune system? *Science*. 330: 1768–1773.

Ley, R.E. 2010. Obesity and the human microbiome. *Curr. Opin. Gastroenterol.* 26: 5–11.

Ley, R.E., Ba ckhed, F., Turnbaugh, P., Lozupone, C.A., Knight, R.D., and Gordon, J.I. 2005. Obesity alters gut microbial ecology. Proc. Natl. Acad. Sci. USA. 102: 11070–11075

Ley, R.E., Lozupone, C.A., Hamady, M., Knight, R., and Gordon, J.I. 2008. Worlds within worlds: evolution of the vertebrate gut microbiota. *Nat. Rev. Microbiol.* 6: 776–788.

Linz, B., Balloux, F., Moodley, Y., Manica, A., Liu, H., Roumagnac, P., Falush, D., Stamer, C., Prugnolle, F., van der Merwe, S.W., *et al.*, 2007. An African origin for the intimate association between humans and Helicobacter pylori. *Nature.* 445: 915–918.

Loftus, E.V., Jr. 2004. Clinical epidemiology of inflammatory bowel disease: Incidence, prevalence, and environmental influences. *Gastroenterology*. 126: 1504–1517.

Marteau, P.R., Flourie, B., Pochart, P., Chastang, C., Desjeux, J.F. and Rambaud, J.C. 1990. Effect of the microbial lactase (EC 3.2.1.23) activity in yoghurt on the intestinal absorption of lactose: an in vivo study in lactase defi cient humans. *British Journal of Nutrition*. 64: 71–79.

McFarland, L.V., Surawicz, C.M., Greenberg, R.N., Elmer, G.W., Moyer, K.A., Melcher, S.A., Bowen, K.E. and Cox, J.L. 1995. Prevention of β-lactam-associated diarrhea by *Saccharomyces boulardii* compared with placebo. *American Journal of Gastroenterology*. 90: 439–448.

McFarland, L.V., Surawicz, C.M., Greenberg, R.N., Fekety, R., Elmer, G.W., Moyer, K.A., Melcher, S.A., Bowen, K.E., Cox, J.L. and Noorani, Z. 1994. A randomized placebo-controlled trial of *Saccharomyces boulardii* in combination with standard antibiotics for *Clostridium diffi cile* disease. *Journal of American Medical Association*. 271: 1913–1918.

Nasidze, I., Li, J., Quinque, D., Tang, K., and Stoneking, M. 2009. Global diversity in the human salivary microbiome. *Genome Res.* 19: 636–643.

O'Hara, A.M., and Shanahan, F. 2006. The gut flora as a forgotten organ. *EMBO Rep.* 7: 688–693.

Palmer, C., Bik, E.M., DiGiulio, D.B., Relman, D.A., and Brown, P.O. 2007. Development of the human infant intestinal microbiota. PLoS Biol. 5, e177.

Perry, S., de Jong, B.C., Solnick, J.V., de la Luz Sanchez, M., Yang, S., Lin, P.L., Hansen, L.M., Talat, N., Hill, P.C., Hussain, R., *et al.*, 2010. Infection with Helicobacter pylori is associated with protection against tuberculosis. *PLoS ONE*. 5: e8804.

Perry, S., de la Luz Sanchez, M., Yang, S., Haggerty, T.D., Hurst, P., Perez- Perez, G., and Parsonnet, J. 2006. Gastroenteritis and transmission of Helicobacter pylori infection in households. *Emerg. Infect. Dis.* 12: 1701–1708.

Peterson, J., Garges, S., Giovanni, M., McInnes, P., Wang, L., Schloss, J.A., Bonazzi, V., McEwen, J.E., Wetterstrand, K.A., Deal, C., *et al.*, NIH HMP Working Group. 2009. *The NIH Human Microbiome Project. Genome Res.* 19: 2317–2323.

Plummer, S., Weaver, M.A., Harris, J.C., Dee, P., and Hunter, J. 2004. *Clostridium difficile* pilot study: effects of probiotic supplementation on the incidence of *C. difficile* diarrhoea. *International Microbiology*. 7: 59–62.

Qin, J., Li, R., Raes, J., Arumugam, M., Burgdorf, K.S., Manichanh, C., Nielsen,T., Pons, N., Levenez, F., Yamada, T., *et al.*, MetaHIT Consortium. 2010. A human gut microbial gene catalogue established by metagenomic sequencing. *Nature*. 464: 59–65.

Ravel, J., Gajer, P., Abdo, Z., Schneider, G.M., Koenig, S.S., McCulle, S.L., Karlebach, S., Gorle, R., Russell, J., Tacket, C.O., *et al.*, 2011. Vaginal micromicrobiome of reproductive-age women. Proc. Natl. Acad. Sci. USA. 108 (Suppl 1): 4680–4687.

Reyes, A., Haynes, M., Hanson, N., Angly, F.E., Heath, A.C., Rohwer, F., and Gordon, J.I. 2010. Viruses in the faecal microbiota of monozygotic twins and their mothers. *Nature*. 466: 334–338.

Robinson, C.J., Bohannan, B.J., and Young, V.B. 2010. From structure to function: the ecology of host-associated microbial communities. *Microbiol. Mol. Biol. Rev.* 74: 453–476.

Saavedra, J.M., Bauman, N.A., Oung, I., Perman, J.A., and Yolken, R.H. 1994. Feeding of *Bifidobacterium bifidum* and *Streptococcus thermophilus* to infants in hospital for prevention of diarrhoea and shedding of rotavirus. *Lancet*. 344: 1046–1049.

Sartor, R.B. 2006. Mechanisms of disease: pathogenesis of Crohn's disease and ulcerative colitis. *Nat. Clin. Pract. Gastroenterol. Hepatol.* 3: 390–407.

Scanlan, P.D., and Marchesi, J.R. 2008. Micro-eukaryotic diversity of the human distal gut microbiota: qualitative assessment using culture-dependent and -independent analysis of faeces. *ISME J.* 2: 1183–1193.

Schmidt, T.S.B., Raes, J., Bork, P. 2018. The Human Gut Microbiome: From Association to Modulation. *Cell.* 172: 1198-1214.

Seekatz, A.M., Aas, J., Gessert, C.E., Rubin, T.A., Saman, D.M., Bakken, J.S., and Young, V.B. 2014. Recovery of the gut microbiome following fecal microbiota transplantation. *MBio*. 5: e00893–e14.

Shreiner, A.B., Kao, J.Y. and Vincent, B.Y. 2015. The gut microbiome in health and in disease. *Curr Opin Gastroenterol*. 31(1): 69–75.

Sokol, H., *et al.*, 2016. Fungal microbiota dysbiosis in IBD. Gut gutjnl–2015-310746.

Sommer, F., Anderson, J.M., Bharti, R., Raes, J., and Rosenstiel, P. 2017. The resilience of the intestinal microbiota influences health and disease. *Nat. Rev. Microbiol.* 15: 630–638.

Sommer, M.O. 2015. Advancing gut microbiome research using cultivation. *Curr Opin Microbiol.* 27: 127-32. doi: 10.1016/j.mib.2015.08.004.

Sommer, M.O., Dantas, G., and Church, G.M. 2009. Functional characterization of the antibiotic resistance reservoir in the human microflora. *Science*. 325: 1128–1131.

Sullivan, A., Edlund, C., and Nord, C.E. 2001. Effect of antimicrobial agents on the ecological balance of human microflora. *Lancet Infect. Dis*. 1: 101–114.

Swiatczak, B., and Cohen, I.R. 2015. Gut feelings of safety: tolerance to the microbiota mediated by innate immune receptors. *Microbiol. Immunol*. 59 (10): 573–585.

Turnbaugh, P.J., Ba¨ ckhed, F., Fulton, L., and Gordon, J.I. 2008. Diet-induced obesity is linked to marked but reversible alterations in the mouse distal gut microbiome. *Cell Host Microbe.* 3: 213–223.

Turnbaugh, P.J., Hamady, M., Yatsunenko, T., Cantarel, B.L., Duncan, A., Ley, R.E., Sogin, M.L., Jones, W.J., Roe, B.A., Affourtit, J.P., *et al.*, 2009a. A core gut microbiome in obese and lean twins. *Nature*. 457: 480–484.

Turnbaugh, P.J., Ley, R.E., Hamady, M., Fraser-Liggett, C.M., Knight, R., and Gordon, J.I. 2007. The human microbiome project. *Nature.* 449: 804–810.

Turnbaugh, P.J., Ley, R.E., Mahowald, M.A., Magrini, V., Mardis, E.R., and Gordon, J.I. 2006. An obesity-associated gut microbiome with increased capacity for energy harvest. *Nature*. 444: 1027–1031.

Turnbaugh, P.J., Ridaura, V.K., Faith, J.J., Rey, F.E., Knight, R., and Gordon, J.I. 2009b. The effect of diet on the human gut microbiome: a metagenomic analysis in humanized gnotobiotic mice. *Sci. Transl. Med*. 1: 6ra14.

Vaishampayan, P.A., Kuehl, J.V., Froula, J.L., Morgan, J.L., Ochman, H., and Francino, M.P. 2010. Comparative metagenomics and population dynamics of the gut microbiota in mother and infant. *Genome Biol. Evol.* 2: 53–66.

Vandeputte, D., Falony, G., Vieira-Silva, S., Wang, J., Sailer, M., Theis, S., Verbeke, K., and Raes, J. 2017. Prebiotic inulin-type fructans induce specific changes in the human gut microbiota. Gut 66, 1968–1974.

Vanderhoof, J.A., Whitney, D.B., Antonson, D.L., Hanner, T.L., Lupo, J.V. and Young, R.J. 1999 *Lactobacillus* GG in the prevention of antibiotic-associated diarrhea in children. *Journal of Pediatrics*. 135: 564–568.

Vijay-Kumar, M., Aitken, J.D., Carvalho, F.A., Cullender, T.C., Mwangi, S., Srinivasan, S., Sitaraman, S.V., Knight, R., Ley, R.E., and Gewirtz, A.T. 2010. Metabolic syndrome and altered gut microbiota in mice lacking Toll like receptor 5. *Science*. 328: 228–231.

Walker, A.W. 2016. Studying the human microbiota. *Adv Exp Med Biol.* 902: 5- 32. doi: 10.1007/978-3-319-31248-4_2.

Wen, L., Ley, R.E., Volchkov, P.Y., Stranges, P.B., Avanesyan, L., Stonebraker, A.C., Hu, C.Y., Wong, F.S., Szot, G.L., Bluestone, J.A., *et al.*, 2008. Innate immunity and intestinal microbiota in the development of Type 1 diabetes. *Nature*. 455: 1109–1113.

Wu, G.D., Chen, J., Hoffmann, C., Bittinger, K., Chen, Y.Y., Keilbaugh, S.A., Bewtra, M., Knights, D., Walters, W.A., Knight, R., *et al.*, 2011. Linking long term dietary patterns with gut microbial enterotypes. *Science*. 334: 105–108.

Wunderlich, P.F., Braun, L., Fumagalli, I., D'Apuzzo, V., Heim, F., Karly, M., Lodi, R., Politta, G., Vonbank, F. and Zeltner, L. 1989. Double-blind report on the efficacy of lactic acid-producing *Enterococcus* SF68 in the prevention of antibiotic associated diarrhoea and in the treatment of acute diarrhoea. *Journal of Internal Medicine and Research*. 17: 333–338.

Zhang, C., Yin, A., Li, H., Wang, R., Wu, G., Shen, J., Zhang, M., Wang, L., Hou, Y., Ouyang, H., *et al.*, 2015. Dietary modulation of gut microbiota contributes to alleviation of both genetic and simple obesity in children. *EBioMedicine*. 2: 968–984.

Zhang, H., DiBaise, J.K., Zuccolo, A., Kudrna, D., Braidotti, M., Yu, Y., Parameswaran, P., Crowell, M.D., Wing, R., Rittmann, B.E., and Krajmalnik-Brown, R. 2009. Human gut microbiota in obesity and after gastric bypass. Proc. Natl. Acad. Sci. USA. 106: 2365–2370.

Chapter 13

Endophytic Microbes as Biological Factories of Bioactive Secondary Metabolites: An Integrative View

Priyanka Sharma[1] *and Pranay Jain*[2]*

[1]*Department of Biotechnology, Goswami Ganesh Dutta S.D. College, Sector 32-C, Chandigarh*
[2]*Department of Biotechnology, UIET, Kurukshetra University, Kurukshetra, Haryana*
**E-mail: drpranayjain@gmail.com*

ABSTRACT

A long history exists pointing to the exploitation of microorganisms by human civilization. Production of secondary metabolitesby the microorganisms represents a vast array ofactive compounds which are both structurally and functionally diversified and that are known to promote its survivability through various ways. Endophytes represent most abundant, poorly investigated and probably the most reliable source of industrially important compounds. They have been recognized as a potential candidate for bioactive metabolite production. The endophytes (both bacterial and fungal) have been explored from the medicinal plants for their therapeutic applications.

Keywords: *Endophytes, Microbial biotechnology, Secondary metabolites, Bioactive compounds.*

Introduction: Secondary Metabolites and its Significance

Secondary metabolites are generally defined as the heterogeneous group of natural metabolic products produced by an organism that are not essential for their growth, development and reproduction (Vining, 1992). These metabolites are produced in the later stages of growth of the organism and are functionally and structurally diversified. The pattern of distribution of secondary metabolites is unique, some metabolites are found in a related group of microorganisms, while some metabolites are found in only one or a few species.

The functions or importance of these compounds to the organism is usually of an ecological nature as they include antibiotics, enzyme inhibitors, antitumour agents, pigments, toxins, effectors of ecological competition and symbiosis, pheromones, immune modulating agents, receptor antagonists, pesticides and growth promoters of animals and plants. They have a key effect on the health as well as commercialization of our society. They are unique for having unusual discrete structures and their synthesis is regulated by amount of available nutrients, growth rate of the microbe, feedback mechanism, enzyme inactivation and stimulation (Stone and Williams, 1992).

Microorganisms as Source of Secondary Metabolites

Microorganisms are relevant in multitude of ways. History of microbiology reveals that many industrially valuable products like foods, beverages, enzymes, vitamins, pesticides and antibiotics were obtained from microorganisms. In eighteenth century, Louis Pasteur's finding that fermentation is brought about by microbes and also that bacteria responsible for wine spoilage during fermentation can be destroyed without ruining the taste of the product (by pasteurization) helped to save the wine and beer manufacturing industry of France. The relevance of microorganisms was becoming evident. The exploitation of microorganisms for valuable metabolites began after the production of first wonder drug 'penicillin' in the nineteenth century. In 1940's, penicillin was discovered by Sir Alexander Fleming but the work went unnoticed until Sir Howard W. Florey and E.B. Chain developed the methods for industrial production of penicillin in England. This led researchers to explore microorganisms for the production of secondary metabolites and marks the beginning of the era of antibiotic research (Demain and Fang, 2000; Aneja, 2003). Another antibiotic, streptomycin was discovered in 1944 by S.A. Waksman which was used in treatment of tuberculosis. It is a broad spectrum microbicidal antibiotic. By 1950, three more antibiotics, chloramphenicol, aureomycin, terramycin were discovered from *Streptomyces* species. The number of antibiotics continued to increase and by the year 1985, there were 7650 antibiotics described. Until 1970s, only two antibiotics, penicillin and cephalosporin, were known. With the introduction of new approaches for screening and isolation of microorganisms, a large number of β-lactam-antibioticswere identified (Wells *et al.*, 1992).

The production of new therapeutic agentsfrom microorganisms has rapidly increased during the last few years with the emergence of new infectious diseases. The prime focus of research for production of antibiotics has been from filamentous microorganisms such as fungi and actinomycetes, which produce secondary metabolites with widely chemically discrete structures. Most of them were isolated from soil. Demain and Fang (2000) discovered that some isolates of *Streptomyces* species were found to produce more than 180 different secondary metabolites.

Among bacteria, genus *Bacillus* produces large number of secondary metabolites. Different strains of *Bacillus* have been found to produce antibiotics such as moenomycins, difficidins, bacillomycins and bacillaenes (Zweerink and Edison, 1987; Patel *et al.*, 1995). *Myxobacterium* is another genus which also produces diverse antibiotics. Dolak *et al* (1983) reported that approximately 80 per cent of the isolated *Myxobacteria* was found to produce secondary metabolites with antibiotic activity, many of them exhibited antifungal activity.

Research efforts have also been focused in finding low molecular weight secondary metabolites with other biological activities from microorganisms parallel to the screening for new antibiotics. Secondary metabolites from microorganisms with activity as enzyme inhibitors, plant growth stimulators, herbicides, insecticides, antihelminthics and immunosuppressants have been obtained. There are two main strategies which have been used during the screening of microbes for the isolation of secondary metabolites by the researchers. The first strategy is to isolate the known secondary metabolites to evaluate the biological activities not identified earlierand the second one is to isolate novel secondary metabolites with biological potential.

Majority of secondary metabolites isolated from the microorganisms have been used in the clinical practice as antibiotics like mithramycin, bleomycin, daunomycin and adriamycin (Table 13.1) (Kieslich, 1986). Other therapeutic applications of secondary metabolites include anabolics, anesthetics, anticoagulants, anti-inflammatories, immunosuppressants (Cyclosporine-A and Tacrolimus), hemolytics, hypocholesterolemics (Statin), and vasodilatories (Bentley, 1997).

Bioactive Compounds from Endophytic Microorganisms

Endophytes are the microorganisms that live asymptomatically inside the plant tissues. Approximately, thirty thousand plants exists on the earth hosting one or more endophytes. Only few plant species have been explored for their endophytic microflora and their potential for human use. Since, all the plant species examined so far are known to harbour at least one endophyte, endophytes represent a rich and diverse source of micro flora left unexplored (Strobel and Daisy, 2003). Extensive colonization of host plants with endophytes out-competes pathogenic organisms and prevents pathogenic organisms from taking hold. Endophytes produce a wide range of antimicrobial and antitumour agents. Endophytes also produce antidiabetic, immunomodulatory, herbicidal, plant protective agents and plant growth promoting agents (Singh *et al.*, 2011).

Current Status of Research on Endophytic Microorganismsin India

Interest in studying endophytes for their medicinal potential has grown tremendously since the discovery of the world's first billion-dollar anticancer compound - Paclitaxel (Taxol) which could be biosynthesized by *Pestalotiopsis microspora,* an endophytic fungus of Himalayan yew (Metz *et al.,* 2000). Several species of genus *Ocimum* have been found to be effective in treating catarrhal bronchitis and bronchial asthma to cancers and diabetes. *Ocimum tenuiflorum* L. expressed both hypoglycemic and antihyperglycemic properties. (Agarwal *et al.,* 1996; Grover *et al.,* 2002).

Table 13.1: Natural Compounds Obtained from Endophytes (Strobel and Daisy, 2003; Rukachaisirikul *et al.*, 2008; Cheng *et al.*, 2012).

Endophytic Microorganism	*Source/Host Plant*	*Natural Product Obtained*
Antibacterial Compounds		
Phomopsis sp.	Garcinia dulcis	Alkaloids
Nodulisporium sp.	Juniperus cedre	Flavonoids
Penicillium sp.	Cerbera manghas	Phenols
Phoma species	*Saurauia scaberrinae*	Phenols
Nodulisporium sp.	*Juniperus cedre*	Steroids
Daphnopsis Americana	Not mentioned	Terpenoids
Daphnopsis Americana	Not mentioned	Terpenoids
Taxus cuspidate	*Periconia* sp.	Terpenoids
Antiviral Compounds		
Cytonaema sp.	Unknown	Cytonic Acids A And B (Novel Human Cytomegalovirus Protease Inhibitors)(*p*-Tridepside Isomers)
Antifungal Compounds		
Phomopsis sp.	*Excoecaria agallocha*	Aliphatic Compound
Chaetomium globosum	*Ginkgo biloba*	Aliphatic Compound
Cladosporium sp	*Quercus variabilis*	Aliphatic Compound
Cryptosporiopsis quercina	*Tripterigeum wilfordii*, a medicinal plant native to Eurasia	Cryptocandin (Peptide) And Cryptocin (Tetramic Acid)
Pseudomonas viridiflava	leaves of many grass species	Ecomycins (Lipopeptides)
Plant-Associated Pseudomonad	Unknown	Pseudomycins (Lipopeptides)
Pestalotiopsis microspora	common rainforest endophyte	Ambuic Acid (Cyclohexenone)
Pestalotiopsis microspora	*Torreya taxifolia*	Pestaloside (An Aromatic DIPPA Glucoside) And Two Pyrones: Pestalopyrone And Hydroxypestalopyrone
Pestalotiopsis microspora	*Taxus brevifolia*	Pestalotiopsins A And B (Two New Caryophyllene Sesquiterpene)
Pestalotiopsis jester	Sepik River area of Papua New Guinea	Jesterone (Cyclohexenone Epoxide) And Hydroxy-Jesterone
Phomopsis sp.	Unknown	Phomopsichalasin (Cytochalasin-Type Compound)
Fusarium sp.	*Selaginella pallescens*	CR377 (A New Pentaketide Antifungal Agent)
Antimalarial Compounds		
Colletotrichum sp.	*Artemisia annua*	Artemisinin
Antiprotozoal Compounds		
Streptomyces sp. strain NRRL 30562	*Kennedia nigricans*	Munumbicins (Peptides)

Endophytic Microorganism	Source/Host Plant	Natural Product Obtained
	Broad Spectrum Antibiotics	
Colletotrichum gloeosporioides	*Artemisia mongolica*	Colletotric Acid
Streptomycete (NRRL 30566)	*Grevillea pteridifolia* (Northern Territory of Australia)	Kakadumycin A
Chaetomium globosum	*Ginkgo biloba*	Alkaloids
Acremonium zeae	Maize	Alkaloids
Cryptosporiopsis sp., *Pezicula* sp.	*Pinus sylvestris* and *Fagus sylvatica*	Peptides
Cryptosporiopsis quercina	*Tripterigium wiflordii*	Peptides
Muscodor albus	Tropical tree and vine sp.	Peptides
Edenia gomezpompae	*Callicarpa acuminata*	Quinones
Pestalotiopsis adusta	Unidentified	Quinones
Ampelomyces sp.	*Urospermum picroides*	Quinines
Cassia spectabilis	*Phomopis cassiae*	Terpenoids
	Anticancer Compounds	
Pestalotiopsismicrospora	*Taxus taxifolia*	Torreyanic acid (cytotoxic quinone dimer)
Rhinocladiella sp.	*Tripterygium wilfordii*	22-oxa-(12)-cytochalasins
Xylaria, *Phoma*, *Hypoxylon*, and *Chalara*	*Various sources*	Cytochalasins (alkaloids)
	Anticancer and Fungicidal Compounds	
Taxomyces andreanae	*Taxus brevifolia*	Paclitaxel (Diterpenoid)
Pestalotiopsis microspora	*Taxus wallichiana*	Paclitaxel (Diterpenoid)
Pestalotiopsis guepini	*Wollemia nobilis*	Paclitaxel (Diterpenoid)
Seimatoantlerium tepuiense	*Maguireothamnus speciosus*	Paclitaxel (Diterpenoid)
Periconia sp. and *Seimatoantlerium nepalense*	*Unknown*	Paclitaxel (Diterpenoid)
Tubercularia sp.	*Taxus mairei*	Paclitaxel (Diterpenoid)
Sporormia minima and *Trichothecium* sp.	*Taxus wallichiana*	Paclitaxel (Diterpenoid)
	Antioxidants	
Pestalotiopsis microspora	*Terminalia morobensis*	Pestacin and Isopestacin
	Insecticidal Compounds	
Nodulisporium sp.	*Bontia daphnoides*	Nodulisporic Acids (Novel Indole Diterpenes)
Unidentified fungus	*Gaultheria procumbens*	5-Hydroxy-2-(1'-Hydroxy-5'-Methyl-4'-Hexenyl)Benzofuran and 5-Hydroxy-2-(1'-Oxo-5'-Methyl-4'-Hexenyl)Benzofuran
Paullina paullinioides	*Muscodor vitigenus*	Naphthalene
	Antidiabetic Compounds	
Pseudomassaria sp.	African rainforest	L-783,281 (Insulin Mimetic)

Endophytic Microorganism	*Source/Host Plant*	*Natural Product Obtained*
Antitubercular Compounds		
Phomopsis sp. PSU-D15	Unknown	Phomoenamide (Amide)
Biscogniauxia formosana	*Cinnamomum* sp.	Biscogniazaphilone A (Polyketide)
Biscogniauxia formosana	*Cinnamomum* sp.	Biscogniazaphilone B (Polyketide)

Eugenia jambolana Lam. Syn. *Syzygium cumini* seeds, kernel, and seed coat have been extensively explored for its hypoglycaemic and antihyperglycaemic potential (Grover *et al.*, 2002). Murali *et al.*, 2007 isolated endophytic fungi *Colletotrichum gloeosporiodes* and *Pestalotiopsis* sp. from *Syzygium cumini* leaves. A napthoquinone (called Javanicin) exhibiting high antibacterial activity against *Pseudomonas aeruginosa* and *Pseudomonas fluorescens* at 2 μg/ml concentration was produced by *Chloridium* sp. isolated from *Azadirachta indica* A. Juss obtained in Varanasi district (Kharwar *et al.*, 2009). Biswas *et al.* (2002) extracted Nimbidin from the oil of seed kernels of *Azadirachta indica* which exhibited anti-inflammatory, antiarthritic, antipyretic, hypoglycaemic, antigastric ulcer, spermicidal, antifungal, antibacterial and diuretic effects.

The researchers at Orchidology division of State Forest Research Institute (SFRI), Itanagar, Arunachal Pradesh had the knowledge from the previous literature that for uptake of sugar, vitamins and other nutrients, orchids require specific mycorrhizal association during seed germination stage. During the non-photosynthetic period of orchid,fungal hyphae supply nutrients to undifferentiated mass of cells (embryo). Experiments were conducted for applying endophytic mycorrhiza as a biofertilizer, it was inoculated into the *in vitro* cultured seedlings of rare, endangered and commercially viable orchids for successful establishment in the field with better survival rate. Majority of fungi isolated from the orchids belong to the genus *Rhizoctonia* (Warcup and Talbot, 1966) and less frequently members of the *Agaricales* (Saha and Rao, 2006).

Mahesh and Satish (2008) reported that methanolic extracts of *Acacia nilotica, Sida cordifolia, Tinospora crdifolia, Withania somnifer* and *Ziziphus mauritiana* exhibited antibacterial property against *Bacillus subtilis, Escherichia coli, Pseudomonas fluorescens, Staphylococcus aureus* and *Xanthomonas axonopodis* pv. *Malvacearum* and antifungal activity against *Aspergillus flavus, Dreschlera turcica* and *Fusarium verticilloides.*

Banu and Kumar (2009) carried out the screening of antimicrobial activity on isopropanol extracts prepared from 121 isolates of endophytic fungi against Gram-positive bacteria, Gram-negative bacteria, yeast and molds. Their study revealed that crude extracts from the culture broth of endophytic fungi grown, aerobically displayed anti-bacterial, anti-fungal, anti-yeast and anti-tumour activity.

Rajagopal *et al.* (2010) studied the diversity of endophytes in four medicinal plants named, *Coleus aromaticus, Leucas aspera, Ocimum basilicum, Ocimum sanctum* and *Tridax procumbens* of South India. From each herb, five hundred lamina and petiole segments were evaluated. Eighteen species of endophytic fungi were obtained from five different species of medicinal plants. Of these eighteen endophytic fungi, six were from Ascomycetous fungi; four were Coelomycetous; three belonged to Hyphomycetes class and rest were sterile forms. Some endophytic fungi were common to all the medicinal plants whereas few appeared to be host specific.

Jalagaonwala *et al.* (2010) screened various parts of medicinal plants and their endophytes for their antimicrobial potential. The medicinal plants included *Azadirachta indica* A. Juss, *Curcuma longa* L., *Eucalyptus globulusc* Dehnh, *Musa paradiasica, Pongamia glabra* vent *Aloe vera* L., *Murraya koenigii* L. *Spengel, Ocimum sanctum*L. Mant. A total of seventy eight bacterial endophytes and one hundred forty two fungal endophytes were obtained from the aerial and underground parts of medicinal plants, out of which fifteen bacterial and fourteen fungal isolates possessed antifungal and antibacterial activity respectively. The antimicrobial activity of endophytes varied greatly different depending upon isolates which suggested that several bioactive constituents contributed in antimicrobial activity.

Gayathri *et al.* (2010) evaluated the bioprospecting potential of endophytic bacteria isolated from the leaves of 5 species of mangroves (*Rhizophora apiculata* Blume, *Avicennia marina* (Forssk.) Vierth., *Excoecaria agallocha* L., *Ceriops decandra* (Griff.) Ding Hou and *Aegiceras corniculatum* (L.) Blanco) and two species of salt marshplants (*Suaeda maritime* Forssk. ex J.F. Gmel. and *Sesuvium portulacastrum* (L.). Thirty six fast growing isolates were selected and screened for biological activities out of 104 bacterial endophytes that were isolated from the leaf samples of mangrove plants,. Of 36 isolates, some isolates showed antimicrobial activity and the remaining isolates exhibited pectinase, protease, inulinase and invertase activities respectively. Plant growth promoting activity, such as, ammonia and acetoin production, was shown by someendophyte isolates, whereas few isolated showed nitrogen fixing activity. Very few isolates exhibited phosphate solubilization activity, indole acetic acid (IAA) production, and malachite green and phenol degrading activities were observed in some endophytic bacterial isolates, respectively. Some of the endophytic bacteria were sensitive to antibiotics like streptomycin and trimethoprim, while few were resistant to vancomycin and bacitracin. The study proved that the mangroves are the sources of endophytic bacteria with bioprospecting potential.

Prabavathy and Nachiyar (2011) evaluated the leaves of *Baliospermum montanum, Plumbagozeylanica, Aegle marmelos* and *Ficus carica* for production of bioactive secondary metabolites. The ethyl acetate extract of *Aspergillus* sp. isolated from *Ficus carica*was found to possess good antimicrobial activity against *Pseudomonas aeruginosa*.

Rehman *et al.* (2011) evaluated the antimicrobial activity of *Nodulisporium* sp. an endophytic fungus isolated from *Nothapodytes foetida*, a medicinal plant. Ethyl acetate and methanol fractions of *Nodulisporium* exhibited antimicrobial activity against pathogenic gram positive bacteria. Dual culture study was carried out for evaluating antifungal activity where maximum antagonism was exhibited against *Alternaria alternata* and *Colletotrichum gleosporoides*.

Verma *et al.* (2011) isolated an alkaloid, piperine (5-(3,4-methylenedioxyphenyl)-1-pipereridinopent-2,4-diene-1-one) from the endophytic fungus *Periconia* sp. which was found to be associated with plant *Piper longum*. Piperine is a potent cancer preventive agent which exhibited strong antimicrobial activity against *Mycobacterium tuberculosis* and *Mycobacterium smegmetis* with minimum inhibitory concentrations (MIC) of 1.74 and 2.64 µg/ml respectively.

Kannan *et al.* (2012) carried out a study forexploring the defense mechanisms of endophytic microbial isolates from medicinal plants. Traditional medicinal plants such as *Andrographis paniculata, Aloe barbadensis, Cynodon dactylon, Ocimum tenuiflorum, Phyllanthus amarus* and *Zingiber officinale* were selected for the study. A total of 44 different bacterial isolates were isolated from the leaves and roots of these medicinal plants. Selected six medicinal plants were found to have symbiosis with arbuscular mycorrhizal fungus (AMF). The colonization percentage of AMF in these plants was above 60 per cent. A total of 209 of the endophytes were Gram-positive and 164 were Gram-negative bacteria. *Bacillus subtilis* strains isolated from the xylem sap of healthy chestnut trees was found to exhibit antifungal activity against *Cryphonectria parasitica* which is known to cause chestnut blight.

Ahmed *et al.* (2012) screened endophytes from four medicinal plants of Jammu and Kashmir. The ethnomedicinal angiospermic plants included *Digitalis lanata* (Wooly Foxglove), *Digitalis purpurea* (Purple Foxglove), *Plantago ovata* (Psyllium/ Isabgol) and *Dioscorea bulbifera* (Air Potato). A total of one hundred and thirty two isolates of microbial endophytes were isolated having enormous applications in the medicinal/pharmaceutical sectors.

Pal *et al.* (2012) carried out a research on a perennial shrub *Paederia foetida* L. (*Rubiaceae*) which is native to temperate and tropical Asia for its endophytic bacterial diversity as well as screening for antagonistic activity. The antimicrobial activity of the bacterial endophytes isolated from surface sterilized stem and leaf tissues of *Passiflora foetida* showed the antagonistic activity against *Escherichia coli, Staphylococcus aureus* and *Klebsiella pneumoniae*.

Prabavathy and Nachiyar (2012) screened endophytic *Aspergillus* sp. from the leaf samples of *Justicia adathoda* for antimicrobial activity. Ethyl acetate extracts of fungal metabolites were screened for their antimicrobial activity against *Candida albicans, Escherichia coli, Staphylococcus aureus, Klebsiella pneumoniae* and *Pseudomonas aeruginosa*. Crude mycelial extract inhibited all the strains significantly.

Elavarasi *et al.* (2012) studied a taxol producing mangrove endophytic fungi *Fusarium oxysporum* from *Rhizophora annamalayana*.The amount of taxol produced by *Fusarium oxysporum* was found to be 172.3 µg/l.

A study on Identification and bioactive potential of endophytic fungi isolated from selected plants of the Western Himalayas was carried out by Riyaz-Ul-Hassan *et al.* (2013). In this study, several extracts prepared from the fermented broth of these fungi demonstrated strong antagonistic activity gainst *Eschericia coli* and *Staphylococcus aureus*. The fungal extracts also possessed immunomodulatory activities which exhibited significant immune suppression as demonstrated by the *In vitro* lymphocyte proliferation assay.

Devi and Prabakaran (2014) prepared the ethyl acetate extract of endophytic *Penicillium* sp. isolated from medicinal plant *Centella asiatica* which exhibited high antioxidant capacity of 325.76 ± 0.14 mg which was equivalent to ascorbic acid and also exhibited promising cytotoxic activity against HeLa, A431 and human breast cancer (MCF7) cell line.

Roy and Banerjee (2015) studied endophytic actinomycete *Cinnamomum* sp. from Indian rainforest, Cherapunji which was found to produce significant antibacterial activity against *Aeromonas caviae, Vibrio parahemolyticus, Pseudomonas aeruginosa, Proteus vulgaris, Shigella flexnerii, Escherichia coli, Bacillus subtilis* and *Bacillus cereus*. The extracted antibacterial metabolites were purified by silica gel column chromatography.

Ananda *et al.* (2012) carried out a study on the enzyme production and antimicrobial potential of forty endophytic fungi isolated from leaves and branches of *Ocimum*. 50 per cent of fungal isolates were found positive for amylase and protease enzymes production and 27.5 per cent showed positive assay for tyrosinase.

Desale and Bodhankar (2013) isolated *Phomopsis* sp. from *Vitex negundo* L. and screened for its antimicrobial potential. The crude extract showed significant antimicrobial activity against *Escherichia coli, Salmonella typhimurium, Bacillus cereus, Bacillus subtilis, Klebsiella pneumoniae* and *Staphylococcus aureus*. The antimicrobial activity was highest against *Escherichia coli*, followed by *Salmonella typhimurium* and *Bacillus cereus*. The study justified the traditional use of *Vitex negundo* L. against human pathogenic bacteria and also suggested the possible role of endophytic fungi in its antimicrobial activity.

In an attempt to evaluate the antimicrobial and antioxidant activity, Joshi *et al.* (2014) isolated two endophytic fungi *Aspergillus oryzae*and *Colletotrichum gloeosporioides* from *Centella asiatica* and *Murraya koenigii* respectively. The DPPH free radical assay revealed that the ethanolic extract of endophytic fungi *Aspergillus oryzae* had higher radical scavenging activity than *Colletotrichum gloeosporioides*. The crude extract of *Aspergillus oryzae* inhibited the growth of *Candida albicans* but had no effect on *Streptococcus pyogenes* and *Enterococcus faecalis*. In contrast, the crude extract of *Colletotrichum gloeosporioides* had no inhibitory effect on *Candida albicans* and *Staphylococcus aureus*.

A similar study on fungal endophytes was conducted by Devi and Singh (2014) on *Phlogacanthus thyrsiflorus* Nees which was collected from Imphal West, Manipur. *Colletotrichum gloeosporioides* extract was used for analyzing antioxidant activity, total phenolic content and for evaluating antibacterial activity. The crude *Colletotrichum* extract exhibited an effective antibacterial activity against *Staphylococcus aureus* with a zone of growth inhibition of 13 mm. Free radical scavenging activity of the extract was shown 90 per cent. Total phenolic content was found to be 18.3 ± 0.36 mg gallic acid/g of extract.

Sandhu *et al.* (2014) isolated fifteen fungi from the plants *Bombex cebia* and *Argemone Mexicana* for their antimicrobial activity against six bacterial strains viz *Escherichia coli, Klebsiella pneumoniae, Bacillus subtilis, Salmonella typhimurium, Enterococcus* sp. and *Streptococcus pyogenes*. The fungal isolate SR2 isolated from leaf of *Bombex cebia*, AMS2 and AMS3 isolated from stem of *Argemone mexicana* showed excellent antibacterial activity. SR2 showed maximum zone of growth inhibition

(27mm) against *Bacillus subtilis*, AMS2 showed maximum zone of growth inhibition against *Klebsiella pneumoniae* (21mm), AMS3 showed maximum zone of inhibition against *Klebsiella pneumoniae* and *Salmonella typhimurium* (19 mm).

Devi and Wahab (2012) carried out a study on *Camellia sinesis* plant to evaluate the antimicrobial potential of endophytic fungi isolated from its leaves. The crude *Penicillium* extract identified through 18S ribosome RNA sequence analysis, showed growth inhibition of all the test pathogens. The *Penicillium* sp. showed 100 per cent gene sequence similarities with *Penicillium funiculosum* strain TS08 and *Penicillium* sp.12-2018S. The bioactive compounds responsible for antibacterial activity were identified (by GC-MS and NMR) as 3-(3-azidopropyl)-1H-indene (15.39 per cent) and 3-Cyano-1,2-dimethylindole.

Sati and Singh (2014) isolated *Anguillospora longissimi*, an endophytic freshwater hyphomycetes from roots of aquatic and riparian plants in the streams and catchment areas of Nainital, Uttarakhand, India. The endophyte was evaluated for antagonistic activity against five bacterial pathogenic strains. In the preliminary antimicrobial assay, maximum zone of growth inhibition was recorded against *Escherichia coli* (23mm) followed by *Erwinia chrysanthemi* (22mm), *Agrobacterium tumefaciens* (21mm), and *Xanthomonas phaseoli* (21mm), while the minimum zone of inhibition was observed against *Bacillus subtilis* (20mm).

Sandhu and his coworkers (2014) isolated 10 species of endophytic fungi from *Ricinus communis* including *Aspergillus fumigates, Aspergillus japonicas, Aspergillus niger, Fusarium semitectum, Curvularia pallescens, Phoma hedericola, Alternaria tenuissima, Fusarium solani, Drechslera australien, Aspergillus repens* and assessed their antibacterial activity against six human pathogenic bacterial strains: *Bacillus subtilis, Enterococcus sp., Klebsiella pneumoniae, Escherichia coli, Salmonella typhimurium* and *Staphylococcus aureus*. Most of the fungal extracts were found to exhibit *In vitro* growth inhibition of pathogenic bacteria.

In another study, Srinivas and Ramesha (2014) surface sterilized different parts of *Plumeria acuminata* and *Plumeria obtusifolia* and then isolated endophytic fungi which were then screened for their antagonistic activity against *Staphylococcus aureus, Bacillus cereus, Pseudomonas aeruginosa, Escherichia coli, Salmonella typhimurium* and *Candida albicans*. The *Colletotrichum gloeosporioides* extract exhibited effective inhibition to the pathogens evaluated except *Candida albicans*; whereas *Fusarium oxysporum* extract exhibited effective inhibition against pathogens except *Pseudomonas aeruginosa*. Phytochemical screening of the extracts revealed the presence of alkaloids, flavonoids, steroids, phenol and phenolic compounds.

Debnath *et al.*(2014) carried out a study to screen and isolate endophytic fungi from root and stem tissues of *Argemone maxicana*exhibiting antimicrobial activity. A total of fourteen endophytic fungi were isolated (assigned as AMS1- AMS8 for stem isolates and AMR9-AMR14 for root isolates). Among these fourteen isolates, AMS5 revealed maximum antagonistic activity against *Escherichia coli, Pseudomonas aeroginosa, Bacillus subtilis, Staphylococcus aureus, Ralstonia solanacearum, Xanthomonas oryzae Penicillium chrysogenum, Candida albicans, Phoma exigua, Sclerotium rolfsii* and *Sclerotinia scleratiourum*. It was revealed from the study that isolates AMS1, AMS2

AMS7, AMS8, AMS9 AMS11 and AMR14 showed antibacterial activity against at least one bacterium while AMS3, AMS4 AMS6,AMR10 AMR12 and AMR13 did not show any antibacterial activity.

Current Status of Research on Bioactive Compounds from Endophytic Bacteria and Fungi: Global View

The use of endophytes in pharmacy, agriculture, industry, and biofuel production has been thoroughly investigated on a global level. Endophytes synthesize the substances which protect and promote their host plants. On isolation and characterization, many of these substances have been developed into novel antimicrobial compounds, immunosuppressive and anticancer agents (Strobel *et al.*, 2004).

The production of novel antimicrobial metabolites from endophytes is an important alternative to the ineffective antibiotics against human and plant pathogens. Antimicrobial compounds could also be used as food preservatives. The increasing number of antibiotic resistant strains of microorganisms has raised the concern of health officials. This has lead to an ever rising need for new antibiotics to replace the old and now ineffective ones (Walsh, 2000; Wang *et al.*, 2000)

Apart from pathogenic microorganisms, some endophytic fungi have been shown to confer protection of plants from herbivore animals (Schardl *et al.*, 2004). Strobel *et al.* (1996) isolated *Pestalotiopsis microspore* which is a common rainforest endophyte. Enormous biochemical diversity does not exist in this endophytic fungus, and many secondary metabolites produced by a myriad of strains of this widely dispersed fungus (Strobel, 2002). Li *et al.* (2001) isolated an antifungal agent, Ambuic acid from the isolates of *Pestalotiopsis microspora*. Lee *et al.* (1996) isolated *Pestalotiopsis microspora* from an endangered tree *Torreya taxifoli* which produced metabolites with antifungal activity, including pestaloside, an aromatic beta-glucoside, and two pyrones: pestalopyrone and hydroxypestalopyrone. These metbolites also had phytotoxic properties. Pulici *et al.* (1996) obtained two new caryophyllene sesquiterpenes- pestalotiopsin A and B and two novel sesquterpenes: 2-α-hyroxydimeninol and humulane.

Li and Strobel (2001) isolated two bioactive compounds, jesterone and hydroxyjesterone from *Pestalotiopsis jesteri*which exhibited antifungal activity against large number of plant pathogenic fungi. This fungal endophyte was isolated from the Sepik river area of Papua, New Guinea. In another study, a secondary metabolite, phomopsichalasin, was isolated from an endophytic *Phomopsis* sp. which was found to exhibit antibacterial activity against *Bacillus subtilis, Salmonella enteric* serovar Gallinarum and *Staphylococcus aureus*. It also showed antagonistic activity against yeast *Candida tropicalis* (Horn *et al.*, 1995).

Strobel *et al.* (1999) isolated Crptocandin, a unique peptide from *Cryptosporiopsis cf. quercina*, an endophytic fungus of *Tripterigeum wilfordii* (a medicinal plant native to Eurasia). Cryptocandin and its related compounds are effective against pathogenic fungi causing skin and nail diseases. It has also been found to have antagonistic activity against phytopathogenic fungi, *Sclerotinia sclerotiorum* and *Botrytis cinerea*.

Li *et al.* (2000) isolated a unique tetramic acid, Cryptocin, from *Cryptosporiopsis quercina*. This unique metabolite exhibited potent activity against *Pyricularia oryzae*and other plant-pathogenic fungi. The compound was found to be ineffective against human-pathogenic fungi. Nevertheless, this compound is being evaluated as a natural chemical control agent for rice blast and is being used as a base model for synthesis of antifungal agents.

Zou *et al.* (2000) isolated endophytic fungus *Colletotrichum gloeosporioides* from *Artemisia mongolica* and was found to produce colletotric acid as secondary metabolite. Colletotric acid was found to exhibit antagonistic activity against bacteria and a fungus *Helminthsporium sativum*. Another strain of *Colletotrichum* sp. isolated from a traditional Chinese herb, *Artemisia annua* produced metabolites active against not only human pathogenic bacteria and fungi but also antagonistic to plant fungal pathogens (Lu *et al.*, 2000). This Chinese herb, *Artemisia annua* is well known for its ability to inhabit different geographically areas and to synthesize artemisinin, an antimalarial drug.

The population density of endophytes is variable and depend not only on the bacterial species and host genotypes but also on the host developmental stage, inoculum density, and environmental conditions (Pillay and Nowak, 1997; Tan *et al.*, 2003). It has been suggested that fungal colonization could affect colonization by endophytic bacteria (Araujo *et al.*, 2001) or the reverse could also happen. It was revealed in a study that plant genotype, the plant age, the tissue sampled, and also on the season of isolation affectedthe presence of endophytic population in soybean (Kuklinsky-Sobral *et al.*, 2004). It was observed that the type of soil determined to a large extent endophytes in wheat (Conn and Franco, 2004).

Tian *et al.* (2004) studied the endophytic fungi and actinomycetes diversity from four rice cultivars in the Panyu district and Wushan district (Site 2) in Guangdong province, South China. The observation that endophytic fungi isolated from rice in Wushan district were more diverse than that in Panyu district which indicated that the endophytic diversity varies in geographical site even within the same host plant. Dual culture and the metabolites activity detection assays showed that 41.2 per cent of all endophytic fungal isolates showed antagonism towards rice pathogens. It was observed that fifty per cent of all the isolated endophytic actinomycetes were antagonistic to those pathogens.

Liu *et al.* (2004) reported twelve secondary metabolites obtained from endophytic *Aspergillus fumigates* CY018 which was isolated from *Cynodon dactylon*leaves. These metabolites were screened for their antagonistic potential. It was revealed from the study that metabolites 'asperfumoid', 'fumigaclavine C', 'fumitremorgin C', 'physcion' and 'helvolic acid' were found to inhibit the growth of *Candida albicans.*

Kumar *et al.* (2011) isolated a root-colonizing endophytic fungus, named *Piriformospora indica*in a study. It was identified as a basidiomycete that possesses highly prospective agricultural potential by conferring biotic and abiotic resistance to its host plants. *Piriformospora indica* is also found to increase biomass and grain yield of crop plants (Varma *et al.*, 1999). In barley; *Piriformospora indica*has been found to confer resistance against blight fungus, *Fusarium culmorum* and systemic resistance to *Blumeria graminis* which causebarley powdery mildew *P. indica* also protected barley plants from abiotic stresses (Waller *et al.*, 2005). *Piriformospora indica* is also popularly known as plant prebiotic as it is responsible for mobilization of insoluble phosphates and translocation of the phosphorous by using phosphate transporter.

Tayung and Jha (2010) analyzed the fungal endophytes in the bark of *Taxus baccata* L., which is an important source of taxol, an anticancer drug. The bark samples were collected from Arunachal Pradesh and Indo-Burma mega hotspot during monsoon and winter which showed the colonizing frequency of 38.5 per cent with higher species diversity during monsoon than winter seasons irrespective of the locations. *Fusarium, Penicillium* and *Aspergillus* were the most common isolated genera which displayed antimicrobial activity against *Bacillus subtilis, Staphylococcus aureus, Staphylococcus epidermis, Klebsiella pneumonia, Shigella flexneri, Escherichia coli, Candida albicans* and *Candida tropicalis*.

A gas-producing endophyte, *Phomopsis* spp. was discovered from *Odontoglossum* sp. obtained from a cloud forest in northern Ecuador by (Singh *et al.*, 2011). This endophytic fungus was found to produce a unique mixture of volatile organic compounds (VOCs) including sabinene, 1-butanol, 3-methyl; benzene ethanol; 1-propanol,2-methyl and 2-propanone. These VOCs exhibited greatest bioactivity against a wide range of phytopathogenic fungi including *Pythium, Phytophthora, Sclerotinia, Rhizoctonia, Fusarium, Botrytis, Verticillium, and Colletotrichum.*

Moussa *et al.* (2011) characterized endophytic actinomycetes from wild egyptian plants which were found to be an antagonists to some phytopathogenic fungi. The ability of endophytic actinomycetes to inhibit phytopathogenic fungi was found mainly to be the result of bioactive compounds, such as antibiotics and cell wall degrading enzymes produced by these organisms. In addition, endophytes compete with phytopathogens for nutrients (El-Tarabily and Sivasithamparam, 2006). Endophytic actinomycetes also trigger plant induced systemic resistance (ISR) (Haas and Defago, 2005). This study focused on the antifungal activities of 65 endophytic actinomycetes that were recovered from 12 wild medicinal plants in Egypt. *In vitro* antifungal activity was determined using the dual culture bioassay against *Fusarium solani, Phytophthora infestans,Macrophomina phaseolina* and *Botrytis cinerea*, which are the causative agents of tomato's wilt, potato's late blight, soybean's charcoal rot and eggplant fruit rot, respectively. The results showed that out of the various isolates screened, *Streptomyces* sp. and *Kitasatosporia* sp. werefound to be the prolific producers of fungal inhibitory compounds. Both isolates were investigated for the production of antifungal diffusible compounds, volatile compounds and enzymes (chitinase, cellulase and Carboxy Methyl Cellulase). They exhibited *In vitro* activity against the fungal pathogens as well as another pathogen (*Fusarium oxysporum*), with a probability of $p<0.05$. The competitive growth between pathogens and antagonists on solid media revealed lesser suppression, comparing to that recorded for the diffusates on agar. However, there was no production of enzymes by both actinomycete isolates.

In another study carried out by Daoud *et al.*(2013), endophytic fungi actinomycetes were isolated from wheat (*Triticum durum*) from Bordj Bou Arreridj region (Algeria). The isolates were screened and evaluated for their antimicrobial activity against several pathogenic bacteria, yeast, and phytopathogenic fungi and were also evaluated for their antioxidant activities. All the crude ethyl acetate extracts showed antagonistic activity on at least some pathogenic microorganisms. Results showed that 60 per cent of these extracts had antioxidant activity. The inhibitory activity was found to be less than 50 per cent.

In a study conducted by Rosa *et al.* (2012) on evaluating the diversity of microbial community associated with healthy *Echinacea purpurea* clones and their ability to produce defense compounds, thirty-nine fungal endophytes were recovered and identified through the molecular methods in 15 distinct phylotypes, which were closely related to species of the genera; *Ceratobasidium, Cladosporium,Colletotrichum, Fusarium, Glomerella,* and *Mycole* which were previously reported as decomposer and phytopathogenic fungi. Crude extracts of fungal isolates were evaluated for antifungal and insecticidal activities. A total of 16 extracts (41 per cent) exhibited antifungal properties and the extract of *Mucor indicus* exhibited larvicidal activity against *Aedes aegypti*. The results clearly suggested that the symbiosis between the endophytic fungal community and micro propagated clones of *Echinacea purpurea* resulted in the production of antimicrobial compounds against phytopathogenic fungi.

An endophytic fungus *Fusarium proliferatum* isolated from a South African plant *Syzygium cordatum*was found to exhibit 100 per cent cytotoxicity against the brine shrimp *Artemia salina* at 100 µg/ml. Several bioactive metabolites, including ergosta- 5,7,22-trien-3β-ol, nectriafurone-8-methyl ether, 9-O-methyl fusarubin, bostrycoidin, bostrycoidin-9-methyl ether, and 8-hydroxy- 5,6-dimethoxy-2-methyl-3-(2-oxo-propyl)-1,4-naphthoquinone, were isolated from the extract. The crude extract also showed a significant antimicrobial activity (Dame *et al.*, 2015).

Powthong and his coworkers in 2012 evaluated the antimicrobial potential of the ethyl acetate extracts of endophytic fungi isolated from *Sesbania grandiflora* (L.) Pers. They exhibited high antibacterial activity against *Bacillus subtilis* with the high susceptibility, followed by *Psedomonas aeruginosa, Staphylococcus aureus*and *Escherichia coli*respectively. The MIC and MBC of this group were in the range of 0.49-250µg/ml and 1.95≥250µg/ml, respectively (Powthong *et al.*, 2012).

Liang and his colleges (2012) investigated the antimicrobial activities of endophytic fungi isolated from *Ophiopogon japonicus*, a traditional Chinese medicinal plant (collected from Chongqing City, Southwestern China). 53.33 per cent of the mycelial extracts and 33.33 per cent of the fermentation broths displayed potent inhibition of peptide deformylase inhibition fluorescence assays. 80 per cent of mycelia extracts and 33.33 per cent of fermentation broths significantly inhibited *Staphylococcus aureus*. 70 per cent of mycelia extract and 36.67 per cent of fermentation broths showed strong activities against *Cryptococcus neoformans*. There was no effect of any strain on *Escherichia coli*.

Another traditional medicinal plant, *Sapindus saponaria* L. (commonly known in Brazil as "sabãode-soldado") is well known for its therapeutic uses such as neutralizing bleeding, antifungal, larvicidal, produces substances which have anxiolytic, astringent, diuretic and expectorant properties, they are also used as tonic, blood depurative and cough medicine. The crude extracts of fungal endophytes (*Cochliobolus intermedius*, non-identified strain, *Phomopsis* sp. and other identified strain) isolated from *S. saponaria*, were evaluated against pathogenic bacteria; *Escherichia coli*, *Staphylococcus aureus*, *Salmonella typhi*, *Micrococcus luteus* and *Enterococcus hirae*. The results revealed that the antibacterial activities were effective, as all of them inhibited the growth of at least one of the evaluated bacteria. One metabolite, extracted *Cochliobolus intermedius* exhibited activity against all the bacteria evaluated (Garcia *et al.*, 2012).

Higginbotham and coworkers (2013) studied foliar endophytes isolated from plant samples collected from nine geographically distinct regions of Panama. More than 2700 isolates were evaluated for their bioactivity against a human breast cancer cell line (MCF-7) and the causative agents of malaria, leishmaniasis and Chagas disease. The results revealed that up to 17 per cent of isolates isolated from plants in cloud forestswere very effective. The study made it possible to conclude that the fungal strains varied in degree of bioactivity.

Endophytic microorganisms are alternative and sustainable source of these compounds. However, the commercial implications of production of desirable compounds by endophytic microorganisms still remain a challenge. Another challenge is that the endophytic microorganisms always remain in inconstant, adaptable and multifaceted interactions with their host plants and other surrounding endophytes. Therefore, even a slight variation in the *In vitro* cultivation conditions is likely to affect the kind and the range of secondary metabolites produced by the endophytes (Scherlach and Hertweck, 2009; Winter *et al.*, 2011).

Some of the major obstacles in the sustainable production of pharmaceutically important compounds from endophytes are the reduction of secondary metabolite production on repeated sub-culturing of endophytes under axenic monoculture conditions and the inability to establish, restore and sustain the *in vitro* biosynthetic capabilities of endophytes (Kusari and Spiteller, 2011).

Another major concern is the rapidly declining world's biodiversity. Endemic, endangered and the valuable medicinal plants are on the verge of extinction. The Rainforests, which are regarded as the 'hot spots of biodiversity', are diminishing at an alarming rate and so are the endophytic microorganisms associated with them. Multistep processes involving efforts for biodiversity conservations, establishment of national collections of microorganisms for the immediate preservation of life forms, their molecular and genetic information are required for production of bioactive metabolites against the present and emerging diseases (Strobel and Daisy, 2003).

Further investigations on the ecological aspects of endophytic microorganisms such as host range, evolutionary relatedness, infection, colonization, transmission patterns, tissue specificity, mutualistic fitness benefits and genetic improvements of endophytic microbes to increase their productivity are required to promote the biotechnological applications of endophytic microorganisms. Researches along these directions are highly desirable to overcome the obstacles in the path of natural product discovery process from endophytic source and its commercialization (Kusari and Spiteller, 2011).

Conclusion

Secondary metabolites obtained from endophytic microorganisms have been found to originate from various biosynthetic pathways and belongs to divergent structural groups such as alkaloids, amines, peptides, *etc.* Several studies conducted on the bioactive metabolites of endophytes obtained from medicinal plants have marked the huge diversity of its secondary metabolites. There exists an extensive literature stating that they possess pharmacological properties such as anticancer, antimicrobial, antifungal, antiviral and antimalarial property (Puri *et al.*, 2018).

Acknowledgement

The authors are grateful to Hon'ble Vice-Chancellor, Kurukshetra University, Kurukshetra for providing the necessary infrastructure to carry out the research on endophytic microorganisms.

REFERENCES

Agrawal, P., Rai, V. and Singh, R.B. 1996. Randomized Placebocontrolled Single Blind Trial of Holy Basil Leaves With Noninsulin-Dependent Diabetes Mellitus. *Int.J. Clin. Pharmacol. Therapeut.* 34: 406–409.

Ahmed, M., Hussain, M., Dhar, M.K. and Kaul, S., 2012. Isolation of Microbial Endophytes from Some Ethnomedicinal Plants of Jammu and Kashmir *Journal of Natural Products. Plant Res.*2(2): 215-220.

Ananda, K., Pavithra, N. and Sathish, L. 2012. Antimicrobial and Enzyme Activity of Endophytic Fungi Isolated from Tulsi. *J. Pharma. Biomed. Sci.* 16(12): 1-6.

Aneja, K.R. 2003. Experiments in Microbiology Plant Pathology and Biotechnology. 4th Ed. *New Age International Publishers*. New Delhi. 1-632.

Araujo, W.L., Maccheroni, W., Jr. Aguilar-Vildoso, C.I., Barroso, P.A.V., Saridakis, H.O. and Azevedo, J.L. 2001. Variability and Interactions Between Endophytic Bacteria And Fungi Isolated From Leaf Tissues Of Citrus Rootstocks. *Canad. J. Microbiol.*47: 229-236.

Banu, G.S. and Kumar, G. 2009. Preliminary Screening of Endophytic Fungi from Medicinal Plants in India for Antimicrobial and Antitumor Activity. *Int. J. Pharm. Sci. Nanotechnol.*2(2): 566-571.

Bentley, R. 1997. Microbial Secondary Metabolites Play Important Roles in Medicine; Prospects for Discovery of New Drugs. *Persp. Biol. Med*40: 364-394.

Biswas, K., Chattopadhyay, I., Banerjee, R.K. and Bandyopadhyay, U. 2002.Biological Activities and Medicinal Properties of Neem (*Azadirachta indica*) *Curr. Sci,* 82(11): 1336-1345.

Cheng, M.J., Wu, M.D., Yanai, H., Su, Y.S., Chen, I.S., Yuang, G.F., Hsieh, S.Y. and Chen, J.J. 2012. Secondary metabolites from the endophytic fungus Biscogniauxia formosana and their antimycobacterial activity. *Phytochem. Lett.* 5: 467–472.

Conn, V.M. and Franco, C.M.M. 2004. Analysis of the Endophytic Actinobacterial Population in the Roots of Wheat (*Triticum aestivum* L.) by Terminal Restriction Fragment Length Polymorphism and Sequencing of 16s rRNA Clones. *Appl. Environ. Microbiol.* 70: 1787-1794.

Dame, Z.T., Silima, B., Gryzenhout, M. and Ree, T.V. 2015. Bioactive Compounds from the Endophytic Fungus *Fusarium proliferatum*. *Nat. Prod. Res.* **7**(12): 53-57.

Daoud,H., Sadrati,N., Zerroug,A., Dahamna,S. and Saddek Bouharati,S. 2013. Screening of Antimicrobial and Antioxidant Secondary Metabolites from Endophytic Fungi Isolated from Wheat (*Triticum durum*). *J. Plant Protect. Res.* 53(2): 128–136.

Debnath(das), M., Verma, S.V. and Kumar, A. 2014. Antimicrobial Activity of Endophytic Fungal Isolate in Argemone Maxicana; A Traditional Indian Medicinal Plant. *Int. J. Innov. Res. Sci. Engg. Technol.*3(3): 10151-10162.

Demain, A.L. and Fang, A. 2000. The Natural Function Of Secondary Metabolites Sheper TIn *Advances In Biochemical Engineering/Biotechnology*. Berlin, Germany: *Springer*. 69: 1–39.

Desale, M.G. and Bodhankar, M.G. 2013. Antimicrobial Activity of Endophytic Fungi Isolated from *Vitex negundo* Linn. *Int. J. Curr. Microbiol. Appl. Sci.*2(12): 389-395.

Devi, N.N. and Prabakaran, J.J. 2014. Bioactive Metabolites from an Endophytic Fungus *Penicillium* sp. Isolated From *Centella asiatica*. *Curr Res. Environ. Appl. Mycol.*4(1): 34-43.

Devi, N.N. and Singh, M.S. 2014. Bioactivity of Endophytic Fungus *Colletotrichum gloeosporioides* Isolated from *Phlogacanthus thyrsiflorus* Nees. *Int. Res. J. Biol. Sci.* **3**(10): 97-99.

Devi, N.N. and Wahab, F. 2012. Antimicrobial Properties of Endophytic Fungi Isolated from Medicinal Plant *Camellia sinesis*. *Int. J. Pharma Bio Sci.*3(3): (P) 420-427.

Dolak, L.A., Reusser, F., Baczynskyj, L., Mizsak, S.A., Hannon, B.R. and Castle T.M. 1983. Desertomycin: Purification And Physical-Chemical properties. *J.Antibiot.* 1: 13-9.

Elavarasi, A., Rathna, G.S. and Kalaiselvam, M. 2012. Taxol Producing Mangrove Endophytic Fungi *Fusarium oxysporum* from *Rhizophora Annamalayana*.*Asian Pacific J. Trop. Biomed.* S1081-S1085.

Garcia, A.1, Rhoden, S.A., Bernardi-Wenzel, J., Orlandelli, R.C., Azevedo, J.L. and Pamphile, J.A. 2012. Antimicrobial Activity of Crude Extracts of Endophytic Fungi Isolated from Medicinal Plant *Sapindus saponaria* L. *J. Appl. Pharmaceut. Sci.* 2(10): 35-40.

Gayathri, S., Saravanan, D., Radhakrishnan, M., Balagurunathan, R. and Kathiresan, K. 2010. Bioprospecting Potential of Fast Growing Endophytic Bacteria from Leaves of Mangrove and Salt-Marsh Plant Species. *Ind. J. Biotechnol.* 9: 397-402.

Grover, J.K., Yadav, S. and Vats, V. 2002. Medicinal Plants of India With Anti-Diabetic Potential. *J. Ethnopharmacol.* 81: 81–100.

Higginbotham, S.J., Arnold, A.E., Ibañez, A., Spadafora, C., Coley, P.D. and Kursar, T.A. 2013. Bioactivity of Fungal Endophytes as a Function of Endophyte Taxonomy and the Taxonomy and Distribution of Their Host Plants. *PLoS ONE*.8(9): e73192.

Horn, W.S., Simmonds, M.S.J., Schwart, R.E. and Blaney, W.M. 1995.Phomopsichalasin, a Novel Antimicrobial Agent from an Endophytic *Phomopsis* sp.*Tetrahedron*. 51(14): 3969-3978.

Jalgaonwala, R.E., Mohite, B.V. and Mahajan, R.T. 2010Evaluation of Endophytes for their Antimicrobial activity from Indigenous Medicinal Plants belonging to North Maharashtra region India. *Int. J. Pharm. Biomed. Res.* 1(5): 136-141.

Joshi, S.R., Nath, A. and Pathak, J. 2014. Bioactivity Assessment of Endophytic Fungi Associated with *Centella asiatica* and *Murraya koengii*. *J. Appl. Biol. Biotechnol.* 2 (5): 6-11.

Kannan, V.R., Sumathi, C.S., Balasubramanian, V., Ramesh, N. and Rajesh, P. 2012. Exploration of Defense Mechanisms on Endophytic Microbial Isolates from Selected Traditional Medicinal Plants. *Int. J. Medicobiol. Res.*1(6): 315-320.

Kharwar, R.N., Verma, V.C., Kumar, A., Gond, S.K., Harper, J.K., Hess, W.M., Lobkovosky, E., Ma C., Ren, Y. and Strobel, G.A. 2009. Javanicin, an Antibacterial Naphthaquinone from an Endophytic Fungus of Neem, *Chloridium* sp. *Curr. Microbiol.*58: 233–238.

Kieslich, K. 1986. Production of Drugs by Microbial Biosynthesis and Biotransformation Possibilities, Limits and Future Development. *Drug Res.* 36: 888–892.

Kuklinsky-Sobral, J., Araujo, W.L., Mendes, R., Geraldi, I.O., Pizzirani- Kleiner, A.A. and Azevedo, J.L. 2004. Isolation and Characterization of Soybean-Associated Bacteria and Their Potential for Plant Growth Promotion. *Environ. Microbiol.* 6: 1244-1251.

Kumar, M., Yadav, V., Kumar, H., Sharma, R., Singh, A., Tuteja, N. and Johri, A.K. 2011.*Piriformospora Indica* enhances Plant Growth by Transferring Phosphate. *Plant Signal. Behav.* 6(5): 723–725.

Kusari, S. and Spiteller, M. 2011. Are We Ready for Industrial Production of Bioactive Plant Secondary Metabolites Utilizing Endophytes? *Nat. Prod. Rep.*28: 1203-1207.

Lee, J.C., Strobel, G.A., Lobkovsky, E. and Clardy, J.C. 1996. Torreyanic Acid a Selectively Cytotoxic Quinone Dimer from the Endophytic Fungus *Pestalotiopsis microspora*. *J. Org. Chem.* 61: 3232–3233.

Li, J.Y., Harper, J.K., Grant, D.M., Tombe, B.O., Bashyal, B., Hess, W.M. and Strobel, G.A. 2001. Ambuic Acid, a Highly Functionalized Cyclohexanone with Antifungal Activity from *Pestalotiopsis* sp. and *Monochaetia* sp. *Phytochem.* 56: 463-468.

Li, J.Y. and Strobel, G.A. 2001. Jesterone And Hydroxy-Jesterone Antioomycete Cyclo-Hexenenone Epoxides from the Endo-Phytic Fungus *Pestalotiopsis Jesteri. Phytochem.*57: 261–265.

Li, J.Y., Strobel, G.A., Harper, J.K., Lobkovsky, E. and Clardy, J. 2000. Cryptocin, a Potent Tetramic Acid Antimycotic from the Endophytic Fungus *Cryptosporiopsis Cf. quercina. Org. Lett.***2**: 767–770.

Liang, H., Xing, Y., Chen, J., Zhang, D., Guo, S. and Wang, C. 2012. Antimicrobial Activities of Endophytic Fungi Isolated from *Ophiopogon japonicus* (Liliaceae). *BMC Compl. Alt. Med.* 12(238): 1-6.

Liu, J.Y., Song, Y.C., Zhang, Z., Wang, L., Guo, Z.J., Zou, W.X. and Tan, R.X. 2004. *Aspergillus fumigatus* CY018, an Endophytic Fungus in *Cynodon dactylon* as a Versatile Producer of New and Bioactive Metabolites. *J. Biotechnol.*114(3): 279–287.

Lu, H., Zou, W.X., Meng, J.C., Hu, J. and Tan, R.X. 2000. New Bioactive Metabolites Produced by *Colletotrichum* sp., an Endophytic Fungus in *Artemisia annua*. *Plant Sci.* 151: 61–73.

Mahesh, B. and Satish, S. 2008.Antimicrobial Activity of Some Important Medicinal Plant Against Plant And Human Pathogens. *World J. Agric. Sci.*4 (5): 839-843.

Metz, A.M., Haddad, A., Worapong, J., Long, D.M., Ford, E.J., Hess, W.M. and Strobel, G.A. 2000. Induction of the Sexual Stage of *Pestalotiopsis microspora*, a Taxol-Producing Fungus. *Microbiol.* 146: 2079–2089.

Moussa, H.E., El-Shatoury,S.A., Wahid, O.A.A. and Dewedar, A. 2011. Characterization of Endophytic *Actinomycetes* from Wild Egyptian Plants as Antagonists to Some Phytopathogenic Fungi. *Egyptian J. Nat. Toxins.* 8(1, 2): 32-48.

Murali, T.S. and Venkatesan, G. 2007. Fungal Endophyte Communities in Two Tropical Forests of Southern India: Diversity and Host Affiliation. *Mycol. Prog.* 6(3): 191–199.

Pal, A. and Paul, A.C.A.K.2012. Diversity and Antimicrobial Spectrum of Endophytic Bacteria Isolated from *Paederia foetida* L.*Int. J. Curr. Pharm. Res.* 4(3): 123-127.

Patel, P.S., Huang, S., Fisher, D., Pirnik, C., Aklonis, C., Dean, L., Meyers, E., Fernandes, P. and Mayerl, F. 1995. Bacillaene, a Novel Inhibitor of Prokaryotic Protein Synthesis Produced by *Bacillus Subtilis*: Production, Taxonomy, Isolation, Physicochemical Characterization and Biological Activity. *J. Antibiot.* 48: 997-1003.

Pillay, V.K. and Nowak, J. 1997. Inoculum Density, Temperature, and Genotype Effects on *In Vitro* Growth Promotion and Epiphytic and Endophytic Colonization of Tomato (*Lycopersicon Esculentum* L.) Seedlings Inoculated with a *Pseudomonad* Bacterium. *Canad.J.Microbiol.* 43: 354- 361.

Powthong, P., Jantrapanukorn, B., Thongmee, A. and Suntornthiticharoen, P. 2012. Evaluation of Endophytic Fungi Extract for Their Antimicrobial Activity from *Sesbania grandiflora* (L.) Pers. *Int. J. Pharm. Biomed. Res.*3(2): 132-136.

Prabavathy, D. and Nachiyar, C.V. 2012. Study on the Antimicrobial Activity of *Aspergillus* sp. Isolated from *Justicia adathoda*. *Ind.J. Sci. Technol*.5(9): 3317-3320.

Prabavathy, D. and Nachiyar, C.V. 2011. Screening and Characterisation of Antimicrobial Compound from Endophytic *Aspergillus* Sp. Isolated from *Ficus Carica*. *J. Pharm. Res*. **4**(6): 1935-1936.

Pulici, M., Sugawara, F., Koshino, H., Uzawa, J., Yoshida, S., Lobkovsky, E. and Clardy, J. 1996. Pestalotiopsin-A and Pestalopsin-B- New Caryophylenes from an Endophytic Fungus of *Taxus brevifolia*. *J. Org. Chem*.61: 2122–2124.

Puri, S.K., Habbu, P.V., Kulkarni, P.V. and Kulkarni, V.H. 2018. Nitrogen Containing Secondary Metabolites from Endophytes of Medicinal Plants and their Biological/Pharmacological Activities- A Review. *Systemat. Rev. Pharm*. 9(1): 22-30.

Rajagopal, K., Kalavathy, S., Kokila, S., Karthikeyan, S., Kathiravan, G., Prasad, R. and Balasubraminan, P. 2010. Diversity of Fungal Endophytes in few medicinal herbs of South India. *Asian J. Exp. Biol Sci*.1(2): 415- 418.

Rehman, S., Mir, T., Kour, A., Qazi1, P.H., Sultan, P. and Shawl, A.S. 2011.*In vitro* antimicrobial studies of *Nodulisporium* specie: An endophytic fungus.*J. Yeast Fungal Res*. 2(4): 53- 58.

Riyaz-Ul-Hassan, S., Qadri, M., Johri, S., Shah, B.A., Khajuria, A., Sidiq, T., Lattoo, S.K. and Abdin, M.Z. 2013. Identification and Bioactive Potential of Endophytic Fungi Isolated from Selected Plants of the Western Himalayas. *Springerplus*. 2(8): 1-14.

Rosa, L., Tabanca, N., Techen, N., Wedge, D., Pan, Z., Bernier, U., Becnel, J., Agramonte, N., Walker, L. and Moraes, R. 2012. Diversity and Biological Activities of Endophytic Fungi Associated with Micropropagated Medicinal Plant *Echinacea purpurea* (L.) Moench. *Am. J.Plant Sci*. 3(8): 1105-1114.

Roy, S. and Banerjee, D. 2015. Bioactive Endophytic Actinomycetes of *Cinnamomum* Sp.; Isolation, Identification, Activity Guided Purification and Process Optimization of Active Metabolite. *Am. J. Microbiol*. 6(1): 4-13.

Rukachaisirikul, V., Sommart, U., Phongpaichit, S., Sakayaroj, J. and Kirtikara, K. 2008. Metabolites from the endophytic fungus Phomopsis sp. PSU-D15. *Phytochem*. 69: 783–787.

Saha, D. and Rao, A.N. 2006. Studies on Endophytic Mycorhiza of Some Selected Orchids of Arunachal Pradesh - 1. Isolation andIdentification.*Bull. Arunachal Forest Res*. **22**(1 and 2): 9-16.

Sandhu, S.S., Kumar, S., Aharwal, R.P. and Kumar, S. 2014. Isolation and Detection of Anti-Bacterial Activity of Endophytic Fungi from *Bombex cebia* and *Argemone mexicana*. *J. Chem.Pharm Res*. **6**(11): 95-100.

Sandhu, S.S., Kumar, S., Aharwal, R.P., Shukla, H. and Rjak, R.C. 2014. Endophytic Fungi as a Source of Antimicrobials Bioactive Compounds. *World J. Pharm. Pharmaceut. Sci*.3(2): 1179-1197.

Sati, S.C. and Singh, L. 2014. Bioactivity of Root Endophytic Freshwater Hyphomycetes *Anguillospora longissima* (Sacc. and Syd.) Ingold. *The Scient World J*. 2014: 1-5.

Schardl, C.L., Leuchtmann, A. and Spiering, M.J. 2004. Symbioses of Grasses with Seedborne Fungal Endophytes. *Ann. Rev.Plant Biol*. 55: 315-340.

Scherlach, K., Hertweck, C. 2009. Triggering Cryptic Natural Product Biosynthesis in Microorganisms. *Org. Biomol. Chem*.7: 1753-1760.

Singh, S.K., Strobel, G.A., Knighton, B., Geary, B., Sears, J. and Ezra, D. 2011. An endophytic *Phomopsis* sp. Possessing Bioactivity and Fuel Potential with its Volatile Organic Compounds. *Microb. Ecol*.61: 729–739.

Srinivas, C. and Ramesha, A. 2014. Antimicrobial Activity and Phytochemical Analysis of Crude Extracts of Endophytic Fungi Isolated from *Plumeria acuminata* L. and *Plumeriaobtusifolia* L. *EuropJ. Exp. Biol*. 4(2): 35-43.

Stone,M.J. and Williams, D. H. 1992.On the Evolution of Functional Secondary Metabolites (Natural Products). *Mol. Microbiol*. 6(1): 29–34.

Strobel, G. and Daisy, B. 2003. Bioprospecting for Microbial Endophytes and their Natural Products. *Microbiol. Mol. Biol. Rev*.67: 491–502.

Strobel, G., Daisy, B., Castillo, U. and Harper, J. 2004. Natural Products from Endophytic Microorganisms. *J. Nat. Prod*.67: 257-268.

Strobel, G., Ford, E., Worapong, J., Harper, J.K., Arif, A.M., Grant, D.M., Peter Fung, P.C.W. and Chau, R.M.W. 2002. Isopestacin, an Isobenzofuranone from *Pestalotiopsis Microspora* Possessing Antifungal and Antioxidant Activities. *Phytochem*. 60: 179- 183.

Strobel, G., Miller, R.V., Miller, C., Condron, M., Teplow, D.B., Hess, W.M. 1999. Cryptocandin, a Potent Antimycotic from the Endophytic Fungus *Crypto-sporiopsis Cf. quercina*. *Microbiol*.145: 1919–1926.

Strobel, G., Yang, X., Sears, J., Kramer, R., Sidhu, R.S. and Hess, W.M. 1996. Taxol from *pestalotiopsis microspora*, an Endophytic Fungus of *Taxus wallachiana*. *Microbiol*. 142: 435–440.

Tan, Z., Hurek, T. and Reinhold-Hurek, B. 2003. Effect of N-Fertilization, Plant Genotype and Environmental Conditions on *nifh* Gene Pools in Roots of Rice. *Environ. Microbiol*. 5: 1009-1015.

Tayung, K. and Jha, D. K.2010.Antimicrobial Endophytic Fungal Assemblages Inhabiting Bark of *Taxus Baccata* L. of Indo-Burma Mega Biodiversity Hotspot*Ind. J. Microbiol*.50(1): S74–S81.

Tian, X.L., Cao, L.X., Tan, H.M., Zeng, Q.G., Jia,Y.Y., Han, W.Q. and Zhou, S.N. 2004. Study on the Communities Of Endophytic Fungi and Endophytic Actinomycetes from Rice and Their Antipathogenic Activities *In Vitro.World J. Microbiol. Biotechnol*.20: 303–309.

Verma, V.C., Lobkovsky, E., A.C., Singh, S.K. and Prakash, S. 2011. Piperine Production by Endophytic Fungus *Periconia* sp. Isolated from *Piper longum* L. *J. Antibiot.* 64: 427-431.

Vining, L.C. 1992. Secondary Metabolism, Inventive Evolution and Biochemical Diversity -A Review. *Gene.* 115: 135-140.

Walsh, C. 2000. Molecular Mechanisms That Confer Antibacterial Drug Resistance. *Nature.* 406: 775-781.

Wang, W., Mukhopadhyay, B., Cho, L., Hu, F. and Hui, L. 2000. High Prevalence of *Helicobacter Pylori* Infection with Dual Resistance to Metronidazole and Clarithromycin in Hong Kong. *Aliment. Pharmacol.Therapeut.* 14(7): 901–910.

Warcup, J.H. and Talbot, P.H.B. 1966. Perfect States of Some Rhizoctonias. *Transactions of the Brit.Mycol. Soc.* 49(3): 427-435.

Wells, S.J., O'Sullivan, J., Aklonis, C., Ax, H.A. and Tymiak, A.A. 1992. Dactylocyclines, Novel Tetracycline Derivatives Produced by a *Dactylosporangium* sp. *J. Antibiot.* 45: 1892-1898.

Winter, J.M., Behnken, S. and Hertweck, C. 2011. Genomics-Inspired Discovery of Natural Products. *Curr. Opin. Chem. Biol.*15: 22-31.

Zou, W.X., Meng, J.C., Lu, H., Chen, G.X., Shi, G.X., Zhang, T.Y. and Tan, R.X. 2000. Metabolites of *Colletotrichum gloeosporioides*, an Endophytic Fungus in *Artemisia mongolica*. *J. Nat. Prod.* 63: 1529–1530.

Zweerink, M.M. and Edison, A. 1987. Difficidin and Oxydifficidin: Novel Broad Spectrum Antibacterial Antibiotics Produced by *Bacillus subtilis*. III. Mode of Action of Difficidin. *J. Antibiot.* 40: 1692-1697.

Chapter 14

Metagenomics: Mining the Untapped Environments

*Mandeep S. Multani[1], P.K. Sharma[2] and Ranjana Bhatia[1]**

[1]Department of Biotechnology, University Institute of Engineering and Technology, Panjab University, Chandigarh
[2]Ex. Professor, Dept. of Microbiology, CCS Haryana Agricultural University, Hisar, Haryana
**E-mail: ranjanabhatia20@gmail.com*

ABSTRACT

Microorganisms are abundantly found and actively involved in almost all the natural processes and ecological phenomena. But currently only a portion of this natural resource is known. The actual number, diversity and function of these microorganisms is yet undiscovered. Metagenomics has emerged as a powerful and effective technology to have access to the so called "unculturable" species of the microbiome. It is a culture independent approach that involves cloning and analysis of the genetic material isolated directly from the environment. The evolution of next generation sequencing techniques has further expanded the purview of this technology. This in turn has led to widespread applications. Although, metagenomics has been put to use to explain the yet unapprehended phenomena, still its true potential remains untapped.

***Keywords**: Metagenomics, Microorganisms, Soil, Screening, Analysis, Phylogeny.*

Introduction

Microorganisms are regarded to be one of the most abundant natural resources found not only on earth but in every portion of the biosphere including soil, oceans, space and even deep inside the earth's crust. Being ubiquitous, they play an important role in supporting and maintaining various biological processes (Alexander, 1977). Microorganisms are actively involved in biogeochemistry (Madsen, 2011), nutrient cycling (Sparling, 1997), degradation of natural wastes (Watanabe, 2001) and in the production of primary and secondary metabolites (Drew and Demain, 1977). Soil is regarded as a living organ of Nature. Experimental evidences and analytical results regard soil as the most potential model to study the diversity and functions of microorganisms. Stating simply, soil is one of the leading frontiers of discovery and bio-prospecting.

Shift to the Uncultured/Unknown World

The association of microorganisms with various biogeochemical phenomena and human metabolism generated a keen interest to study their diversity. Microbiologists earlier relied on traditional microscopic analysis which though not very accurate, revealed morphological characteristics of the microorganisms. The petriplate continues to provide a crude yet essential glimpse of the microbial world. However the physiological traits could not be successfully uncovered. Studies reveal that only 1 per cent of the viable prokaryotes are culturable (Torsvik *et al.*, 1998, Davis *et al.*, 2005) and hence the known so called "culturable microorganisms" represent only the tip of the actual microbiota iceberg, the range of which (the uncultured world) could easily outnumber the inhabitants of the cultured world (Dunbar *et al.*, 1999; Smit *et al.*, 2001)

All this induced a new zeal in the microbiologists to conquer the "unknown world". With time, the use of non-traditional techniques of classification and identification became popular. 16 sRNA gene analyses provided an insight into the diversity analysis (Hugenholtz *et al.*, 1998, Hill *et al.*, 2000, Kirk *et al.*, 2004) and functional aspects of the unculturable species while the high throughput sequencing of culturable microorganisms could lead to the identification of novel group of microorganisms. This paved the way to 'metagenomics'.

Metagenomics: Road to the "Unknown" Microbiome

With the advent of molecular biology and arising awareness about the unculturable species, scientists started adopting the non-traditional, culture-independent approaches to study the diversity among the microbial species. Pace *et al.* (1985) coined the idea of cloning DNA directly from environmental samples. This technique was formally named "metagenomics" and is now quoted as the 'the study of collective genomes of the members of a microbiome' or 'the study of genetic material isolated directly from environmental samples (Environmental Metagenomics)' (Tringe and Rubin, 2005). So, the analysis assesses the community (of microorganisms) as a single unit instead of individual studies conducted on every member of the community. Metagenomics is thus regarded as 'Community Genomics' and 'Population Genomics'. Metagenomics involves cloning and analyzing genomes without culturing the organisms in the community (*i.e.* it is a culture-independent approach). Metagenomic analysis enables us to understand the whole spectrum of microbial diversity, population structure, their ecological roles and establishing a relationship between functional and phylogenetic diversity.

The shift of focus towards the application of metagenomics was accelerated by techniques such as 5s RNA and 16s RNA analysis and PCR. Using PCR 52 phyla of uncultured microorganisms have been delineated (Handelsman, 2004) and thus, this new field has helped to widen the horizon of "the unknown" world in terms of physiology and genetics.

Methodology followed for Metagenomic Studies

Metagenomic analysis involves the isolation of DNA from an environmental sample, cloning the DNA into a suitable vector and transformation of the clones into a host. This is followed by screening of the transformants for the desired characteristics. Metagenomics approach can be generalized as represented in six steps:

1. Isolation of DNA from environmental samples (soil, water, *etc.*)
2. Treatment with digestive restriction enzymes resulting in DNA fragments of varying lengths.
3. Integration of these DNA fragments into suitable vectors.
4. Transformation of these vector-DNA complexes into a suitable host.
5. Cloning to form METAGENOMIC LIBRARIES.
6. Screening

The salient features of the above mentioned steps include:

Extraction and Isolation of DNA

Distribution of microbial species is non-uniform in the environmental samples such as soil due to its heterogeneous nature. Non-uniformity may also arise due to variations in the population size of different species. This may lead to biased extraction of DNA.

Whatever maybe the size of the DNA insert, DNA preparation presents a critical step in the preparation of a metagenomic library. An important aspect in DNA preparation is the use of digestive restriction enzymes employed for this step, necessitating to start with DNA at least three times longer than the desired insert size (Gillespie *et al.*, 2005). Thus, if an insert size of 100 kb is required, then the pre-digested DNA should be ≥300 kb.

In order to overcome the problem of heterogeneity, different methods of DNA extraction are employed whose results are then overlapped and combined to gain access to majority of the microbiome spectrum flanking the sample (Delmont *et al.*, 2011). DNA extraction can be done in two ways (Shmeisser *et al.*, 2007):

1. Directly: direct lysis of cells in the soil sample; and
2. Indirectly: cells are first separated and then subjected to lysis.

DNA extraction and isolation can be divided into for most basic steps. There are many available and abundantly used protocols for the above (Surzycki, 2000):

1. Cell breakage
2. Removal of protein and RNA
3. Concentration of DNA
4. Determination of the purity and quantity of DNA

Purification of DNA Sample

Various factors govern the amount of DNA that can be extracted from the sample. For instance, in order to isolate DNA sample from soil, it is important to reduce the interference by inorganic soil components like positively and negatively charged clay particles, and biochemical contaminants which include humic acid and DNases. This demands physical separation of microbes from the soil matrix which results in a pellet from which DNA can be isolated. Further purification can be done by using techniques like agarose gel electrophoresis which significantly separates high molecular weight DNA from humic acids and DNases.

Choice of Vector

Choice of the right cloning vector is also very significant. Cloning vectors govern various features of metagenomic libraries such as the size of insert and copy number. The size of the DNA fragment to be studied affects the type of vector to be used for its metagenomic analysis. For instance:

1. For small DNA insert libraries, a high copy number vector with an inducible promoter is desired to maximize the chances of gene expression. Thus for small DNA fragments where the need is to study a single unknown gene, vectors like pUC, pBluescript *etc.* may be used.
2. For large DNA fragments, vectors like cosmids, fosmids, BACs (bacterial artificial chromosomes), *etc.* may be used. Large DNA fragments are analysed to study functions encoded by multiple genes, to capture complete biosynthetic pathway, or to study complete genome organisation. Among these mentioned vectors, BACs are most widely used due to their ability to accommodate inserts of over 300 kb size (Schizuya *et al.*, 1992).

Host Cell Selection

Generally *E.coli* is used as a host organism in metagenomic analysis (Henne *et al.*, 2000). For instance, the strain DH10B is employed for large insert BAC libraries. Method of transformation used is electroporation due to its high transformation efficiency.

16 sRNA is considered as an authentic phylogenetic marker because of its highly conserved nature and its involvement in polypeptide synthesis and is often used in phylogenetic analysis. This fact is taken into consideration while screening.

Screening

Metagenomic screening is often carried out with the aim of either reconstructing the genome of an organism or to know the function encoded by the cloned DNA. For this very basis, screening is driven by two approaches:

1. Sequence-based screening (sequence analysis)
2. Function-based screening (functional analysis)

Sequence Based Analysis

It was proposed by DeLong group (Stein *et al.,* 1996) and employs rRNA or other genes as "anchors" to identify clones of interest in the metagenomic libraries. PCR is done to amplify the anchor linked sequence to create a genomic library. Using sequence based analysis scientists can find a relationship between function and phylogeny. Sequence based analysis reveals physiological and ecological capacity of the transformant. However, there are certain limitations associated with this approach. Phylogenetic markers are not available for all types of genes and so it becomes difficult to authentically place all the accessed genomic fragments in the "Tree of Life" (Handelsman, 2004). As a result, sequencing of such genomic fragments does not help in determining the organisms of origin.

It is of two types:

1. Complete sequencing- this sequencing involves identification of clones based on methods that identify a particular sequence. For this, phylogenetic markers like RecA or 16sRNA gene are exploited to retrace the source of DNA fragment. Hence it aids to identify the taxonomic group of the organism and study the genomic fragments within a taxon. Although it is difficult to reconstruct the entire genome of an organiosm, still genomic data sheds light on the physiological and ecological features of organisms in the sample.
2. Random sequencing- In this method a gene of interest is identified in the sequenced genome and a search for phylogenetic markers is made in the flanking region of the gene of interest. These phylogenetic markers are then used for determining the phylogeny of the gene of interest. It provides knowledge about the distribution of functions among the members of the community, genomic organisation and the linkage of traits of different members. Random sequencing allows studying the behaviour of biofilms, mutualism and symbiosis phenomena.

Sequencing thus helps to reconstruct the genomes of the organisms that cannot be cultured.

Function Based Analysis

This type of analysis is used to assess a community's functions. It involves identification of clones that express certain functions of the metagenomic DNA. In order to identify an organism by its gene's protein, the protein must first be made. This depends on the transcription of the cloned DNA fragment followed by translation of the m-RNA so formed. This screening is further categorized into two wings depending upon how the protein (the translated gene product) is expressed:

1. Extracellular functional screening- It is done for the proteins that are secreted extracellularly by the transformant. This may involve selective enrichment of clones and subsequent amplification of the expressed gene

product. One limitation of this type of screening is that the cloned gene may not be able to express itself fully in the host or there may be a low expression of the inserted gene. The latter can be curbed by using proper amplification and selection methods. Also, high throughput sequencing can be used for selection (Zhou *et al.*, 2015).

2. Intracellular functional screening- (Sabree *et al.*, 2009) Sometimes the protein encoded by gene may not be secreted in the extracellular region, but rather remains intracellular. This calls for intracellular screening which is more rapid and sensitive as compared to extracellular screening and hence allows the identification of even poorly expressed gene products. Intracellular screening can be categorised further as:

   - ✰ METREX (Metabolite Regulated Expression) - This requires the recombinant vector (containing the metagenomic DNA) and the reporter plasmid to be present within the host itself. This allows for detection of low concentration of poorly expressed metagenomic gene product. The transcription of metagenomic DNA leads to intracellular expression of gene product. The reporter plasmid has three components:

     1. luxR gene
     2. lux I (promoter for gfp gene)
     3. Reporter gene- gfp gene (gene encoding green fluorescent protein) lying downstream of lux I promoter.

     The intracellular gene product (of metagenomic DNA) stimulates the dimerisation of the luxR proteins (metabolite produced by the luxR gene on reporter plasmid) and the complex so formed induces the transcription of the gfp gene by attaching to its promoter site (the luxI promoter). The gfp gene product produces a green coloured fluorescence on exposure to UV light and thus helps in the screening of such host cells.

   - ✰ SIGEX (Substrate Induced Gene Expression) - This screening involves the exposure of the clones to a substrate that induces gene expression. Here, the metagenomic DNA is entrapped in a vector which has a promoterless gfp gene present downstream of the cloning site. A suitable substrate is introduced which attaches to the promoter region in the metagenomic DNA and allows its transcription. Since the gfp gene is present downstream of metagenomic DNA, it also gets transcribed. This is now followed by translation of the gfp gene and the production of green fluorescent protein (responsible for giving fluorescence under UV light). Thus substrate induces the expression of gfp which can be easily screened.

Functional screening provides for the identification of new genes encoding known or unknown proteins. If many clones with a common function are identified they can be sequenced to find phylogenetic anchors and hence establish the genome structure flanking the DNA. This can be used as a method of selection for diverse genomes possessing the same function.

Molecular Biology Assays and Metagenomics

Various assays used in cell and molecular biology are being used for metagenomic analysis. For instance, the knowledge of diversity of rRNA gene sequences in the total metagenomic DNA (extracted from sample) is used to study structural diversity with the aid of techniques such as Denaturing Gradient Gel Electrophoresis (DGGE) (Muyzer *et al.*, 1993), Temperature Gradient Gel Electrophoresis (TGGE) (Muyzer, 1999), Amplified rDNA Restriction Analysis (ARDRA) (Liu *et al.*, 1997; Tiedje *et al.*, 1999) Terminal-Restriction Fragment Length Polymorphism (T-RFLP) (Liu *et al.*, 1997; Tiedje *et al.*, 1999) and Automated Ribosomal Intergenic spacer analysis (ARISA) (Fisher and Triplett, 1999). MAR (microautoradiography) coupled with FISH (Fluorescent in-situ Hybridisation) is also used for phylogenetic identification of members of a community (Rogers *et al.*, 2007).

Various techniques like Community Level Physiological Profiling (CLPP) (Preston-Mafham *et al.*, 2002), Stable Isotope Probing (SIP) (Manefield *et al.*, 2006), Immunocapture of Bromodeoxyuridine (Bru) DNA (Artursson and Jansson, 2003) and MAR-FISH are being widely used to assess functional diversity.

Microarrays: A Boon for Metagenomic Studies

Soil microarrays technology provides rapid, simple, reliable, and cost effective tool for detection of microorganisms. Microarrays are also a powerful tool for the simultaneous high throughput identification of many organisms in the soil. Microbial diagnostic microarrays (MDMs) have several thousand probes which are specific for different strains, species, genera or higher taxa and help in their identification (Bodrossy and Sessitsch, 2004).

On the basis of the probe and target molecules employed, MDMs can be classified as follows (Sessitsch *et al.*, 2006):

1. Phylogenetic Oligonucleotide Microarrays (phylochips)-These MDMs employ short oligonucleotides which detect a particular phylogenetic marker gene (like the 16SrRNA gene).
2. Functional Gene Arrays (FGAs)-These have oligonucleotides that target or identify genes of the function of interest. For example, genes like *amo*A is used to identify genes expressing nitrification, *nir*K and *nir*S for denitrification, *nif*H for nitrogen fixation, *nod*C for symbiotic nitrogen fixation, *dsr*AB for sulphite reduction, *etc.*
3. Community Genome Arrays (CGAs)- CGAs are based on Reverse Sample Genome Probing (RSGP), wherein the total DNA is extracted from an environmental sample, labelled and then allowed to hybridize on membranes on which denatured chromosomal DNA from relevant bacterial standards have been spotted (Greene and Voordouw, 2003). These arrays employ whole bacterial genomes as probes. With the metagenomic approach on the scene, these arrays can be used for genome reconstruction of uncultutarble members in a particular sample (community).

Thus, microarrays offer advantages of rapid detection, low cost, automation and low background levels.

Preponderant Applications of Metagenomics

Metagenomic analysis answers various questions and explains various phenomena. The following illustrations throw a light on some of the classic examples on advancements in the field of science which has been possible largely due to the development in metagenomics.

Identification of Unknown Microorganisms

As is evident from the "Great Plate Anomaly", the actual microbial world lies undiscovered. Metagenomic assays have revealed certain species which were earlier regarded as 'unculturable' and couldn't be discovered using usual cultivation techniques. Metagenomic studies have shown that Acidobacteria, Verrucomicrobia, Green non-sulphur bacteria and OP-11 divisions are most abundant species in unculutrable soil bacteria (Harris *et al.*, 2004). These species contribute 1-10 per cent of the total microbiome and constitute 10^7 to 10^8 cells per gram of soil sample. Sequencing can be used to reconstruct genomes of unculturable organisms in crude environments. Examples of successful sequence based analysis include studies of Acid Mine Drainage biofilm communities (Tyson *et al.*, 2004) and Sargossa Sea Analysis (Venter *et al.*, 2004).

Relationship between Phylogeny and Function

Metagenomics helps to draw a relation between phylogeny and known/ unknown functions of genes. This can be done by identifying an appropriate anchor marker (phylogenetic/functional) on a transformant that expresses a certain function or contains a sequence which is similar to the genes whose functions are known. This helps to establish a link between phylogeny and function.

Small Molecules and Metagenomics

Metagenomics has also uncovered the knowledge of proteins and small molecules encoded by genomes of microorganisms in soil. Several antibiotic resistance determinants have been found in soil. For example, *Mycobacterium* inhibiting antibiotic Terragine was discovered by metagenomic analysis of soil sample (Wang *et al.*, 2000). Metagenomic libraries have also shown the production of a set of diverse small molecules like violacein (Brady *et al.*, 2001), indirubin (Mac Neil *et al.*, 2001), acyl tyrosine (Brady and Clardy, 2000) and fatty dienic alcohols (Courtois *et al.*, 2000). Some of these molecules are found to be encoded by a single gene while some require collaborative expression of multiple genes for their production. *e.g.* Violacein is encoded collectively by 7 different genes (Brady *et al.*, 2001).

Sequence based analysis employing PCR amplification can be used to identify sequences encoding different antibiotics. For instance, polyketide synthase, the enzyme involved in synthesis of a class of antibiotics called polyketides, was first cloned from soil in a PCR based approach using primers that hybridised with the genes of polyketide synthase, thus allowing selective amplification of that gene (Ginolhac *et al.*, 2004).

This discovery of new molecules synthesised by microorganisms has led to the application of metagenomics to BIOPROSPECTING (such as discovery of novel antibiotics like acyl tyrosine, terragine and turbomycin). Metagenomics analysis has also discovered new enzymes (Uchiyama and Miyazaki, 2009) and antiporters (Xiang *et al.*, 2010).

Ecological Application

Metagenomics has been well exploited to study ecological phenomena such as symbiosis. Symbiotic relationship involves inter-dependence of microorganisms and hosts. Since the symbiotic microflora cannot be cultured outside the host, therefore these are widely studied using metagenomic analysis. For instance, Buchnera-aphid symbiosis was analysed by using bacterial DNA isolated from the insect. This symbiosis involves the uncultured organism *Buchnera aphidicola* and aphids (Moran and Mira, 2001; Abbot *et al.*, 2001). Sequenced data revealed a reduced genome size of *Buchnera*. Comparison with a reconstructed ancestral genome revealed that the functions encoded by the lost genes were being performed by the host insect indicating that this genome shrinkage is a result of the symbiotic lifestyle.

Metagenomics has also provided an insight into antagonism and mutualism among different microbial species. Genes responsible for microbial competition have been explored but their relevance is very ambiguous due to non-expression of these genes outside their actual environment owing to their dependence on limiting substrate and other species.

Metagenomics and Virology

Metagenomics has also been employed to characterise viruses (that are difficult to cultivate *in vitro* due to their obligatory parasitic nature) (Prabha *et al.*, 2013). This application faces various limitations like:

1. Separation of viral DNA from the free DNA sample.
2. Inactivation of the viral DNA as it can be infectious.
3. This inactivation may leave the virus unclonable.

However, since most of the hosts have not been cultured, metagenomics provides the only method to access these virus infected hosts. For instance in plant virology, this is done by coupling next generation sequencing techniques (*e.g.* ELISA, PCR, Microarrays) with metagenomic analysis in order to produce large quantity of cDNA sequences in a model system of tomato plants infected with Pepino-mosaic virus, and this helped in identification of these viruses in the infected tomato plants (Adams *et al.*, 2009).

Challenges to Metagenomics

Metagenomics has been a boon to microbiology in terms of explaining the yet unanswered questions. But since every assay has a dark side, same goes here.

Metagenomics faces various challenges that need to be taken care of, for successful implementation of this very beneficial analysis. These challenges include:

1. Samples (from which metagenomic DNA is to be extracted) like soil possess a high degree of heterogeneity both spatially and temporally owing to the non-uniform distribution and prokaryotic diversity. Due to this, a uniform sample cannot be obtained and hence replication of samples is not possible.
2. Since various communities are highly complex in terms of the number of species they harbour, so complete sequencing and genome reconstruction of each member of the community is a difficult task.
3. Another challenge is the low expression of certain clones or masking by over expressing clones. This calls for fast and effective selection and enrichment techniques which add to the cost of analysis. Also, some bacteria are reluctant to lysis and do not show up in clones or their genes or gene products may be toxic to the host and would not be maintained in the libraries.

Conclusion

The world of microorganisms invades almost every physically accessible and non-accessible region of Earth. They are the most useful resource gifted by nature and play a crucial role in sustaining life. However, the current access to such a potentially useful resource is very limited. Amidst this, metagenomics paves its way towards accessing knowledge about the yet unknown microorganisms which may help open new areas of industrial use and scientific research. Although the advent in Next Generation Sequencing Techniques gives a boost to the application of this new field, but the plausible potential of this tool still remains uncovered and unused. Nevertheless, this scientific field comes as a blessing in understanding various natural phenomena and the involvement of microorganisms. Metagenomics is undoubtedly a very promising tool which can aid in exploring the currently untouched and unfathomable aspects of the nature.

REFERENCES

Abbot, P., Withgott, J.H. and Moran, N.A. 2001. Genetic conflict and conditional altruism in social aphid csolonies. Proc. Natl. Acad. Sci. USA 98: 12068-12071.

Adams, I.P., Glover, R.H., Monger, W.A., Mumford, R., Jackeviciene, E., Navalinskiene, M., Samuitiene, M. and Boonham, N. 2009. Next-generation sequencing and metagenomic analysis: a universal diagnostic tool in plant virology. Mol Plant Pathol 10 (4): 537-545.

Alexander, M. 1977. Introduction to soil microbiology, 2nd edn. *New York, NY, USA: Wiley.*

Artursson, V. and Jansson, J.K. 2003. Use of bromodeoxyuridine immunocapture to identify active bacteria associated with arbuscular mycorrhizal hyphae. Appl Environ Microbiol 69: 6208–6215.

Bodrossy, L. and Sessitsch, A. 2004. Oligonucleotide microarrays in microbial diagnostics. Curr. Opin. Microbiol. 7*:* 245–254.

Brady S. F. and Clardy J. 2000. Long-chain *N*-acyl amino acid antibiotics isolated from heterologously expressed environmental DNA. J. Am. Chem. Soc. 122: 12903–12904.

Brady, S.F., Chao, C.J., Handelsman, J. and Clardy, J. 2001. Cloning and heterologous expression of a natural product biosynthetic gene cluster from eDNA. Org. Lett. 3 (13): 1981-1984.

Courtois, S., Cappellano, C. M., Ball, M., Francou, F. X., Normand, P., Helynck, G., Martinez, A., Kolvek, S. J., Hopke, J., Osburne, M. S., August, P. R., Nalin, R., Guerineau, M., Jeannin, P., Simonet, P. and Pernodet, J. L. 2003. Recombinant environmental libraries provide access to microbial diversity for drug discovery from natural products. Appl. Environ. Microbiol. 69**:** 49-55.

Davis, K.E.R., Joseph, S.J. and Janssen, P.H., 2005. Effects of growth medium, inoculum size, and incubation time on culturability and isolation of soil bacteria Appl. Environ. Microbiol. 71 (2): 826–834.

Delmont, T.O., Robe, P., Cecillon, S., Clark, I. M., Constancias, F., Simonet, P., Hirsch, P.R. and Vogel, T. M. 2011. Accessing the soil metagenome for studies of microbial diversity. Appl. Environ. Microbiol. 77(4): 1315-1324.

Drew, S. W. and Demain, A.L., 1977. Effect of primary metabolites on secondary metabolism. Ann. Rev. Microbiol., 31: 343-356.

Dunbar, J., Takala, S., Barns, S.M., Davis, J.A. *and* Kuske, C.R. 1999. Levels of bacterial community diversity in four arid soils compared by cultivation and 16S rRNA gene cloning. Appl. Environ. Microbiol. 65*:* 1662–1669.

Fisher, M.M. and Triplett, E.W. 1999. Automated approach for ribosomal intergenic spacer analysis of microbial diversity and its application to freshwater bacterial communities. Appl. Environ. Microbiol. 65: 4630- 4636.

Gillespie, D.E., Rondon, M.R., Williamson, L.L. and Handelsman, J. 2005. Metagenomic libraries from uncultured microorganisms. *In:* A.M. Osborn and C. J. Smith (eds.). Molecular microbial ecology. Francis and Taylor group. pp. 261-280.

Ginolhac A., Jarrin C., Gillet B., Robe P., Pujic P., Tuphile K., Bertrand H., Vogel T.M., Perrière G., Simonet P., Nalin R. 2004. Phylogenetic analysis of polyketide synthase I domains from soil metagenomic libraries allows selection of promising clones. Appl. Environ. Microbiol. 70: 5522–5527.

Greene, E.A. and Voordouw, G. 2003. Analysis of environmental microbial communities by reverse sample genome probing. J. Microbiol. Meth. 53*:* 211–219.

Handelsman, J. 2004. Metagenomics: application of genomics to unculutred microorganisms. Microbiol. Mol. Biol. Rev.68 (4): 669-685.

Harris, J.K., Kelley, S.T. and Pace, N.R. 2004. New perspective on uncultured bacterial phylogenetic division OP11. Appl. Environ. Microbiol. 70: 845-849.

Henne, A., Schmitz, R.A., Bomeke, M., Gottschalk, G. and Daniel, R. 2000. Screening of environmental DNA libraries for the presence of genes conferring lipolytic activity on Escherichia coli. Appl. Environ. Microbiol. 66: 3113-3116.

Hill, G.T., Mitkowski, N.A., Aldrich-Wolfe, L., Emele, L.R., Jurkonie, D.D., Ficke, A., Maldonado-Ramirez, S. Lynch, S.T. and Nelson, E.B., 2000.Methods for assessing th composition and diversity of soil microbial communities. Appl. Soil Ecol. 15: 25–36.

Hugenholtz, P., Goebel, B.M. and Pace, N.R., 1998.Impact of culture-independent studies on the emerging phylogenetic diversity J. Bacteriol. 180: 4765-4774.

Kirk, J.L., Beaudette, L.A., Hart, M., Moutoglis, P., Klironomos, J.N., Lee, H., and Trevorsa, J.T., 2004. J. Microbiol. Meth. 58: 169-188.

Liu, W.-t., Marsh, T.L., Cheng, H. and Forney, L.J. 1997.Characterization of microbial diversity by determining terminal restriction fragment length polymorphisms of genes encoding 16S rRNA. Appl. Environ. Microbiol. 63: 4516– 4522.

MacNeil, I.A., Tiong, C.L., Minor, C., August, P.R., Grossman, T.H., Loiacono, K.A., Lynch, B.A., Phillips, T., Narula, S., Sundaramoorthi, R., Tyler, A., Aldredge, T., Long, H., Gilman, M., Holt, D. and Osburne, M.S. 2001. Expression and isolation of antimicrobial small molecules from soil DNA libraries. J Mol. Microbiol. Biotechnol. 3 (2): 301-308.

Madsen, E.L. 2011. Microorganisms and their roles in fundamental biogeochemical cycles. Curr. Opin. Biotechnol., 22(3): 456-464.

Manefield, M., Griffiths, R.I., Bailey, M.J. and Whiteley, A. S. 2006. Stable isotope probing: a critique of its role in linking phylogeny and function. *In:* P. Nannipieri, K. Smalla (eds.). Nucleic acids and proteins in soil. Springer, Heidelberg, Germany, pp 205–216.

Moran, N. A. and Mira, A. 2001. The process of genome shrinkage in the obligate symbiont *Buchnera aphidicola.* Genome Biology, 2(12): research0054.1–0054.12.

Muyzer, G., 1999. DGGE/TGGE a method for identifying genes from natural ecosystems. Curr. Opin. Microbiol. 2: 317–322.

Muyzer, G., Waal, E.C.D., Uitterlinden, A.G. 1993. Profiling of complex microbial populations by denaturing gradient gel electrophoresis analysis of polymerase chain reaction-amplified genes coding for 16S rRNA. Appl. Environ. Microbiol. 59: 695– 700.

Pace, N.R., Stahl, D.A., Lane, D.J. and Olsen, G.J. 1985. Analyzing natural microbial populations by r-RNA sequences. ASM News 51: 4-12.

Prabha, K., Baranwal, V. K. and R. K. Jain. 2013. Applications of next generation high throughput sequencing technologies in characterization, discovery and molecular interaction of plant viruses. Indian J. Virol. 24 (2): 157-165.

Preston-Mafham, J., Boddy L. and Randerson, P.F. 2002. Analysis of microbial community functional diversity using sole-carbon-source utilisation profiles – a critique. FEMS Microbiol. Ecol., 42(1): 1–14.

Rogers, S.W., Moorman, T.B. and Ong, S.K. 2007. Fluorescent in situ hybridization and microautoradiography applied to ecophysiology in soil. Soil Sci. Soc. Am. J. 71: 620–631.

Sabree, Z. L, Rondon, M.R. and Handelsman, J. 2009.Metagenomics. pp. 622-632. Elsevier Inc.

Schmeisser C., Helen S. and Wolfgang R.S. 2007. Metagenomics, biotechnology with non-culturable microbes. Appl Microbiol Biotechnol 75: 955- 962.

Sessitsch, A., Hackl, E., Wenzl, P., Kilian, A., Kostic, T, Stralis-Pavese, N., Tankouo Sandjong, B. and Bodrossy, L. 2006.New Phytologist 171 : 719-736.

Shizuya, H., Birren,B., Kim, U.J., Mancino, V., Slepak, T., Tachiiri, Y. and Simon, M. 1992. Cloning and stable maintenance of 300-kilobase-pair fragments of human DNA in Escherichia coli using an F-factor-based vector. Proc Natl Acad Sci USA. 89(18): 8794–8797.

Smit, E., Leeflang, P., Gommans, S., Van Den Broek, J., Van Mil, S. and Wernars, K. 2001. Diversity and seasonal fluctuations of the dominant members of the bacterial soil community in a wheat field as determined by cultivation and molecular methods. Appl. Environ. Microbiol. 67: 2284–2291.

Sparling, G.P. 1997. Soil microbial biomass, activity and nutrient cycling as indicators of soil health. *In*: Biological indicators of soil health. C.E. Pankhurst, B.M. Doube and V.V.S.R. Gupta (eds.) pp. 97-119. CAB International, Wallingford, Oxon. UK.

Stein, J. L., Marsh, T. L., Wu, K. Y., Shizuya, H. and DeLong, E. F. 1996. Characterization of uncultivated prokaryotes: isolation and analysis of a 40-kilobase-pair genome fragment from a planktonic marine archaeon. J. Bacteriol. 178: 591–599.

Surzycki, S. 2000. General aspects of DNA isolation and purification. *In*: Basic techniques in molecular biology, Springer Lab Manuals, S. Surzycki (ed.), pp. 1-32. Springer-Verlag, Berlin Heidelberg.

Tiedje, J.M., Asuming-Brempong, S., Nusslein, K., Marsh, T.L. and Flynn, S.J. 1999. Opening the black box of soil microbial diversity. Appl. Soil Ecol. 13, 109– 122.

Torsvik, V., Daae, F.L., Sandaa, R.-A., Ovreas, L., 1998. Review article: novel techniques for analysing microbial diversity in natural and perturbed environments. J. Biotechnol. 64: 53–62.

Tringe S.G. and Rubin, E.M., 2005. Metagenomics: DNA sequencing of environmental samples. Nat. Rev. Genet., 6(11): 805-814.

Tyson, G.W., Chapman, J., Hugenholtz, P., Allen, E.E., Ram, R.J., Richardson, P.M., Solovyev V.V., Rubin E.M., Rokhsar, D.S. and Banfield, J.F. 2004. Community structureand metabolism through reconstruction of microbial genomes from the environment. Nature, 428: 37–43.

Uchiyama, T. and Miyazaki, K. 2009. Functional metagenomics for enzyme discovery: challenges to efficient screening. Curr. Opin. Biotechnol., 20: 616–622.

Venter, J.C., Remington, K., Heidelberg, J.F., Halpern, A.L., Rusch, D., Eisen, J.A., *et al.*, 2004. Environmental genome shotgun sequencing of the Sargasso Sea. Science 304: 66–74.

Wang, G.Y., Graziani, E., Waters, B., Pan, W., Li, X., McDermott, J., Meurer, G., Saxena, G., Andersen, R.J. and Davies, J. 2000. Novel natural products from soil DNA libraries in a Streptomycete host. Org. Lett. 2(16): 2401-2404.

Watanabe, K., 2001. Microorganisms relevant to bioremediation. Curr. Opin. Biotechnol., 12: 237-241.

Xiang, W., Zhang, J., Li, L., Liang, H., Luo, H., Zhao, J., Yang, Z., Sun, Q. 2010. Screening a novel Na^+/H^+ antiporter gene from a metagenomic library of halophiles colonizing in the Dagong Ancient Brine Well in China. FEMS Microbiol. Lett. 306: 22-29.

Zhou, J., He, Z., Yang, Y., Deng, Y., Tringe, S.G., Alvarez-Cohen, L. 2015. High-throughput metagenomic technologies for complex microbial community analysis: open and closed formats. mBio 6(1): e02288-14.

Chapter 15

Effects of Probiotics in Diabetes Mellitus: A Review

Shikha Jaggi

Department of Zoology, Gandhi Memorial National College, Ambala Cantt. – 133001, Haryana
E-mail: shikha.jaggi22@gmail.com

ABSTRACT

Diabetes mellitus is a multifactorial disorder of a major public health concern. Diabetes mellitus is a metabolic disease that leads to high blood sugar level due to either insulin insufficiency, insulin resistance or both. Interest in studying the human gut microbiota has only recently been emerged and it has been observed that human microbiota might play a major role in maintaining the homeostasis of human metabolism. Application of probiotics serves as a useful approach for modulating human microbiota. Present study aimed to investigate the effects of probiotics on oxidative stress, immune system, inflammatory response and expression of genes in diabetes mellitus. This review summarizes the potential mechanisms for the effect of probiotics in diabetes mellitus. It elucidates the capability of probiotics as a novel approach towards controlling diabetes. Probiotics can be efficiently used a valuable tool for the prevention and management of diabetes and its use has dramatically increased over the last decade. Based on the understanding of the advantages of probiotics in diabetes mellitus at the molecular level, use of probiotics should be investigated for their application in clinical sciences.

Keywords: *Gut microbiota, Oxidative stress, Probiotics, T1DM, and T2DM.*

Introduction

Diabetes mellitus is the most common complex metabolic disorder among the world's population. Diabetes mellitus which was long considered a disease of minor significance to world health is now one of the major threats in the 21st century. The past two decades have an explosive increase in the number of diabetics' worldwide. The number of people (20-79 years) with diabetes worldwide was 425 million in 2017. It is expected to rise to 629 million by the year 2045 (IDF, 2017).The epidemic of diabetes mellitus is accompanied by an increase in the incidence of diabetic complications including cardiovascular disease, retinopathy, nephropathy, microangiopathy, neuropathy and several other complications (Gale and Anderson, 2005).

Type 2 Diabetes mellitus (T2DM) is characterized by disorders of insulin action and insulin secretion, either of which may be predominant feature. Such patients are at increased risk of developing microvascular and macrovascular complications and do not need insulin treatment to survive. T2DM possesses strong genetic predisposition (Valle *et al.*, 1997). The risk of developing T2DM increases with age, obesity and lack of physical activity (Harris *et al.*, 1995). Risk factors associated with T2DM are family history of diabetes, overweight, unhealthy diet, physical inactivity, increasing age, high blood pressure, smoking, ethnicity, impaired glucose tolerance,history of gestational diabetes and poor nutrition during pregnancy. Type 1 Diabetes mellitus (T1DM) is an autoimmune condition, where the body's defense system attacks the insulin producing β-cells in pancreas. As a result, insulin is required by patients in order to control the blood glucose levels (IDF, 2017). T1DM is usually characterized by the presence of anti-GAD65, islet-cell or insulin antibodies (Humphrey *et al.*, 1998). Auto-antibodies to the tyrosine phosphatases IA-2 and IA-2β are also the markers of immune destruction of β-cells of pancreas. T1DM develops suddenly and its symptoms include abnormal thirst, frequent urination, lack of energy, extreme tiredness, constant hunger, sudden weight loss, slow healing wounds, recurrent infection, blurred vision, ketoacidosis, prone to autoimmune disorders such as Grave's disease, Hashimoto's thyroiditis, Addison's disease, Vitiligo, Celiac sprue, Autoimmune hepatitis, Myasthenia gravis and Pernicious anaemia (ADA, 2004).

Gut flora consists of microorganisms living in the digestive tracts of humans and is the largest reservoir of human flora with concentration of 1014 cells/ml and features over 5000 bacterial species with approximate weight of 1.5 kg (Larsen *et al.*, 2010).Among bacterial species, Bacteroidetes phylum (composed mainly of Gram– bacteria) and the Firmicutes phylum (composed mainly of Gram + bacteria) represents approximately 90 per cent of them.Other phyla among gut microbes include the Gram-positive Actinobacteria as well as the Gram-negative Verrucomicrobia and Proteobacteria. More than 90 per cent of the bacterial populations are obligate anaerobes (Eubacterium, Bifidobacterium, Bacteroides, Fusobacterium, Peptostreptococcu) (Tsukumo *et al.*, 2009). Our gut microbiota carries more than 3 million genes, which provides a broad range of functions and abilities for dynamical changes according to the environmental factors(Burcelin*et al.*, 2011). Medical conditions such as intestinal inflammatory diseases, diabetes, cancer, allergies, cardiovascular diseases and dyslipidaemia have been found to be associated with composition of intestinal microbiota. Altered intestinal microbiota contributes to the development of diabetes by increasing intestinal permeability as well as mucosal immune response.

Probiotics are live microorganisms that may employ advantageous impact on the microbial balance of the gut. Probiotics can modulate intestinal microbiota by facilitating the management of a number of clinical conditions. Probiotics have also been recognized as efficacious adjuvants in insulin resistance therapies and might be helpful in maintaining healthier gut microbiota (Floch and Montrose, 2005).They have anti-diabetes effect and delay the progression of hyperglycemia, hyperinsulinemia, dyslipidemia and oxidative stress (Naydenov *et al.*, 2012). Probiotic intake influences the composition of the gut flora, improves the strength of the intestinal epithelium, weakens the immune responses and diminishes the toll-like receptor 4 pathways, which leads to decreased pro-inflammatory signaling and increased insulin sensitivity (Miraghajani *et al.*, 2017). The aim of this chapter is to focus on the currently described effects of probiotics in the prevention and management of T1DM and T2DM.

The Effects of Probiotics on Oxidative Stress

Oxidative stress is defined as imbalance between the production of pro-oxidants (reactive oxygen species or free radicals) and antioxidant defences (Betteridge, 2000). Oxidative stress is a regular characteristic of diabetic complications when additional production of reactive oxygen species overburdens the action of antioxidant systems. In the hyperglycemic state, glucose toxicity, glucose auto-oxidation, oxidative phosphorylation leads to elevation of reactive oxygen species (Singh *et al.*, 2001). Antioxidant plays an important role to prevent diabetic complications because these fight or destroys free radicals or oxidants and also enhances the insulin secretion and insulin sensitization. Tanaka and co-workers (2002) reported that glucose toxicity leads to oxidative stress in pancreatic β-cells and glutathione peroxidise protects from oxidative stress. Free radicals are formed disproportionately in diabetes by glucose oxidation, nonenzymatic glycation of proteins, and the subsequent oxidative degradation of glycated proteins. Abnormally high levels of free radicals and the simultaneous decline of antioxidant defense mechanisms can lead to damage of cellular organelles and enzymes, increased lipid peroxidation, and development of insulin resistance. Changes in oxidative stress biomarkers, including catalase, superoxide dismutase, glutathione reductase, glutathione peroxidase, glutathione levels, vitamins, nitrite concentration, lipid peroxidation, non-enzymatic glycosylated proteins, and hyperglycemia occur in diabetes (Maritim *et al.*, 2003). Therefore, reduction of oxidative stress status may contribute towards management of diabetes.

Lactobacillus GG was observed to lower the level of MDA/gm of liver weight, which helps in repression of oxidative stress and ameliorates glucose tolerance (Tabuchi *et al.*, 2003). Matsuzaki and co-workers (1997) studied that *Lactobacillus casei* had inhibitory effect on the incidence of diabetes was partially dependent on nitric oxide production prevention. High fat diets leads to production of reactive oxygen species and reduces mucus production in intestine. Due to weakened intestinal barrier,intestinal bacteria translocate. Furthermore, as a result of oxidation of polyunsaturated fatty acids malondialdehyde is producted which damages epitheial cell membranes, increasing tight junction permeability. In patients with

T2DM, the strain *L. acidophilus La5 è B. lactis Bb12* reduced blood glucose levels, glycated hemoglobin and remarkably improved antioxidant status and total serum antioxidant capacity along with elevation of erythrocyte superoxide dismutase and glutathione peroxidase levels (Ejtahed *et al.*, 2011).

On the other side, probiotic foods have shown to protect against occurrence of diabetes by repressing oxidative stress. Probiotics act as scavengers of reactive oxygen species, chelate metal ion, inhibit pro-oxidant enzyme and reduce and inhibit ascorbate oxidation. In one study, probiotic yogurt consumption was observed to increase the levels and activity of glutathione peroxidase and erythrocyte superoxide dismutase enzymes, scavenging free radicals, and improving the total antioxidant status (Ejtahed *et al.*, 2012).The fermented milk products containing probiotic bacteria administration decreased thiobarbituric acid reactive substances and increased glutathione levels in the liver and pancreatic tissues of diabetic rats. These observations suggested the good antioxidant properties of the drink (Yadav *et al.*, 2007). Probiotic milk inhibits lipid peroxidation and exerts antioxidant effects through preserving the activity of antioxidant enzymes (Yadav *et al.*, 2008). The consumption of dahi, a traditional Indian milk form, containing *L. acidophilus, L. casei and L. lactis* has been reported to reduce the glycemic curve, lipid peroxidation and HbA1c.After consumption of yogurt containing 7,23 x 10^6 CFU of *L. acidophilus* La5 and 6,04 x 10^6 CFU of *B. lactis* Bb12 for 6 weeks, T2DM patients showed decreased HbA1c levels and fasting glucose, followed by increased activity of superoxide dismutase and glutathione peroxidase than control group (Gomes *et al.*, 2014).

The Effects of Probiotics on Immune System and Inflammatory Response

The gut microbiota is capable of regulating the immune response and therefore autoimmunity. The influence of intestinal bacteria in the origination and development of T1DM has been demonstrated by Atkinson and Chervonsky (2012). Interaction between colonizing the gastrointestinal mucosa bacteria and the underlying epithelial and mucosal lymphoid elements stimulates the host defensive mechanism in the gut (Shi and Walker, 2004). *L.casei* was found to suppress the effector functions of $CD4^+$ T cells, accompanied by reducing the pro-inflammatory molecules (So *et al.*, 2008).It was observed that in response to *Clostridium difficile* toxin A, the probiotic fungus *Saccharomyces boulardii* enhances the specific IgG and IgA antibodies. *Lactobacillus rhamnosus* strain GG along with rotaviral gastroenteritis resulted in increased IgA response in infants (Qamar *et al.*, 2001).In streptozotocin induced diabetic rats, the strain *L. casei Shirota* significantly decreased the blood levels of pro-inflammatory cytokine IL-6 and acute-phase protein CRP and concentration of neutrophils was lowered (Zarfeshani*et al.*, 2011). Studies have revealed that in type 2 diabetes mice, reduction in *Bacteroides* and *Prevotella* is associated with significant decrease of metabolic endotoxemia and inflammation (Cani *et al.*, 2009). Lipopolysaccharides present in outer membranes of gram-negative bacteria act as potent stimulators of inflammation and inflammatory response and play an important role in the development of diabetes (Allcock *et al.*, 2001).

Consumption of probiotics might stimulate immune system, decrease cholesterol blood levels, protect against respiratory and intestinal diseases, and reduce inflammatory responses and anti-tumorigenic effects. These purported health claims originated from the ability of probiotics to secrete antimicrobial substances, compete with other pathogens, strengthen the intestinal barrier and modulate the immune system. *Bifidobacteria* and *lactobacilli* are the most frequently used strains in dietary supplements and functional foods. VSL#3, a commercially available mixture of probiotics containing 3 x 10^{11} CFU/g of *Bifido bacterium longum, B. infantis and B. breve,* has been observed to enhance insulin signaling and lower inflammation in the adipose tissue of ApoE–/– rats (Gomes *et al.*, 2014). Oral administration of *Bifido bacterium spp*. was observed to reduce blood glucose levels and elevated the protein expressions of insulin receptor beta, insulin receptor substrate 1, protein kinase B, IKKα, IκBα and extracellular-signal-regulated kinase 2. *Bifidobacterium* spp. was also capable of inducing the expression of adiponectin and decreasing the expression of both macrophage chemoattractant protein-1 and interleukin-6. Thus, *Bifidobacterium spp*. may play a promising role in treatment of diabetes (Le *et al.*, 2015).

Lactobacillus kefiranofaciens M and *Lactobacillus kefiri K* administration alleviated the progression of T1DM in streptozotocin-induced T1D mice. This effect might be due to stimulating effect on the secretion of GLP-1 and inhibitory effect on the proinflammatory and inflammatory cytokines. These strains might elevate the production of IL-10 and modify the intestinal microbiota (Wei*et al.*,2015). *Lactobacillus johnsonii* (Ljo) N6.2 strain could potentially inhibit T1DM through mediating T-helper 17 bias within the mesenteric lymph nodes. Retention in the Th17 differentiation state and not converting to a Th1 state is critical to diabetogenesis and prevents or delays the beginning of type 1 diabetes mellitus (Lau *et al.*,2011). Toll-like receptors (TLRs) have immunoregulatory effects on anti-inflammatory cytokines. Stimulation of TLRs through systemic administration of TLR ligands is potentially capable of suppressing the onset of both allergic and autoimmune diseases. Effects of probiotics mediated through TLR pathway involves immune-regulatory cytokines like transforming growth factor (TGF)-β and interleukin IL-10 as well as some regulatory T cells in diabetes (Aumeunier *et al.*, 2010).

The Effects of Probiotics on Gene Expressions

Lactobacillus strains are well known probiotics which may stimulate up-regulation of mucous genes in intestinal goblet cells (Mack *et al.*, 1999).The gut microbiota composition of T1DM children showed higher occurrence of virulence factors, phage, prophage, motility genes and higher response to stress (Brown *et al.*, 2011). Experimental studies have shown the administration of *Lactobacillus johnsonii* may increase the gene expression of claudin-1 and decrease oxidative stress in diabetes-prone rats (Gomes *et al.*, 2014). A study with the help of DNA microarray indicated that *lactobacilli* have the capability to modulate the regulation of genes encoding particular adherence junction proteins such as E-cadherin and β-catenin (Hummel *et al.*, 2012). Balakumar and co-workers (2018) reported the anti-diabetic effect of probiotic strains of Indian gut origin (MTCC 5690 and MTCC 5689) in high-fat diet-fed C57BL/6J mice by regulating the gene expression patterns of intestinal tight junction markers (occludin and ZO-1), proinflammatory gene markers (TNFα and IL6) and mRNA expression of FIAF and adiponectin. Probiotics were observed to have similar effects to that of standard drugs such as metformin and vildagliptin. *Lactobacillus rhamnosus* GG intervention alleviated ER stress gene expression in skeletal muscle and suppressed macrophage activation, which resulted in improvement of glucose tolerance in db/db mice (Park *et al.*, 2015).

Certain studies on interplay between gene expression and probiotics have suggested that carrot juice fermented with *Lactobacillus plantarum NCU116* was capable of ameliorating T2DM through up regulation of glucose transporter-4 (GLUT-4)mRNA (Li *et al.*, 2014). Moreover, NCU can also maintain insulin sensitivity and glucosehomeostasis in diabetic rats by regulating gene expression of PPAR-α and PPAR-γ. These genes play important roles in glucose homeostasis and inflammation (Soares *et al.*, 2013).Oral administration of *Bifidobacterium* spp. enhanced protein expression involved in the insulin-signaling pathway, such as IR-β, IRS-1 and Akt and reduced blood glucose levels (Le *et al.*, 2015). *Bifidobacterium lactis* HY8101 increased the mRNA levels of GLUT4 and several insulin sensitivity-related genes (PPAR-γ). It upregulated the glycogen synthesis-related gene pp-1 and down-regulated the hepatic gluconeogenesis-regulated genes (PCK1 and G6PC) in mice. *Bifidobacterium lactis* HY8101 has focussed on the mechanisms of probiotics action and their importance in regulating diabetes consequences (Kim*et al.*, 2014). Zhang and co-workers(2014) reported the beneficial effects of *Lactobacillus casei* in protecting against T2DM onset by suppressing hyperglycaemia. It occurs through microbiota-mediated body chloride ion influx by upregulating of chloride ion-dependent genes (ClC1-7, GlyRα1, SLC26A3, SLC26A6, GABAAα1, Bestrophin-3 and CFTR). In another study, multispecies *Lactobacillus*- and *Bifidobacterium*-containing probiotic mixture was found potent against metabolic disorders like T2DM. It significantly down-regulated expression of several genes encoding pro-inflammatory adipokines and leukotriene pathway enzymes like CCL-2, IL-6 and leukotriene C4 synthase in adipose tissue while leukotriene C4 synthase and leukotriene A4 hydrolase in the gut (Holowacz *et al.*, 2015).

Conclusion

Diabetes might be associated with significant changes in the composition and metabolic functions of gut microbiota. So, it is especially necessary to develop the strategy that modifies the gut microbiota. Hence, the intestinal microflora may be an important new target for treatment of metabolic disorders. Probiotics have numerous beneficial roles in health and diseases. It reduces extrinsic nuisances, inhibits inflammation and has antioxidant capacity via reducing and preventing the formation of reactive free radicals with improvement of antioxidant defense in diabetic patients. The evidence available from experimental studies and clinical trials supports our suggestion that molecular mechanism underlying anti-diabetic effects of probiotics may be due to oxidative stress reduction, weakened inflammatory response, immunomodulation and modification of the intestinal microbiota. Further research is needed to confirm the underlying pathways involved in the advantageous effects from each probiotic strain as well as evaluation of other confounding factors. Need of the hour is to spread awareness among the masses regarding fruitful implications of probiotics in diabetic patients. The findings discussed here give an insight for further hypothesis aiming to elucidate the molecular mechanisms regarding impacts of probiotics on oxidative stress, inflammatory response, immune system and gene expression in diabetes mellitus.

REFERENCES

ADA. 2004. Diagnosis and classification of diabetes mellitus. *Diabetes Care*. 27(1): S5-S10.

Allcock, G.H., Allegra, M., Flower, R.J. and Perretti, M. 2001. Neutrophil accumulation induced by bacterial lipopolysaccharide: effects of dexamethasone and annexin 1. *Clin Exp Immunol*, 123: 62-67.

Atkinson, M.A. and Chervonsky, A. 2012. Does the gut microbiota have a role in type 1 diabetes? Early evidence from humans and animal models of the disease. *Diabetologia*, 55: 2868–28677.

Aumeunier, A., Grela, F., Ramadan, A., Pham Van, L., Bardel, E., Gomez Alcala, A., Jeannin, P., Akira, S., Bach, J.F. and Thieblemont, N. 2010. Systemic Toll-like receptor stimulation suppresses experimental allergic asthma and autoimmune diabetes in NOD mice. *PloS One*, 5(7): e11484.

Balakumar, M., Prabhu, D., Sathishkumar, C., Prabu, P., Rokana, N., Kumar, R., Raghavan, S., Soundarajan, A., Grover, S., Batish, V.K., Mohan, V. and Balasubramanyam, M. 2018. Improvement in glucose tolerance and insulin sensitivity by probiotic strains of Indian gut origin in high-fat diet-fed C57BL/6J mice. *Eur J Nutr*, 57(1): 279-295.

Betteridge, D.J. 2000. What is oxidative stress? *Metabolism*. 49(2): 3-8.

Brown, C.T., Davis-Richardson, A.G., Giongo, A., Mukherjee, N., Novelo, L.L., Casella, G., Drew, J.C., Ilonen, J., Knip, M., Hyöty, H., Veijola, R., Simell, T., Simell, O., Neu, J., Wasserfall, C.H., Schatz, D., Atkinson, M.A. and Triplett, E.W. 2011. Gut microbiome metagenomics analysis suggests a functional model for the development of autoimmunity for type 1 diabetes. *Plos One*, 6: e25792.

Burcelin, R., Serino, M., Chabo, C., Blasco-Baque, V. and Amar, J. 2011. Gut microbiota and diabetes: from pathogenesis to therapeutic perspective. *Acta Diabetol*, 48: 257-273.

Cani, P.D., Possemiers, S., Van de Wiele, T., Guiot, Y., Everard, A., Rottier, O., Geurts, L., Naslain, D., Neyrinck, A., Lambert, D.M., Muccioli, G.G. and Delzenne, N.M. 2009. Changes in gut microbiota control inflammation in obese mice through a mechanism involving GLP-2-driven improvement of gut permeability. *Gut*, 58: 1091-1103.

Ejtahed, H.S., Mohtadi-Nia, J., Homayouni-Rad, A., Niafar, M., Asghari-Jafarabadi, M., Mofid, V. and Akbarian-Moghari, A. 2011. Effect of probiotic yogurt containing *Lactobacillus acidophilus* and *Bifidobacterium lactis* on lipid profile in individuals with type 2 diabetes mellitus. *J Dairy Sci*, 94: 3288-3294.

Ejtahed, H.S., Mohtadi-Nia, J., Homayouni-Rad, A., Niafar, M., Asghari-Jafarabadi, M. and Mofid, V. 2012. Probiotic yogurt improves antioxidant status in type 2 diabetic patients. *Nutrition*, 28(5): 539-543.

Floch, M.H. and Montrose, D.C. 2005. Use of probiotics in humans: an analysis of the literature. *Gastroenterol Clin North Am*, 34: 547–570.

Furet, J.P., Kong, L.C., Tap, J., Poitou, C., Basdevant, A., Bouillot, J.L., Mariat, D., Corthier, G., Doré, J., Henegar, C., Rizkalla, S. and Clément, K. 2010. Differential adaptation of human gut microbiota to bariatric surgery-induced weight loss: links with metabolic and low-grade inflammation markers. *Diabetes*, 59: 3049–3057.

Gale, E.A.M. and Anderson, J.V. 2005. Diabetes mellitus and other disorders of metabolism. In: Kumar P, Clark M (Eds.) 6th edn. Philadelphia, Pa, USA; Saunder Elsevier, pp. 1101-1115.

Goffau, M., Luopajärvi, K., Knip, M., Ilonen, J., Ruohtula, T., Härkönen, T., Orivuori, L., Hakala, S., Welling, G.W., Harmsen, H.J. and Vaarala, O. 2013. Fecal microbiota composition differs between children with beta-cell autoimmunity and those without. *Diabetes*, 62: 1238–1244.

Gomes, A.C., Bueno, A.A., de Souza, R.G. and Mota, J.F. 2014. Gut microbiota, probiotics and diabetes. *Nutrition Journal*, 13: 60. doi: 10.1186/1475-2891-13-60.

Harris MI, Cowie CC, Stern MP, Boyko ES, Reiber GE and Bennett PH (1995) Diabetes in America. 2nd edn. Washington DC: US Government Printing Office.

Holowacz, S., Guigné, C., Chêne, G., Mouysset, S., Guilbot, A., Seyrig, C. and Dubourdeau, M. 2015. A multispecies *Lactobacillus*- and *Bifidobacterium*-containing probiotic mixture attenuates body weight gain and insulin resistance after a short-term challenge with a high-fat diet in C57/BL6J mice. *PharmaNutrition*, 3(3): 101-107.

Hummel, S., Veltman, K., Cichon, C., Sonnenborn, U. and Schmidt, M.A. 2012. Differential targeting of the E-Cadherin/β-Catenin complex by gram-positive probiotic *Lactobacilli* improves epithelial barrier function. *Appl Environ Microbiol*, 78: 1140–1147.

Humphrey, A.R.G., McCarty, D.J., Mackay, I.R., Rowley, M.J., Dwyer, T. and Zimmet, P. 1998. Autoantibodies to glutamic acid decarboxylase and phenotypic features associated with early insulin treatment in individuals with adult-onset diabetes mellitus. *Diabetes Medicine*. 15(2): 113-119.

IDF. 2017. IDF Diabetes Atlas, 8th edn. Brussels, Belgium: International Diabetes Federation. Retrieved on 03-12-2018 from http: //diabetesatlas.org/resources/2017-atlas.html.

Naydenov, K., Anastasov, A., Avramova, M., Mindov, I., Tacheva, T., Tolekova, A. and Vlaykova, T. 2012. Probiotics and diabetes mellitus. *Trakia Journal of Sciences*, 10(1): 300-306.

Kim, S.H., Huh, C.S., Choi, I.D., Jeong, J.W., Ku, H.K., Ra, J.H., Kim, T.Y., Kim, G.B., Sim, J.H. and Ahn, Y.T. 2014. The anti-diabetic activity of *Bifidobacterium lactis* HY8101 *in vitro* and *in vivo*. *J Appl Microbiol*, 117(3): 834-845.

Larsen, N., Vogensen, F.K., Berg, F.W., Nielsen, D.S., Andreasen, A.S., Pedersen, B.K., Al-Soud, W.A., Sørensen, S.J., Hansen, L.H. and Jakobsen, M. 2010. Gut microbiota in human adults with type 2 diabetes differs from non-diabetic adults. *PLoS One*, 5: e9085.

Lau, K., Benitez, P., Ardissone, A., Wilson, T.D., Collins, E.L., Lorca, G., Li, N., Sankar, D., Wasserfall, C., Neu, J., Atkinson, M.A., Shatz, D., Triplett, E.W. and Larkin, J. 2011. Inhibition of type 1 diabetes correlated to a *Lactobacillus johnsonii* N6.2-mediated Th17 bias. *J Immunol*, 186(6): 3538-3546.

Le, T.K., Hosaka, T., Nguyen, T.T., Kassu, A., Dang, T.O., Tran, H.B., Pham, T.P., Tran, Q.B., Le, T.H.and Pham, X.D. 2015. *Bifidobacterium species* lower serum glucose, increase expressions of insulin signaling proteins, and improve adipokine profile in diabetic mice. *Biomed Res*, 36(1): 63-70.

Li, C., Ding, Q., Nie, S.P., Zhang, Y.S., Xiong, T. and Xie, M.Y. 2014. Carrot juice fermented with *Lactobacillus plantarum* NCU116 ameliorates type 2 diabetes in rats. *J Agric Food Chem*, 62(49): 11884-11891.

Mack, D.R., Michail, S., Wei, S., McDougall, L. and Hollingsworth, M.A. 1999. Probiotics inhibit enteropathogenic *E. coli* adherence in vitro by inducing intestinal mucin gene expression. *Am J Physiol*, 276: G941-950.

Maritim, A.C., Sanders, R.A. and Watkins, J.B. 2003. Diabetes, oxidative stress, and antioxidants: a review. *Journal of Biochemical and Molecular Toxicology*. 17(1): 24-38.

Matsuzaki, T., Nagata, Y., Kado, S., Uchida, K., Hashimoto, S. and Yokokura, T. 1997. Effect of oral administration of *Lactobacillus casei* on alloxan-induced diabetes in mice. *APMIS*, 105(8): 637-642.

Miraghajani, M., Dehsoukhteh, S.S., Rafie, N., Hamedani, S.G., Sabihi, S. and Ghiasvand, R. 2017. Potential mechanisms linking probiotics to diabetes: a narrative review of the literature. *Sao Paulo Med J*, 135(2): 169-178.

Murri, M., Leiva, I., Gomez-Zumaquero, J.M., Tinahones, F.J., Cardona, F., Soriguer, F. and Queipo-Ortuño, M.I. 2013. Gut microbiota in children with type 1 diabetes differs from that in healthy children: a case–control study. *BMC Med*, 11: 46. doi: 10.1186/1741-7015-11-46.

Park, K.Y., Kim, B. and Hyun, C.K. 2015. *Lactobacillus rhamnosus* GG improves glucose tolerance through alleviating ER stress and suppressing macrophage activation in db/db mice. *J Clin BiochemNutr*, 56(3): 240-246.

Qamar, A., Aboudola, S., Warny, M., Michetti, P., Pothoulakis, C., LaMont, J.T. and Kelly, C.P. 2001. *Saccharomyces boulardii* stimulates intestinal immunoglobulin A immune response to Clostridium difficile toxin A in mice. *Infect Immun*, 69: 2762-2765.

Shi, H.N. and Walker, A. 2004. Bacterial colonization and the development of intestinal defences. *Can J Gastroenterol*, 18: 493-500.

Singh, R., Barden, A., Mori, T. and Beilin, L. 2001. Advanced glycation end-products: a review. *Dibetologia*. 44: 129-146.

So, J.S., Kwon, H.K., Lee, C.G., Yi, H.J., Park, J.A., Lim, S.Y., Hwang, K.C., Jeon, Y.H. and Im, S.H. 2008. *Lactobacillus casei* suppresses experimental arthritis by down-regulating T helper 1 effector functions. *Mol Immunol*, 45: 2690-2699.

Soares, F.L., de Oliveira Matoso, R., Teixeira, L.G., Menezes, Z., Pereira, S.S., Alves, A.C., Batista, N.V., de Faria, A.M., Cara, D.C., Ferreira, A.V. and Alvarez-Leite, J.I. 2013. Gluten-free diet reduces adiposity, inflammation and insulin resistance associated with the induction of PPAR-alpha and PPAR-gamma expression. *J NutrBiochem*, 24(6): 1105-1111.

Tabuchi, M., Ozaki, M., Tamura, A., Yamada, N., Ishida, T., Hosoda, M. and Hosono, A. 2003. Antidiabetic effect of *Lactobacillus* GG in streptozotocin-induced diabetic rats. *BiosciBiotechnolBiochem*, 67(6): 1421-1424.

Tanaka, Y., Tran, P.O., Harmon, J. and Robertson, R.P. 2002. A role for glutathione peroxidase in protecting pancreatic beta cells against oxidative stress in a model of glucose toxicity. *Proceedings of National Academy of Science*. 99(19): 12363-12368.

Tsukumo, D.M., Carvalho, B.M., Carvalho-Filho, M.A. and Saad, M.J. 2009. Translational research into gut microbiota: new horizons in obesity treatment. *Arq Bras Endocrinol Metabol*, 53: 139-144.

Valle, T., Tuomilehto, J. and Eriksson, J. 1997. Epidemiology of NIDDM in Europoids. In: Alberti K, Zimmet PZ, De Fronzo RA, Keen H (Eds.) International Textbook of Diabetes Mellitus. 2nd edn. Chichester, New York: John Wiley: 125-142.

Wei, S.H., Chen, Y.P. and Chen, M.J. 2015. Selecting probiotics with the abilities of enhancing GLP-1 to mitigate the progression of type 1 diabetes *in vitro* and *in vivo*. *Journal of Functional Foods*, 18(Part A): 473-486.

Yadav, H., Jain, S. and Sinha, P.R. 2007. Antidiabetic effect of probiotic dahi containing *Lactobacillus acidophilus* and *Lactobacillus casei* in high fructose fed rats. *Nutrition*, 23(1): 62-68.

Yadav, H., Jain, S. and Sinha, P.R. 2008. Oral administration of dahi containing probiotic *Lactobacillus acidophilus* and *Lactobacillus casei* delayed the progression of streptozotocin-induced diabetes in rats. *J Dairy Res*, 75(2): 189-195.

Zarfeshani, A., Khaza'ai, H., Mohd Ali, R., Hambali, Z., Wahle, K.W.J. and Mutalib, M.S.A. 2011. Effect of *Lactobacillus casei* on the production of pro-Inflammatory markers in streptozotocin-induced diabetic rats. *Probiotics and AntimicroProt*, 3: 168-174.

Zhang, Y., Guo, X., Guo, J., He, Q., Le, H., Song, Y. and Zhang, H. 2014. *Lactobacillus casei* reduces susceptibility to type 2 diabetes via microbiota-mediated body chloride ion influx. *Sci Rep*, 4: 5654.

Chapter 16

Microbial Urease: Target Molecule in Medicine and Agriculture

*Neha Sikri, Manoj Kumar and Sunita Dalal**

Department of Biotechnology, Kurukshetra University, Kurukshetra – 136 119, Haryana
****E-mail: sdalal@kuk.ac.in***

ABSTRACT

Urease is an important hydrolase that attracted much attention for its impact on organism's health and surrounding atmosphere as it acts on the urea molecule thereby leads to the release of basic molecule viz. ammonia and carbonic acid. It is an important virulence factor for various bacterial species viz. *Helicobacter pylori, Streptococcus salivarius, Staphylococcus capitis* subsps. *urealiticum, Gardnerella vaginalis Lactococcus lactis/cremoris, Enterococcus durans, Lactobacillus fermentum* and fungal species are *Coccidiods immitis, Sporothrix schenckii, Trichosporon* species and *Histoplasma capsulatum, Schizosaccharomyces pombe, Cryptococuus neofomans* as well as atmospheric transformation of different nitrogenous compounds. Diseases caused by urease action are urolithiasis, pyelonephritis, ammonia encephalopathy, hepatic encephalopathy hepatic coma, gastro duodenal infections, infections related to urinary tract, carbonate stones formation. So, there is need to develop novel, selective and nontoxic urease inhibitor. In this article we discussed the microbial origin of urease and various urease inhibitors discovered so far especially in the recent years.

Keywords: *Urease inhibitor, Helicobacter pylori, Peptic ulcers, NBPT (N-(butyl) thiophosphoric triamide).*

Introduction

Urease was the first enzyme to be crystallized by Sumner in 1926. He stated that urease is a pure protein that might function as an enzyme (Mobley *et al.*, 1995). It is a highly efficient catalyst for the hydrolysis of urea with a rate approximately

10^{14} times the rate of non catalyzed reaction. The three dimensional structure of this Ni^{2+} dependent hydrolytic enzyme, from various sources has been elucidated (Dixon *et al.*, 1975). The Enzyme Ureases (EC 3.5.1.5) belong to the superfamily of amidohydrolases and phosphotriestreases. In simplified form reaction can be written as: $(NH_3)_2\ CO + H_2O \rightarrow CO_2 + 2NH_3$

It is an enzyme that catalyses the hydrolysis of urea to produce ammonia and carbamate; the carbamate produced is subsequently degraded by spontaneous hydrolysis to produce another ammonia and carbonic acid (Mobley *et al.*, 1995).

Urease is a significant microbial enzyme playing multifunctional role in microbial physiology. As the urease leads to hydrolyses of its substrate urea to a basic molecule, ammonia, the net effect causes increase in pH of environment surrounding urease producing microorganisms. Urease is also considered as stress response to counteract low environment pH(Cotter and Hill, 2003).

Although Urease is being produced by different microorganisms, invertebrates like *Aplysia californica*, plants like Jack bean(*Canavalia ensiformis*), Mulberry leaves (*Morus alba*), Soyabean (*Glycine max*), *Arabidopsis*,water melon (*Citrullus lanatus*), Yard long bean (*Vigna unguiculata ssp. Sesquipedalis*), Arahar dal (*Cajanus cajan*) and seeds of Sword bean (*Canavalia gladiata*). Urease producing bacteria are *Helicobacter pylori, Streptococcus salivarius, Staphylococcus capitis* subsps. *urealiticum, Gardnerella vaginalis Lactococcus lactis/cremoris, Enterococcus durans, Lactobacillus fermentum.* Fungi producing ureases are *Coccidiods immitis, Sporothrix schenckii, Trichosporon* species and *Histoplasma capsulatum, Schizosaccharomyces pombe, Cryptococuus neofomans etc.* Most of these fungal species are pathogenic to humans (Mobley *et al.*, 1995; Sirko and Brodzik, 2000; Mobley and Hausinger, 1989; Suzuki *et al.*, 1979; Murchan *et al.*, 2004; Jin *et al.*, 2004; Clememns *et al.*, 1995; Dupuy *et al.*, 1997; Lam and Yeo, 1980; Futagmi *et al.*, 1998, Nakano *et al.*, 2001, Orth *et al.*, 2006, Tange and Niwa, 1997, Kaniecza *et al.*, 2012; Yu *et al.*, 1997; Zonia *et al.*, 1995. Urease plays different roles in different organisms like in bacteria it transform urea to ammonia which is used as a nitrogen source by them whereas in plants used to recycle nitrogen and in soil, it (released upon cellular lysis) is bound to organic and inorganic soil components (clays and humic substances) and protected from degradation by extracellular proteolytic enzymes and microorganism that acts as a important enzyme for nitrogen cycle (Pedrozo *et al.*, 1996).

Human Urease Microbiota is Associated with Health

In most of the microorganisms urease is mainly linked with the nitrogen assimilation and recycling of nitrogenous wastes *i.e.* mainly urea. Major nitrogen source for bacteria that colonize human body is ammonium that released from the urea present in the gland secretions. Different oral bacteria viz. *Streptococcus salivarius, Actinomycetes, Naeslumdiii* and many other species of gut produce active urease and causes urea hydrolysis. Human milk also contains its 15 per cent nitrogen in the form of urea (Krajewska, 2009; Chen and Burne, 2003; Bergeron *et al.*, 2000; Wegmann *et al.*, 2014; Yatsunenko *et al.*, 2012). So the infant do not contain much nitrogen. As the urea is being secreted by different parts of the human body, so the urease activity which leads to the urea hydrolysis has positive as well as negative effects. Urease activity associated with plaque decrease the occurrence of dental carries whereas urease activity in the saliva promoted the development of carries (Fuller and Reeds,1998). Out of various urease positive bacterial species, one of the significant bacterial pathogens, *H. pylori* is a gram- negative, helix- shaped bacterium which is able to develop infection in different parts of digestive tract (Moron-Bermudez *et al.*, 2011; Burne and Chen, 2000; Mollenhauer-Rektorschek *et al.*, 2000), it leads to its survivability in human stomach. *H. pylori* infection also leads to the gastric mucosal injury (Skouloubris *et al.*, 1998) as it attaches to the stomach lining, whereby the urea hydrolysis it results in the increase in the pH of the mucosal lining. This increase in the pH prevents the movement of H ions between the gastric gland and gastric lumen. In addition, ammonia produced as a result of urea hydrolysis have an effect on intercellular tight junctions increasing permeability and also disrupting the gastric mucous membrane of the stomach. The other urease positive bacterial species is *Streptococcus thermophilus* which is widely used in synthesis of different dairy products. In this bacterial species urease by its action of urea hydrolysis increase the catabolic efficiency by increasing the activity of β-galactosidases, lactate dehydrogenases and glycolytic enzymes thereby metabolically regulate the biosynthetic pathways involved in the different amino acids (Smoot *et al.*, 1990; Arioli *et al.*, 2009). This action of urease leads to the increase in pH inside as well as outside the cell as the ammonia produced rapidly diffuses out of the cell. The bactera of gastrointestinal tract of human *Bifidobacterium logum subsps. Infantis*, *Klebsiella* and *Proteus* species are the reason of development of hepatic encephalopathy(HE) which occurs due to the high level of gut derived ammonia (Arioli *et al.*, 2010; Locascio *et al.*, 2010). In addition, urease is being identified as a virulence factor for several microbial pathogens which includes different bacteria such as *Clostridium perfringens, Helicobacter pylori, Klebsiella pneumoniae, Proteus mirabilis, Salmonella spp., Staphylococcus saprophyticus, Ureoplasma urealyticum, and Yersinia enterocolitica*, and this leads to the various infections such as hepatic coma, gastroduodenal infections, ammonia encephalopathy and many more (Riordan and William, 2010; Weeks *et al.*, 2000). Different species like *Proteus mirabilis* and *Staphylococuus saprophyticus* act as causative agents of stones formation and urinary tract infections due to the urease activity (Burne and Chen, 2000). More recently (Johnson *et al.*, 1993), the pathogenic role of urease is also being investigated due to different fungal species like *Cryptococcus neoformans* (a basidiomycete) and *Coccidioides posadasii* (an ascomycete).

However, during the past 20 years the first line therapy for eradication of urease positive pathogens of human microbiota consisted of the mixture of two or more antibiotics (amoxicillin, clathromycin and penicillin) and a proton pump inhibitor (Hung *et al.*, 2009). To fight the infection the combination can be given with Bismuth salt (a metal) that have antiurease properties (Follmer, 2010; Megraud, 2012; Zhang *et al.*, 2006; Maltertheiner *et al.*, 2007). But the treatment efficacy can be reduced by acquisition of resistance by the bacterial pathogen to antibiotics (Glupczynsk, 1998). However, these antibiotics also have some side effects like allergic reaction, cramps, nausea, vomiting and diarrhoea. So the activity of urease can be controlled by the application of its inhibitors which seems to be viable strategy to inhibit *H. pylori* infection. So there is a need for the designing of new therapy. So, inhibitors of the urease enzyme may act as better alternative to cure various infections caused by it. Although the substances which are already available such as isatin-thiosemicarbazones, polyhalogenated benzo- and napthoquininoes and imidazoles are toxic and not highly stable and these features prevent their clinical use (Azizian *et al.*, 2012; Ibrar *et al.*, 2013).

Table 16.1: Microbial Ureases in Human Diseases

Urease Role	*Microbial Species*	*Diseases*
Survival in host	*Helicobacter pylori, Mycoplasma tuberculosis, Escherichia coli*	Gastritis, peptic ulcers, tuberculosis, hemorrhagic colitis
Persistence to host	*H. pylori*	Gastritis, peptic ulcers
Polyvalent ions precipitation	*P. mirabilis, U. urealyticum* and *M. morganii*	Urinary tract infections
Stimulation of inflammatory reaction	*Y. enterocolitica, H. pylori*	Gastritis, peptic ulcers, reactive arthritis
Aggregation of platelets	*H. pylori*	Gastritis, cardiovascular disease
Binding to epithelial lining of lung tissues	Various fungal species	Lung tissue damage

Urease in Agriculture

Most widely used nitrogen fertiliser across the world is urea. Various features leads to its maximum use are its low price, 46 per cent N content, easy management and transportation. But urease enzyme presence in the field leads to its lower availability to the soil. The soil-resident microorganisms include urease producing bacteria belonged to Gammaproteobacteria except *Cupriavidus* sp. which has been categorised in Betaproteobacteria. *Cupriavidus* sp. and *Mucor hiemalis*. This list also includes a wide range of microorganisms including saprophytes (*e.g. Gliomastix protea*), a human pathogen (*Trichosporon* sp.), insect and nematode pathogens (*Paecilomyces marquandii; Pochonia bulbillosa*) and plant pathogens (e.g *Fusarium solani*)(Alizadeh at al. 2017).

The high urease activity in agriculture causes significant environment and economical problems by releasing large amount of ammonia in to the atmosphere. This induces plant damage primarily by depriving plants of essential nutrients and secondly through ammonia toxicity. This excessive emission of ammonia to atmosphere gradually will cause an imbalance in nitrogen cycle and adversely affect the natural ecosystem. But, if NH_3 remains in soil solution, NH_3 is converted to ammonium ion (NH_4^+) that, in turn, can undergo nitrification by the action of *Nitrosomonas* and/or *Nitrobacter* species, yielding nitrate (NO_3). Thus the nitrate enriched environment leads to the eutrophication which in turn affect the animal and human health (Kawakami *et al.*, 2012).

However in agriculture, till now NBPT (N-(butyl) thiophosphoric triamide) is commercially available effective soil urease inhibitor. The inhibitors of urease enzyme in soil would lead to following changes in the soil: (i) maintain pH of soil and they control the ammonia volatilization. (ii) increase the efficiency of urea in delivering nitrogen to soil. (iii) also temporarily reduce the activity of the enzyme and slow the rate at which urea is hydrolysed.

In soil, NBPT (synthesized compound) potentially inhibit the urease (Watson, 2000) whereas other chemically synthesized substances, are found to posses not much inhibition potential. So, the great challenge is to find a natural urease inhibitor that is compatible with urea, eco-friendly, non toxic and of low costs.

There is no doubt that we need to start digging up the natural resources as urease enzyme inhibitors. In the past decades, exploration of urease inhibitors from natural products has attracted much attention (Firdous *et al.*, 2012; Ramsay *et al.*, 2012). The isolation of urease inhibitors from some plants and herbs has been revealed in previous literature (Ayaz *et al.*, 2006; Tariq *et al.*, 2011). Typically *Allium sativum* extract is a natural inhibitor of urease (Juszkiewicz *et al.*, 2004). Catechins in green tea extract were reported to inhibit urease strongly (Matsubara *et al.*, 2003). In addition, a *H. pylori* urease inhibitor has been isolated from *Rubus coreanus* (Yang *et al.*, 2004). *Hypericum oblongifolium* has been assayed for antiurease activity (Arfan *et al.*, 2010). Urease inhibition activity of some Pakistani medicinal plants have also been reported recently (Amin *et al.*, 2013). Ismail *et al.* (2015) determined phenolics from hydroalcoholic extracts of pomegranate as significant urease inhibitor. Rasool khan (2015), compared the methanolic and aqueous crude extract of *Cypnostemma digitatum linn.* and observed that methanol crude extract shows good inhibition activity against urease enzyme (Khan *et al.*, 2015).

Aqueous and methanol extract of *Menthe pipertia, Voila odorata, Trachyspermum copticum* and *Foeniculum vulgare* exhibited good antiurease efficacy (Zulfiqar *et al.*, 2017). Hasan *et al.* (2018), depicted that phenolic rich aqueous extract of *Salvia officinalis* possessed the IC50 value against urease of 1.65mg/ml thus can act as curative agent in the treatment of diseases caused by urease producing pathogens. The above quoted references clearly depicts the efficiency of plant extracts belonging to different classes that inhibit the activity of urease enzyme. This also may leads to the discovery of effective substances which exhibit considerable urease inhibitory potential in turn help in fighting different disease and help in improving aminal and human health.

Acknowledgement

The authors greatly acknowledge the Haryana State Council for Science and Technology (HSCST) for financial support for the project entitled "Urease Inhibitors from Medicinal plants of Haryana in context to agricultural soil protection and therapy against clinically significant virulence factor" (Endst.No. HSCST/R and D/2015/2383).

REFERENCES

Alizadeh, H., Kandula, D. R., Hampton, J. G., Stewart, A., Leung, D. W., Edwards, Y., and Smith, C. (2017). Urease producing microorganisms under dairy pasture management in soils across New Zealand. *Geoderma Regional, 11*, 78-85.

Amin, M., Anwar, F., Naz, F., Mehmood, T., and Saari, N. (2013). Anti-Helicobacter pylori and urease inhibition activities of some traditional medicinal plants. *Molecules, 18*(2), 2135-2149.

Arfan, M., Ali, M., Ahmad, H., Anis, I., Khan, A., Choudhary, M. I., and Shah, M. R. (2010). Urease inhibitors from Hypericum oblongifolium WALL. *Journal of enzyme inhibition and medicinal chemistry, 25*(2), 296-299.

Arioli, S., Monnet, C., Guglielmetti, S., Parini, C., De Noni, I., Hogenboom, J.,. and Mora, D. (2007). Aspartate biosynthesis is essential for the growth of Streptococcus thermophilus in milk, and aspartate availability modulates the level of urease activity. *Applied and environmental microbiology, 73*(18), 5789-5796.

Arioli, S., Ragg, E., Scaglioni, L., Fessas, D., Signorelli, M., Karp, M.,. and Guglielmetti, S. (2010). Alkalizing reactions streamline cellular metabolism in acidogenic microorganisms. *PLoS One, 5*(11), e15520.

Arioli, S., Roncada, P., Salzano, A. M., Deriu, F., Corona, S., Guglielmetti, S.,. and Mora, D. (2009). The relevance of carbon dioxide metabolism in Streptococcus thermophilus. *Microbiology, 155*(6), 1953-1965.

Ayaz, M., ARIF LODHI, M. U. H. A. M. M. A. D., Riaz, M., -UL-HAQ, A. Z. H. A. R., Malik, A., and IQBAL CHOUDHARY, M. (2006). Novel urease inhibitors from Daphne oleoids. *Journal of enzyme inhibition and medicinal chemistry, 21*(5), 527-529.

Azizian, H., Nabati, F., Sharifi, A., Siavoshi, F., Mahdavi, M., and Amanlou, M. (2012). Large-scale virtual screening for the identification of new Helicobacter pylori urease inhibitor scaffolds. *Journal of molecular modeling, 18*(7), 2917-2927.

Bergeron, L. J., Morou-Bermudez, E., and Burne, R. A. (2000). Characterization of the fructosyltransferase gene of Actinomyces naeslundii WVU45. *Journal of bacteriology, 182*(13), 3649-3654.

Burne, R. A., and Chen, Y. Y. M. (2000). Bacterial ureases in infectious diseases. *Microbes and Infection, 2*(5), 533-542.

Chen, Y. Y. M., and Burne, R. A. (2003). Identification and characterization of the nickel uptake system for urease biogenesis in Streptococcus salivarius 57. I. *Journal of bacteriology, 185*(23), 6773-6779.

Clemens, D. L., Lee, B. Y., and Horwitz, M. A. (1995). Purification, characterization, and genetic analysis of Mycobacterium tuberculosis urease, a potentially critical determinant of host-pathogen interaction. *Journal of bacteriology, 177*(19), 5644-5652.

Cotter, P. D., and Hill, C. (2003). Surviving the acid test: responses of gram-positive bacteria to low pH. *Microbiology and molecular biology reviews, 67*(3), 429-453.

Dixon, N. E., Gazzola, C., Blakeley, R. L., and Zerner, B. (1975). Jack bean urease (EC 3.5. 1.5). Metalloenzyme. Simple biological role for nickel. *Journal of the American Chemical Society, 97*(14), 4131-4133.

Dupuy, B., Daube, G., Popoff, M. R., and Cole, S. T. (1997). Clostridium perfringens urease genes are plasmid borne. *Infection and immunity, 65*(6), 2313-2320.

Follmer, C. (2010). Ureases as a target for the treatment of gastric and urinary infections. *Journal of Clinical Pathology, 63*(5), 424-430.

Firdous, S., Ansari, N. H., Fatima, I., Malik, A., Afza, N., Iqbal, L., and Lateef, M. (2012). Ophiamides AB, new potent urease inhibitory sphingolipids from Heliotropium ophioglossum. *Archives of pharmacal research, 35*(7), 1133-1137.

Fuller, M. F., and Reeds, P. J. (1998). Nitrogen cycling in the gut. *Annual review of nutrition, 18*(1), 385-411.

Futagami, S., Takahashi, H., Norose, Y., and Kobayashi, M. (1998). Systemic and local immune responses against Helicobacter pylori urease in patients with chronic gastritis: distinct IgA and IgG productive sites. *Gut, 43*(2), 168-175.

Glupczynski, Y. (1998). Antimicrobial resistance in Helicobacter pylori: a global overview. In *Helicobacter pylori*(pp. 398-415). Springer, Dordrecht.

Hassan, S. T., Švajdlenka, E., Rengasamy, K. R., Melichárková, R., and Pandian, S. K. (2018). The metabolic profile of essential oils and assessment of anti-urease activity by ESI-mass spectrometry of Salvia officinalis L. *South African Journal of Botany*.

Hung, I. F., Chan, P., Leung, S., Chan, F. S., Hsu, A., But, D.,. and Tong, T. S. (2009). Clarithromycin–AmoxycillinContaining Triple Therapy: A Valid Empirical FirstLine Treatment for Helicobacter pylori Eradication in Hong Kong?. *Helicobacter, 14*(6), 505-511.

Ibrar, A., Khan, I., and Abbas, N. (2013). Structurally diversified heterocycles and related privileged scaffolds as potential urease inhibitors: a brief overview. *Archiv der Pharmazie, 346*(6), 423-446.

Ismail, T., Akhtar, S., Sestili, P., Riaz, M., Ismail, A., and Labbe, R. G. (2016). Antioxidant, Antimicrobial and Urease Inhibitory Activities of PhenolicsRich Pomegranate Peel HydroAlcoholic Extracts. *Journal of Food Biochemistry, 40*(4), 550-558.

Jin, M., Rosario, W., Watler, E., and Calhoun, D. H. (2004). Development of a large-scale HPLC-based purification for the urease from Staphylococcus leei and determination of subunit structure. *Protein expression and purification, 34*(1), 111-117.

Johnson, D. E., Russell, R. G., Lockatell, C. V., Zulty, J. C., Warren, J. W., and Mobley, H. L. (1993). Contribution of Proteus mirabilis urease to persistence, urolithiasis, and acute pyelonephritis in a mouse model of ascending urinary tract infection. *Infection and immunity, 61*(7), 2748-2754.

Juszkiewicz, A., Zaborska, A., £aptaœ, A., and Olech, Z. (2004). A study of the inhibition of jack bean urease by garlic extract. *Food chemistry, 85*(4), 553-558.

Kawakami, E. M., Oosterhuis, D. M., Snider, J. L., and Mozaffari, M. (2012). Physiological and yield responses of field-grown cotton to application of urea with the urease inhibitor NBPT and the nitrification inhibitor DCD. *European Journal of Agronomy, 43*, 147-154.

Rutherford, J. C. (2014). The emerging role of urease as a general microbial virulence factor. *PLoS pathogens, 10*(5), e1004062.

Khan, R., Saif, A. Q., Quradha, M. M., Ali, J., and Rauf, A. (2015). Phytochemical analysis, antimicrobial, antioxidant and urease inhibitory potential of Cyphostemma digitatum Lam. *Natural product research, 29*(5), 466-468.

Konieczna, I., Zarnowiec, P., Kwinkowski, M., Kolesinska, B., Fraczyk, J., Kaminski, Z., and Kaca, W. (2012). Bacterial urease and its role in long-lasting human diseases. *Current Protein and Peptide Science, 13*(8), 789-806.

Krajewska, B. (2009). Ureases I. Functional, catalytic and kinetic properties: A review. *Journal of Molecular Catalysis B: Enzymatic, 59*(1-3), 9-21.

Lam, S. E. L. E. N. A., and Yeo, M. A. V. I. S. (1980). Urease-positive Vibrio parahaemolyticus strain. *Journal of clinical microbiology, 12*(1), 57-59.

LoCascio, R. G., Desai, P., Sela, D. A., Weimer, B., and Mills, D. A. (2010). Broad conservation of milk utilization genes in Bifidobacterium longum subsp. infantis as revealed by comparative genomic hybridization. *Applied and environmental microbiology, 76*(22), 7373-7381.

Malfertheiner, P., Megraud, F., O'Morain, C., Bazzoli, F., El-Omar, E., Graham, D.,. and Kuipers, E. J. (2007). Current concepts in the management of Helicobacter pylori infection: the Maastricht III Consensus Report. *Gut, 56*(6), 772-781.

Matsubara, S., Shibata, H., Ishikawa, F., Yokokura, T., Takahashi, M., Sugimura, T., and Wakabayashi, K. (2003). Suppression of Helicobacter pylori-induced gastritis by green tea extract in Mongolian gerbils. *Biochemical and biophysical research communications, 310*(3), 715-719.

Mégraud, F. (2012). The challenge of Helicobacter pylori resistance to antibiotics: the comeback of bismuth-based quadruple therapy. *Therapeutic advances in gastroenterology, 5*(2), 103-109.

Mollenhauer-Rektorschek, M., Hanauer, G., Sachs, G., and Melchers, K. (2002). Expression of UreI is required for intragastric transit and colonization of gerbil gastric mucosa by Helicobacter pylori. *Research in microbiology, 153*(10), 659-666.

Mobley, H. L., and Hausinger, R. P. (1989). Microbial ureases: significance, regulation, and molecular characterization. *Microbiological reviews, 53*(1), 85-108.

Mobley, H. L., Island, M. D., and Hausinger, R. P. (1995). Molecular biology of microbial ureases. *Microbiological reviews, 59*(3), 451-480.

Morou-Bermudez, E., Elias-Boneta, A., Billings, R. J., Burne, R. A., Garcia-Rivas, V., Brignoni-Nazario, V., and Suárez-Pérez, E. (2011). Urease activity as a risk factor for caries development in children during a three-year study period: a survival analysis approach. *Archives of oral biology, 56*(12), 1560-1568.

Murchan, S., Aucken, H. M., O'neill, G. L., Ganner, M., and Cookson, B. D. (2004). Emergence, spread, and characterization of phage variants of epidemic methicillin-resistant Staphylococcus aureus 16 in England and Wales. *Journal of clinical microbiology, 42*(11), 5154-5160.

Nakano, M., Iida, T., Ohnishi, M., Kurokawa, K., Takahashi, A., Tsukamoto, T., Yasunaga, T., Hayashi, T., Honda, T. Association of the Urease Gene with Enterohemorrhagic *Escherichia coli* Strains Irrespective of Their Serogroups. *J. Clin. Microbiol.*, **2001**, *39*(12), 4541-4543.

Orth, D., Grif, K., Dierich, M. P., and Würzner, R. (2006). Prevalence, structure and expression of urease genes in Shiga toxin-producing Escherichia coli from humans and the environment. *International journal of hygiene and environmental health*, *209*(6), 513-520.

Pedrozo, H. A., Schwartz, Z., Luther, M., Dean, D. D., Boyan, B. D., and Wiederhold, M. L. (1996). A mechanism of adaptation to hypergravity in the statocyst of Aplysia californica. *Hearing research*, *102*(1-2), 51-62.

Ramsay, K. S. T., Wafo, P., Ali, Z., Khan, A., Oluyemisi, O. O., Marasini, B. P.,. and Choudhary, M. I. (2012). Chemical constituents of Stereospermum acuminatissimum and their urease and α-chymotrypsin inhibitions. *Fitoterapia*, *83*(1), 204-208.

Riordan, S. M., and Williams, R. (2010). Gut flora and hepatic encephalopathy in patients with cirrhosis. *N Engl J Med* 362: 1140–1142.

Sirko, A., and Brodzik, R. (2000). Plant ureases: roles and regulation. *Acta Biochimica Polonica*, *47*(4), 1189-1195.

Skouloubris, S., Thiberge, J. M., Labigne, A., and De Reuse, H. (1998). The Helicobacter pylori UreI protein is not involved in urease activity but is essential for bacterial survival in vivo. *Infection and immunity*, *66*(9), 4517-4521.

Smoot, D. T., Mobley, H. L., Chippendale, G. R., Lewison, J. F., and Resau, J. H. (1990). Helicobacter pylori urease activity is toxic to human gastric epithelial cells. *Infection and immunity*, *58*(6), 1992-1994.

Suzuki, K., Benno, Y., Mitsuoka, T., Takebe, S., Kobashi, K., and Hase, J. (1979). Urease-producing species of intestinal anaerobes and their activities. *Applied and Environmental Microbiology*, *37*(3), 379-382.

Tange, Y., and Niwa, O. (1997). Identification of the ure1+ gene encoding urease in fission yeast. *Current genetics*, *32*(3), 244-246.

Tariq, S. A., Ahmad, M. N., Obaidullah, Khan, A., Choudhary, M. I., Ahmad, W., and Ahmad, M. (2011). Urease inhibitors from Indigofera gerardiana Wall. *Journal of enzyme inhibition and medicinal chemistry*, *26*(4), 480-484.

Watson, C. J. (2000). Urease activity and inhibition-principles and practice. In *Proceedings-International Fertiliser Society*(No. 454, pp. 1-40). International Fertiliser Society.

Weeks, D. L., Eskandari, S., Scott, D. R., and Sachs, G. (2000). A H+-gated urea channel: the link between Helicobacter pylori urease and gastric colonization. *Science*, *287*(5452), 482-485.

Wegmann, U., Louis, P., Goesmann, A., Henrissat, B., Duncan, S. H., and Flint, H. J. (2014). Complete genome of a new F irmicutes species belonging to the dominant human colonic microbiota ('R uminococcus bicirculans') reveals two chromosomes and a selective capacity to utilize plant glucans. *Environmental microbiology, 16*(9), 2879-2890.

Yang, S. W., Ho, J. N., Lee, Y. H., Shin, D. H., Hong, B. S., and Cho, H. Y. (2004). Isolation and characterization of Helicobacter pylori urease inhibitor from Rubus coreanus Miquel. *Journal of the Korean Society of Food Science and Nutrition*.

Yatsunenko, T., Rey, F. E., Manary, M. J., Trehan, I., Dominguez-Bello, M. G., Contreras, M., and Heath, A. C. (2012). Human gut microbiome viewed across age and geography. *nature, 486*(7402), 222.

Yu, J. J., Smithson, S. L., Thomas, P. W., Kirkland, T. N., and Cole, G. T. (1997). Isolation and characterization of the urease gene (URE) from the pathogenic fungus Coccidioides immitis. *Gene, 198*(1), 387-391.

Zhang, L., Mulrooney, S. B., Leung, A. F., Zeng, Y., Ko, B. B., Hausinger, R. P., and Sun, H. (2006). Inhibition of urease by bismuth (III): implications for the mechanism of action of bismuth drugs. *Biometals, 19*(5), 503-511.

Zonia, L. E., Stebbins, N. E., and Polacco, J. C. (1995). Essential role of urease in germination of nitrogen-limited Arabidopsis thaliana seeds. *Plant physiology, 107*(4), 1097-1103.

Zulfiqar, I., Asif, H. M., Sultana, S., Akram, M., and Qayoom, I. (2017). IN-VITRO Antiurease activity of aqueous ethsnol extract of some medicinal plants. *Pakistan Journal of Science*, 69(1).

Chapter 17

Scope of Laccase for the Synthesis of Bioactive Compounds and Functionalization of Biopolymers

*Vikas Sharma, Digvijay Singh and D. Vasanth**

Department of Biotechnology, National Institute of Technology Raipur, Raipur – 492 010, Chhattisgarh
****E–mail: dvasanth.bt@nitrr.ac.in***

ABSTRACT

Today, every major industry is striving to go green in its production technologies due to the increased consciousness of the public towards preservation of nature and natural resources. One of the enzyme groups that have proved effective in realizing this is the oxidoreductase laccase, which is a multi-copper enzyme oxidizing various phenolic and non-phenolic compounds releasing only water as the byproduct. It has the added advantage of having broad specificity and among its substrates are several recalcitrant environmental pollutants. Laccase catalyse a vast range of industrial applications including degradation of hazardous dyes, development of biosensors, degradation of lignin, wine stabilization and hair coloring. Laccase also catalyze a number of synthetic reactions to form cross-coupling, dimeric and polymeric compounds. These compounds are beneficial in the synthesis of medical, pharmaceutical, cosmetic products and their intermediates. Therefore, this chapter majorly focuses on the laccase-catalyzed functionalization of biopolymers and production of bioactive compounds including antioxidants, anticancerous, antiviral, antifungals and antibacterials.

Keywords: *Laccase, Bioactive compounds, Functionalization of polymers, Antioxidants, Antimicrobials.*

Introduction

Laccase (EC 1.10.3.2), an oxidoreductase enzyme, is attracting the attention of scientific community for over two decades because of its wide range of industrial applications and green catalytic mechanism. They belong to a class of multicopper oxidases (MCOs) which possesses the ability to catalyze a large number of phenolic and nonphenolic substrates by transforming an oxygen molecule to water with parallel reduction of four-electrons (Hakulinen and Rouvinen, 2015). These free electrons oxidize various aromatic and non-aromatic substrates including phenolic ring-containing amines that are replaced with different functional groups like amino, diamino, methoxy and hydroxyindols. They can also catalyze various metallic compounds such as $(Mo(CN)_8)_4$-, $(Fe(CN)_6)_4$- and $(Os(CN)_6)_4$- (Chauhan *et al.*, 2017).

Laccase was first isolated a little more than 100 years ago from the Japanese tree *Rhus vernicifera* (Guan *et al.*, 2018). They are glycoproteins that are considered as ubiquitous in nature because of their availability in almost all form of life ranging from various fungi to higher plants. Laccase is majorly isolated from fungi, plants, bacteria and insects but fungal laccases have been studied extensively for the catalysis in various industrial processes because of their higher redox potential (Sharma and Vasanth, 2018). Laccases can be monomeric, dimeric or tetrameric in nature and their molecular mass generally lies in the range of 25-140 kDa. The stability of the enzyme is probably due to a covalently-linked carbohydrate moiety, which constitutes 10 - 45 per cent of total molecular mass of the enzyme (Durán *et al.*, 2002).

The non-specificity of laccases towards their reducing substrates and the variable range for different laccases makes them highly useful, catalyzing a variety of inorganic and organic substrates, including aromatic amines, polyphenols, ascorbate, and methoxy-substituted phenols (Thurston, 1994). Hydroxyl and amine groups are the main targets of laccase, hence it also reacts with the hydroxyl groups present in phenolic structures of lignin (Youn *et al.*, 1995). Laccase has the ability to replace the environmentally hazardous chemical processes, which makes them promising candidate to focus in the emerging trends of new biotechnology. The broad specificity towards substrates and their eco-friendliness (water being the only by-product), laccase is considered as a promising green alternative for a number of industrial applications which includes bioremediation, dye degradation, production of biofuels, development of biosensors, organic synthesis, wine stabilization, synthesis of bioactive compounds and functionalization of biopolymers (Chauhan *et al.*, 2017, Sentivelan *et al.*, 2016, Mogharabi and Faramarzi, 2014, Kudanga *et al.*, 2017).

The value of compounds that possesses therapeutic utility is increasing because of the role of biologically active compounds in industries like pharmaceutical, food and agriculture. Thus researchers around the globe are looking for cost-effective and environment friendly synthesis of bioactive compounds. The process that involves enzymatic catalysis can provide such benefits in the synthesis of useful organic compounds that can replace hazardous chemical processes. Laccase enzyme has the potential to catalyze the synthesis of various bioactive compounds while balancing the production cost and environment risks (Kudanga *et al.*, 2017). Major focus of this chapter lies on laccase-catalyzed synthesis of bioactive compounds with their applications and also discusses the advancement that have been made in the methods involving laccase-based coupling and grafting reactions.

Role of Laccase in Organic Synthesis

Laccases involve a large array of synthetic reactions such as amino acids derivatives and transformation of antibiotics (Piscitelli *et al.*, 2012). That is the reason behind the utilization of laccase in the synthesis of complex pharmaceutical products, like immunosuppressors (*e.g.* cyclosporin A), anti-cancer drugs (for example, mitomycin, vinblastine and actinomycin) and antibiotics such as cephalosporin and penicillin X dimers (Figure 17.1) (Kunamneni *et al.*, 2008). Also, some laccases-mediated derivatives of amino acids are L-tryptophan, L-lysine or L-phenylalanine (Mogharabi and Faramarzi, 2014). Moreover, laccases have been used to generate dimers by oxidation of steroid hormones, stilbenic phytoalexin trans-resveratrol and 17β-estradiol (Nicotra *et al.*, 2004). Another promising application of laccases is the generation of iodine by the oxidation of iodide, a useful antimicrobial product. Oxidation of iodide by laccase was found to inhibit the growth of *Bacillus anthracis* and spores (Mate and Alcalde, 2016).

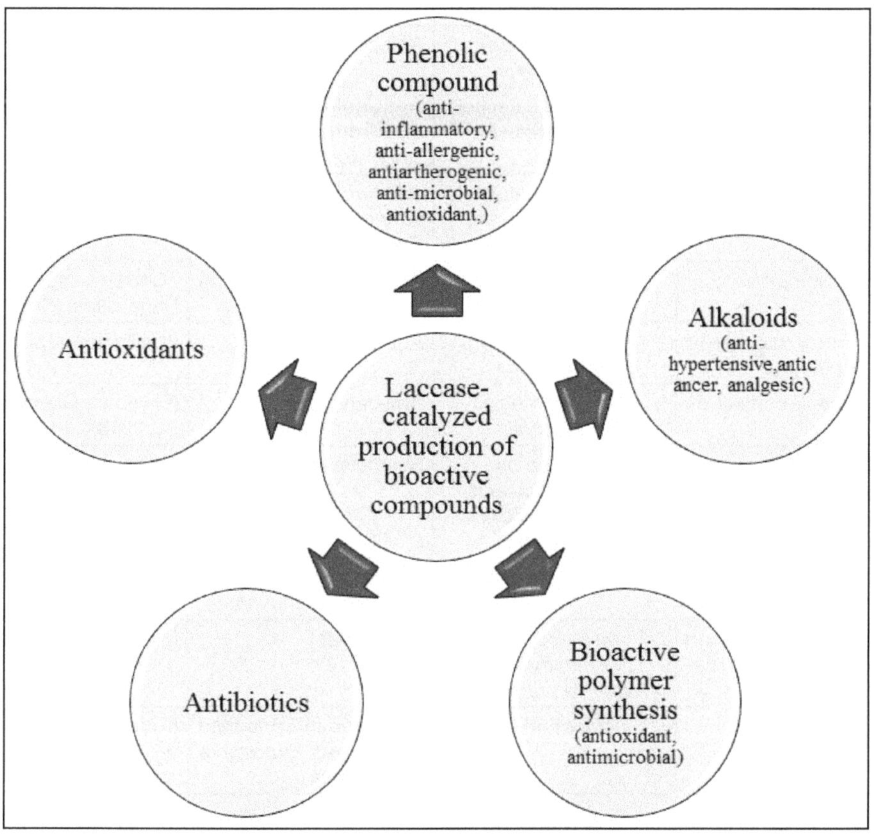

Figure 17.1: Laccase-Catalyzed Synthesis of Bioactive Compounds.

Scope for the Synthesis of Bioactive Compounds

The biologically active compounds were majorly synthesized using commercial laccase obtained from fungal sources. The compounds synthesized are categorized on the basis of their application (Table 17.1). Naphthoquinones and their derivatives are in great demand because of their several applications that include anticancer, anti-inflammatory, wound healing and antibacterial and antiparasitic activities. In a study, laccase-mediated 1, 4-naphthoquinone derivatives were synthesized by using Diels–Alder reaction. During the process, dienes and para-quinones were reacted in an aqueous medium which ultimately provides 1, 4-naphthoquinones. They also investigated the effect of laccase dose, temperature, and substrate sensitivity without

Table 17.1: Laccase-Mediated Synthesis of Bioactive Compounds and their Applications

Compounds	*Source of Laccase*	*Application*	*Reference*
Pyrimido-benzothiazole derivatives	*Agaricus bisporus*	Anti-proliferative	(Abdel-Mohsen *et al.*, 2017)
β-lactam antibiotics	*Myceliophthora thermophila*	Antimicrobial	(Mikolasch *et al.*, 2012)
Amino-naphtoquinones	*Myceliophthora thermophile*	Anti-cancer	(Wellington *et al.*, 2012b)
Novel cephalosporins	Amination of catechols with cefadroxil	Antibacterial against methicillin-resistant *Staphylococcus aureus*	(Mikolasch *et al.*, 2008)
Novel carbacephems	Amination of catechols with carbacephem loracarbef	Antibacterial against vancomycin-resistant *Enterococci*	(Mikolasch *et al.*, 2008)
Benzylic nitriles	*Trametes villosa*	Intermediate of several bioactive compounds	(Cannatelli and Ragauskas, 2015)
5,6-Dihydroxylated benzo(*b*)furans	*Myceliophthora thermophila*	Anti-cancer	(Wellington *et al.*, 2013)
Coumestans derivatives	*Myceliophthora thermophila*	Anti-cancer	(Qwebani-Ogunleye *et al.*, 2016)
Dihydroxybenzoic acid derivatives and aromatic amine compounds	*Pycnoporus cinnabarinus and Myceliophthora thermophila*	Synthetic intermediates	(Hahn *et al.*, 2010)
Benzenediol derivatives	*Tramates versicolor*	Synthetic intermediates of several useful drugs	(Habibi *et al.*, 2017)
stilbenic phytoalexin trans-resveratrol dehydrodimer	*Myceliophtora thermophyla* and *Trametes pubescens*	Anti-oxidant	(Nicotra *et al.*, 2003)
Quinoxalines	*Ganoderma* sp.	Synthetic intermediate of various bioactive compounds	(Kidwai *et al.*, 2012)
Lawsone derivatives	*naphthoquinone*	Antibacterial, antifungal, antiviral, antitumor and antiparasitic	(López *et al.*, 2014)

depending on harmful organic solvents and heavy metal reagents (Witayakran and Ragauskas, 2007). Another derivative 1, 4-naphthoquinone-2, 3-bis-sulfide was synthesized using two different substrates of 1, 4-naphthohydroquinone. The reaction was processed under mild conditions at pH 4.5 while maintaining the aerobic environment and it was found that laccase catalyses the formation of C–S bond (Wellington *et al.*, 2012a). In a separate study, Wellington *et al.*, reported the one-pot synthesis of aminonaphthoquinones with a reaction mixture of 1, 4-naphthohydroquinone and primary amine in the presence of succinate-lactate buffer and dimethylformamide. The reaction conditions were maintained at pH 4.5 and pH 6.0 within an open vessel under mild condition. It was mentioned that aminonaphthoquinones showed anti-cancerous activity against the cell lines including UACC62 (melanoma), TK10 (renal) and MCF7 (breast). Hence, further investigation on naphthoquinone derivatives can make things feasible in the anticancer therapy (Wellington *et al.*, 2012b). The action of bioactive compounds like derivatives of 5, 6-dihydroxylated benzo(b)furans and catechol synthesized by applying commercial laccase was studied against the cancer cell lines cervical (HeLa), melanoma (UACC62), breast (MCF7) and renal (TK10). The results obtained from the screening of compounds showed comparatively better inhibitory activities than the anticancer agent etoposide (Wellington *et al.*, 2013).

Laccase obtained from the fungal strain *Myceliophthora thermophila* (MTL) was utilized for the synthesis of O-carboxymethyl chitosan (CMCS) hydrogel. The chitosan based hydrogel was found to be bioactive when crosslinked with naturally occurring phenolics and showed anti-oxidant and anti-inflammatory activities. Thus, the compound has potential application in wound dressings (Huber *et al.*, 2018). A novel green approach was reported for the synthesis of pyrimidobenzothiazoles and catechol thioethers, and the products obtained showed activity against cancer cell line HepG2 (Abdel-Mohsen *et al.*, 2017). Laccase obtained from *Trametes versicolor* (TvL) catalyses the synthesis of racemic di-hydrobenzofuran dehydrodimers via oxidative coupling. These monomeric stilbenoids that are closely related to resveratrol were found to be active when studied for their anticancer assay against SW480 cancer cells of human colon. It is mentioned that the methoxy group located in the ortho position of C-4 hydroxy group affects the activity when the active compounds were compared with inactive compounds (Bhusainahalli *et al.*, 2012). In another study, 8 novel morpholine-substituted aromatics were successfully produced by the reaction of morpholines with para-dihydroxylated aromatic systems. The cross-linking reaction was carried out by using the laccase isolated from *Myceliophthora thermophila.* It was reported that both the primary and secondary amino groups coupled with para-dihydroxylated aromatic systems when the reaction is catalyzed by laccase. The products synthesized showed antibacterial, antifungal and cytotoxic activity, thus can be utilized for medical and agricultural applications (Hahn *et al.*, 2009). The direct synthesis of anticancerous compounds were reported by Wellington and his co-workers utilizing laccase as a sole biocatalyst, while number of oxidation reactions to produce carpanones, benzopyrans and sesamol trimerisation was reported by Beifuss and co-workers (Habibi *et al.*, 2016).

A laccase isolated from *Trametes sp.* has been reported to synthesize novel compound like cephalosporins, penicillins, and carbacephems which are single aminated products. All these products were reported to have antibacterial activity as they inhibit the growth of various bacterial strains including *Staphylococcus aureus* strains and *Enterococci* (Mikolasch *et al.,* 2008). Many substrates that are used to check the activity of laccase were also studied for their antioxidant property. One of them is 2, 6-Dimethoxyphenol the oxidation product generated by the reaction of laccase with 2, 6-Dimethoxyphenol was studied and found to have higher oxidation capacity than the 2, 6-Dimethoxyphenol before catalysis. The dimer formed after reaction had higher antioxidant activity. This study showed that laccase mediated oxidation of 2, 6-Dimethoxyphenol can lead to the production of useful bioactive compounds (Adelakun *et al.,* 2012). The oxidation of stilbenicphytoalexin trans-resveratrol (3, 5, 4′-trihydroxystilbene, from laccase isolated from *Myceliophtora thermophyla* and *Trametes pubescens* generates dehydrodimers that also poses higher antioxidant activity (Nicotra *et al.,* 2003).

Scope for the Functionalization of Biopolymers

Surface functionalization of biopolymer through laccase demonstrates an eco-friendly approach to enhance the properties of biopolymers. Laccase mediated surface modification offers novel opportunities to impart desired properties. In recent years the interest has been grown in utilizing biological reactions with biopolymer to generate antioxidant and antimicrobial materials. Among various biological agents laccase has been found to be extensively studied oxidative enzyme for functionalization of biopolymers. Recently a novel method was introduced by Ferreira *et al.,* to produce antiviral Kimwipes® made of cellulosic fiber was functionalized with catechin using laccase as a grafting material. These wipes were reported to have potential antiviral activity, thus can be utilized as in wipes or as a face mask in cosmetic products (Catel-Ferreira *et al.,* 2015). Laccase mediated functionalization of wood venner and pulp was also reported to improve the antibacterial resistance against pathogenic *Staphylococcus aureus* (Gram-positive bacterium) and *Escherichia coli* (Gram-negative bacterium) (Widsten *et al.,* 2010).

Laccase has a potential application in the field of pharmaceutical industry. Many studies have been reported that confirms the role of laccase in enhancing the antioxidant and antimicrobial activity of microporous compounds. Tegl *et al.* has reported that the laccase based surface modification of chitosan-zeolite particles with glucose oxidasein and caffeic acid where the caffeic acid was attached on the chitosan moieties by laccase. Further the glucose oxidase as immobilized on both the functional chitosan moieties and caffeic acid. This modification causes the generation of hydrogen peroxide in the system. The system was studied against *E. coli* and *S. aureus* for their oxidative and antimicrobial property and gave positive results against both the strains for up to 24h (Tegl *et al.,* 2018). A naturally occurring polymer chitosan was also reported to be functionalized by both homogeneous and heterogeneous methods using either tannic acid or quercetin with laccase isolated from *Trametes versicolor*. This study reveals the formation of a product with efficient antioxidant, antimicrobial and antibacterial property (Boziè *et al.,* 2012). In his another study Boziè *et al.,* reported the laccase-catalyzed functionalization of caffeic acid and gallic acid on chitosan. The laccase mediated oxidation caused the functionalization of chitosan that enhances its antibacterial activity by incorporating the antioxidant activity of caffeic and gallic acid (Boziè *et al.*, 2013). Chitosan has been reported to functionalize with 4-hexyloxyphenol (HP) and the reaction was catalyzed by laccase. The copolymer generated with this reaction was found to have antioxidant property. The functionalization did not only provide antimicrobial activity but also increase the tensile strength, thermal stability as well as hydrophobic properties (Liu *et al.*, 2018). For the first time the lignin model compound dibenzodioxocin was functionalized with laccase that eventually provides an alternative method of functionalizing biopolymer (Kudanga *et al.*, 2009). Iodine which is well known antimicrobial agent was coupled using laccase to oxidize unreactive iodide to reactive iodine. A phenolic compound vanillin which is another model compound of lignin was ionized by reacting it with laccase and iodide. The resulting compound was reported to have antifungal activity against wood degrading fungal strain (Ihssen *et al.*, 2014). Sharma *et al.* has utilized thermostable laccase isolated from the bacterium *Bacillus* sp. PC-3 for the functionalization of chitosan film using tannic acid. This is the first report where bacterial laccase has been studied for the functionalization of chitsan-based biopolymers (Sharma *et al.*, 2018b). Table 17.2 summarizes the functionalized products catalyzed by laccase.

Table 17.2: Laccase Functionalized Products and their Applications

Products	*Source of Laccase and their Substrates*	*Applications*	*Reference*
Functionalized chitosan film	Tannic acid	Antioxidant and antimicrobial	(Sharma *et al.*, 2018)
Kimwipes	Cellulosic fibers, functionalized with polyphenol catechin	Anti-viral	(Catel-Ferreira *et al.*, 2015)
Catechin polymers	*Trametes versicolor*	Anti-oxidant	(Jadhav and Singhal, 2014)
Polyrutin and polyesculin	*Trametes versicolor*	Anti-oxidant	(Chebil *et al.*, 2015)
Catechin polymers	*Aspergillus oryzae*	Anti-glycation compound	Jeon *et al.*, 2013
Catechin polymers	*Trametes versicolor*	Anti-oxidant	(Ezgi Ünlü *et al.*, 2017)
Catechol thioeters	*Agaricus bisporus*	Anti-oxidant	(Abdel-Mohsen *et al.*, 2014)
Veneers and paper	Wood veneer and pulp treated with tannins	Antibacterial	(Widsten *et al.*, 2010)
Zeolites	Functionalization of chitosan-zeolite particles (CTS-zeolites) with caffeic acid (CA) and glucose oxidase (GOX)	Antioxidant and antimicrobial	(Tegl *et al.*, 2018)

Products	*Source of Laccase and their Substrates*	*Applications*	*Reference*
Polyphenol functionalized chitosan	Functionalization of chitosan with quercetin or tannic acid	Synergistic antioxidant and antimicrobial properties	(Boži *et al.*, 2012)
Iodovanillin	Unreactive iodide and softwood lignin	Wash-out resistant antimicrobial surface	(Ihssen *et al.*, 2014)
Iodo-ethyl vanillin	Unreactive iodide and softwood lignin	Anti-fungal	(Ihssen *et al.*, 2014)
Functionalized chitosan	Chitosan with 4-hexyloxyphenol (HP)	Anti-oxidant and hydrophobic properties	(Liu *et al.*, 2018)

Future Perspectives

Considering the scope of laccase in various applications and its green catalytic mechanism, there is a need to enhance the catalytic properties of this enzyme. It is very well known that industrial processes involves harsh conditions including elevated pressure and temperature, increased concentration of salts, extreme pH and oxidative conditions, whereas in some cases high shear forces or delays were also noticed. Researchers are constantly looking for enzymes with novel characteristics for efficient catalysis of bioprocesses. Major concern is the stability of enzyme in industries is thermotolerancy and thermostability. Production of thermostable laccase can be beneficial in overcoming the limitations of current scenario because of their catalytic robustness and resistant to denaturation characteristic. Isolation of thermophilic bacterial strains can play a major role in the production of thermostable enzymes as they also offers better overexpression facility. The studies that focus on the physico-chemical properties and thermodynamics of the metallo-enzymes that affects their catalytic performance can provide better understanding of thermal stability or thermotolerance of laccases. Laccase-mediated coupling reactions can provide an alternate for hazardous chemical-based processes to many industries. Therefore, there is a need to reduce the cost of production of highly stable and active enzymes, utilization of eco-friendly redox mediators, improvising the cost of scaling up processes, to make the enzyme commercially feaslible. Enzyme production can also be enhanced by the overexpression of laccase producing gene in heterologous hosts, whereas the activity and stability of enzyme can also be improved by the modification of enzyme using chemical and biotechnological methods. Techniques to immobilize enzymes need to be improved to make their catalysis feasible in food industry as the utilization of enzyme as food additive is not yet approved. Gas chromatography/mass spectrometry (GC-MS) can provide more efficient methods for monitoring and analysis of coupling and grafting reactions. Application of laccase-oxidized products can be widened by incorporating the reactive groups to non reactive polymers.

REFERENCES

Abdel-Mohsen, H.T., Conrad, J. and Beifuss, U. 2013. Laccase-catalyzed domino reaction between catechols and 6-substituted 1, 2, 3, 4-tetrahydro-4-oxo-2-

thioxo-5 pyrimidinecarbonitriles for the synthesis of pyrimidobenzothiazole derivatives. *J. org. chem.* 78: 7986-8003.

Abdel-Mohsen, H.T., Conrad, J., Harms, K., Nohr, D. and Beifuss, U. 2017. Laccase-catalyzed green synthesis and cytotoxic activity of novel pyrimidobenzothiazoles and catechol thioethers. *RSC Adv.* 7: 17427-17441.

Adelakun, O.E., Kudanga, T., Green, I.R., le Roes-Hill, M. and Burton, S.G. 2012. Enzymatic modification of 2, 6-dimethoxyphenol for the synthesis of dimers with high antioxidant capacity. *Process Biochem.* 47: 1926-1932.

Bhusainahalli, V.M., Spatafora, C., Chalal, M., Vervandier Fasseur, D., Meunier, P., Latruffe, N. and Tringali, C. 2012. Resveratrol Related Dehydrodimers: Laccase Mediated Biomimetic Synthesis and Antiproliferative Activity. *Eur. J. Org. Chem.* 2012: 5217-5224.

Bozic, M., Gorgieva, S. and Kokol, V. 2012. Homogeneous and heterogeneous methods for laccase-mediated functionalization of chitosan by tannic acid and quercetin. *Carbohydr. Polym.* 89: 854-864.

Bozic, M., Štrancar, J. and Kokol, V. 2013. Laccase-initiated reaction between phenolic acids and chitosan. *React. Funct. Polym.* 73: 1377-1383.

Cannatelli, M.D. and Ragauskas, A.J. 2015. Laccase-catalyzed α-arylation of benzoylacetonitrile with substituted hydroquinones. *Chem. Eng. Res. Des.* 97: 128-134.

Catel-Ferreira, M., Tnani, H., Hellio, C., Cosette, P. and Lebrun, L. 2015. Antiviral effects of polyphenols: Development of bio-based cleaning wipes and filters. *J. Virol. Methods.* 212: 1-7.

Chauhan, P.S., Goradia, B. and Saxena, A. 2017. Bacterial laccase: recent update on production, properties and industrial applications. *3 Biotech.* 7: 323-343. DOI 10.1007/s13205-017-0955-7.

Chebil, L., Rhouma, G.B., Chekir-Ghedira, L. and Ghoul, M. 2015. Enzymatic Polymerization of Rutin and Esculin and Evaluation of the Antioxidant Capacity of Polyrutin and Polyesculin. In *Biotechnology*. InTech. DOI http: //dx.doi.org/10.5772/60413.

Durán, N., Rosa, M.A., D'Annibale, A. and Gianfreda, L. 2002. Applications of laccases and tyrosinases (phenoloxidases) immobilized on different supports: a review. *Enzyme Microb. Technol.* 31: 907-931.

Ezgi Ünlü, A., Prasad, B., Anavekar, K., Bubenheim, P. and Liese, A. 2017. Investigation of a green process for the polymerization of catechin. *Prep. Biochem. Biotechnol.* 47: 918-924.

Guan, Z.B., Luo, Q., Wang, H.R., Chen, Y. and Liao, X.R. 2018. Bacterial laccases: promising biological green tools for industrial applications. *Cell. Mol. Life Sci.* 75: 3569–3592.

Habibi, D., Rahimi, A., Rostami, A. and Moradi, S. 2017. Green and mild laccase-catalyzed aerobic oxidative coupling of benzenediol derivatives with various sodium benzenesulfinates. *Tetrahedron Lett.* 58: 289-293.

Hahn, V., Davids, T., Lalk, M., Schauer, F. and Mikolasch, A. 2010. Enzymatic cyclizations using laccases: Multiple bond formation between dihydroxybenzoic acid derivatives and aromatic amines. *Green Chem.* 12: 879-887.

Hahn, V., Mikolasch, A., Wende, K., Bartrow, H., Lindequist, U. and Schauer, F. 2009. Synthesis of model morpholine derivatives with biological activities by laccasecatalysed reactions. *Biotechnol. Appl. Biochem.* 54: 187-195.

Hakulinen, N. and Rouvinen, J. 2015. Three-dimensional structures of laccases. *Cell. Mol. Life Sci.* 72(5): 857-868.

Huber, D., Grzelak, A., Baumann, M., Borth, N., Schleining, G., Nyanhongo, G.S. and Guebitz, G.M. 2018. Anti-inflammatory and anti-oxidant properties of laccase-synthesized phenolic-O-carboxymethyl chitosan hydrogels. *N. Biotechnol.* 40: 236-244.

Ihssen, J., Schubert, M., Thöny-Meyer, L. and Richter, M. 2014. Laccase catalyzed synthesis of iodinated phenolic compounds with antifungal activity. *PloS one.* 9: p.e89924. DOI 10.1371/journal.pone.0089924.

Jadhav, S.B. and Singhal, R.S. 2014. Laccase–gum Arabic conjugate for preparation of water-soluble oligomer of catechin with enhanced antioxidant activity. *Food Chem.* 150: 9-16.

Jeon, S.Y., Oh, S., Kim, E. and Imm, J.Y. 2013. α-Glucosidase inhibiton and antiglycation activity of laccase-catalyzed catechin polymers. *J. Agric. Food Chem.* 61: 4577-4584.

Kidwai, M., Jain, A., Sharma, A. and Kuhad, R.C. 2012. First time reported enzymatic synthesis of new series of quinoxalines–A green approach. *J. Mol. Catal. B Enzym.* 74: 236-240.

Kidwai, M., Jain, A., Sharma, A. and Kuhad, R.C. 2013. Laccase-catalysed reaction between Meldrum's acid and catechols/hydroquinones–An investigation. *C. R. Chim.* 16: 728-735.

Kudanga, T., Nemadziva, B. and Le Roes-Hill, M. 2017. Laccase catalysis for the synthesis of bioactive compounds. *Appl. Microbiol. Biotechnol.* 101: 13-33.

Kudanga, T., Prasetyo, E.N., Sipilä, J., Nyanhongo, G.S. and Guebitz, G.M. 2010. Enzymatic grafting of functional molecules to the lignin model dibenzodioxocin and lignocellulose material. *Enzyme and microb. technol.* 46: 272-280.

Kunamneni, A., Plou, F.J., Ballesteros, A. and Alcalde, M. 2008. Laccases and their applications: a patent review. *Recent Pat. Biotechnol.* 2: 10-24.

Liu, N., Ni, S., Ragauskas, A.J., Meng, X., Hao, N. and Fu, Y. 2018. Laccase-mediated functionalization of chitosan with 4-hexyloxyphenol enhances antioxidant and hydrophobic properties of copolymer. *J. biotechnol.* 269: 8-15.

López López, L.I., Flores, N., Daniel, S., Silva Belmares, S.Y. and Sáenz Galindo, A. 2014. Naphthoquinones: biological properties and synthesis of lawsone and derivatives-a structured review. *Vitae.* 21: 248-258.

Mate, D.M. and Alcalde, M. 2017. Laccase: a multipurpose biocatalyst at the forefront of biotechnology. *Microb. Biotechnol.* 10: 1457-1467.

Mikolasch, A., Manda, K., Schlüter, R., Lalk, M., Witt, S., Seefeldt, S., Hammer, E., Schauer, F., Jülich, W.D. and Lindequist, U. 2012. Comparative analyses of laccasecatalyzed amination reactions for production of novel βlactam antibiotics. *Biotechnol. Appl. Biochem.* 59: 295-306.

Mikolasch, A., Wurster, M., Lalk, M., Witt, S., Seefeldt, S., Hammer, E., Schauer, F., Jülich, W.D. and Lindequist, U. 2008. Novel β-lactam antibiotics synthesized by amination of catechols using fungal laccase. *Chem. Pharm. Bull.* 56: 902-907.

Mogharabi, M. and Faramarzi, M.A. 2014. Laccase and laccasemediated systems in the synthesis of organic compounds. *Adv. Synth. Catal*, 356: 897-927.

Nicotra, S., Cramarossa, M.R., Mucci, A., Pagnoni, U.M., Riva, S. and Forti, L. 2004. Biotransformation of resveratrol: synthesis of trans-dehydrodimers catalyzed by laccases from *Myceliophtora thermophyla* and from *Trametes pubescens*. *Tetrahedron*. 60: 595-600.

Piscitelli, A., Amore, A. and Faraco, V. 2012. Last advances in synthesis of added value compounds and materials by laccasemediated biocatalysis. *Curr. Org. Chem.*, 16: 2508-2524.

Qwebani-Ogunleye, T., Kolesnikova, N.I., Steenkamp, P., de Koning, C.B., Brady, D. and Wellington, K.W. 2017. A one-pot laccase-catalysed synthesis of coumestan derivatives and their anticancer activity. ý*Bioorganic Med. Chem.* 25: 1172-1182.

Senthivelan, T., Kanagaraj, J. and Panda, R.C. 2016. Recent trends in fungal laccase for various industrial applications: an eco-friendly approach-a review. *Biotechnol. Bioprocess. Eng*, 21: 19-38.

Sharma, V. and Vasanth, D. 2018. Lignocellulolytic enzymes from thermophiles. In: Singh, O.V. and Chandel, A.K. (Eds.). Sustainable Biotechnology-Enzymatic Resources of Renewable Energy, Springer International Publishing AG 2018, pp. 205-217.

Sharma, V., Ayothiraman, S. and Dhakshinamoorthy, V. 2018. Production of highly thermo-tolerant laccase from novel thermophilic bacterium Bacillus sp. PC-3 and its application in functionalization of chitosan film. *J. biosci. Bioeng.* DOI https: //doi.org/10.1016/j.jbiosc.2018.11.008.

Tegl, G., Stagl, V., Mensah, A., Huber, D., Somitsch, W., GrosseKracht, S. and Guebitz, G.M. 2018. The chemo enzymatic functionalization of chitosan zeolite particles provides antioxidant and antimicrobial properties. *Eng. Life Sci.* 18: 334-340.

Thurston, C.F. 1994. The structure and function of fungal laccases. *Microbiology.* 140: 19-26.

Wellington, K.W. and Kolesnikova, N.I. 2012. A laccase-catalysed one-pot synthesis of aminonaphthoquinones and their anticancer activity. *Bioorganic Med. Chem.* 20: 4472-4481.

Wellington, K.W., Bokako, R., Raseroka, N. and Steenkamp, P. 2012. A one-pot synthesis of 1, 4-naphthoquinone-2, 3-bis-sulfides catalysed by a commercial laccase. *Green Chem.* 14: 2567-2576.

Wellington, K.W., Qwebani Ogunleye, T., Kolesnikova, N.I., Brady, D. and de Koning, C.B. 2013. One pot laccase catalysed synthesis of 5, 6- dihydroxylated benzo(b)furans and catechol derivatives, and their anticancer activity. *Arch. Pharm.* 346: 266-277.

Widsten, P., Heathcote, C., Kandelbauer, A., Guebitz, G., Nyanhongo, G.S., Prasetyo, E.N. and Kudanga, T. 2010. Enzymatic surface functionalisation of lignocellulosic materials with tannins for enhancing antibacterial properties. *Process biochem.* 45: 1072-1081.

Witayakran, S. and Ragauskas, A.J., 2007. One-pot synthesis of 1, 4-naphthoquinones and related structures with laccase. *Green Chem.* 9: 475-480.

Youn, H.D., Hah, Y.C. and Kang, S.O. 1995. Role of laccase in lignin degradation by white-rot fungi. FEMS Microbiol. Lett. 132: 183-188.

Index

R

S

U

V

W

X

Z

www.ingramcontent.com/pod-product-compliance
Ingram Content Group UK Ltd.
Pitfield, Milton Keynes, MK11 3LW, UK
UKHW021324190726
13845UKWH00003B/40

9 789390 371211